Lecture Notes in Mechanical Engineering

Lecture Notes in Mechanical Engineering (LNME) publishes the latest developments in Mechanical Engineering—quickly, informally and with high quality. Original research reported in proceedings and post-proceedings represents the core of LNME. Volumes published in LNME embrace all aspects, subfields and new challenges of mechanical engineering. Topics in the series include:

- Engineering Design
- Machinery and Machine Elements
- Mechanical Structures and Stress Analysis
- Automotive Engineering
- Engine Technology
- Aerospace Technology and Astronautics
- Nanotechnology and Microengineering
- Control, Robotics, Mechatronics
- MEMS
- Theoretical and Applied Mechanics
- Dynamical Systems, Control
- Fluid Mechanics
- Engineering Thermodynamics, Heat and Mass Transfer
- Manufacturing
- Precision Engineering, Instrumentation, Measurement
- Materials Engineering
- Tribology and Surface Technology

To submit a proposal or request further information, please contact the Springer Editor of your location:

China: Dr. Mengchu Huang at mengchu.huang@springer.com
India: Priya Vyas at priya.vyas@springer.com
Rest of Asia, Australia, New Zealand: Swati Meherishi at swati.meherishi@springer.com
All other countries: Dr. Leontina Di Cecco at Leontina.dicecco@springer.com

To submit a proposal for a monograph, please check our Springer Tracts in Mechanical Engineering at http://www.springer.com/series/11693 or contact Leontina.dicecco@springer.com

Indexed by SCOPUS. All books published in the series are submitted for consideration in Web of Science.

More information about this series at http://www.springer.com/series/11236

Harshit K. Dave · Dumitru Nedelcu
Editors

Advances in Manufacturing Processes

Select Proceedings of RAM 2020

Springer

Editors
Harshit K. Dave
S. V. National Institute of Technology
Surat, India

Dumitru Nedelcu
Gheorghe Asachi Technical
University of Iasi
Iasi, Romania

ISSN 2195-4356 ISSN 2195-4364 (electronic)
Lecture Notes in Mechanical Engineering
ISBN 978-981-15-9116-7 ISBN 978-981-15-9117-4 (eBook)
https://doi.org/10.1007/978-981-15-9117-4

This Springer imprint is published by the registered company Springer Nature Singapore Pte Ltd.
The registered company address is: 152 Beach Road, #21-01/04 Gateway East, Singapore 189721, Singapore

RAM 2020 Conference Organizing Committee

Chairman

Prof. S. R. Gandhi, Director, SVNIT, Surat

Organizing Secretaries

Dr. Shailendra Kumar, SVNIT, Surat
Dr. Harshit K. Dave, SVNIT, Surat

Coordinators

Dr. Amrut S. Mulay, SVNIT, Surat
Dr. Pawan Sharma, SVNIT, Surat

Steering Committee

Dr. R. V. Rao, SVNIT, Surat
Dr. H. K. Raval, SVNIT, Surat
Dr. K. P. Desai, SVNIT, Surat
Dr. A. A. Shaikh, SVNIT, Surat
Dr. T. N. Desai, SVNIT, Surat

Programme Committee

Dr. D. I. Lalwani, SVNIT, Surat
Dr. J. V. Menghani, SVNIT, Surat
Dr. Ravi Kant, SVNIT, Surat
Dr. Dinesh Singh, SVNIT, Surat
Dr. V. D. Kalyankar, SVNIT, Surat
Dr. S. N. Pandya, SVNIT, Surat
Dr. B. N. Sahoo, SVNIT, Surat
Dr. Sumit Khare, SVNIT, Surat

Advisory Committee

Dr. K. P. Rajurkar, University of Nebraska, USA
Dr. Dumitru Nedelcu, University of Iasi, Romania
Dr. Chintan Vaishnav, MIT, USA
Dr. J. P. Davim, University of Aveiro, Portugal
Dr. M. P. Jahan, Miami University, USA
Dr. H. M. A. Hussein, Helwan University, Egypt
Dr. Andrew Y. C. Nee, NUS, Singapore
Dr. Panagiotis Kyratsis, UWM, Greece
Dr. R. Balasubramaniam, BARC, Bombay
Dr. Suhas Joshi, IIT Bombay
Dr. J. Ram Kumar, IIT Kanpur
Dr. D. Ravi Kumar, IIT Delhi
Dr. P. M. Pandey, IIT Delhi
Dr. Anupam Agrawal, IIT, Ropar
Dr. A. M. Sidpara, IIT Kharagpur
Dr. K. Hariharan, IIT Madras
Dr. G. S. Dangayach, MNIT Jaipur
Dr. Venu Gopal A., NIT, Warangal
Dr. P. K. Patowari, NIT, Silchar
Dr. Rajiv Kumar Garg, NIT, Jalandhar
Dr. P. C. Tewari, NIT, Kurukshetra
Dr. Vishal Sharma, NIT, Jalandhar
Dr. H. S. Mali, MNIT Jaipur
Dr. Prabhat Ranjan, BARC, Bombay
Dr. U. Chandrasekhar, Wipro3d
Dr. P. K. Jain, IITDM Jabalpur
Dr. R. K. Verma, MMMUT, Gorakhpur

List of Reviewers

Dr. H. K. Raval, SVNIT, Surat
Dr. K. P. Desai, SVNIT, Surat
Dr. S. K. Budhwar, SVNIT, Surat
Dr. P. K. Patowari, NIT, Silchar
Dr. T. N. Desai, SVNIT, Surat
Dr. H. K. Dave, SVNIT, Surat
Dr. J. V. Menghani, SVNIT, Surat
Dr. Dinesh Singh, SVNIT, Surat
Dr. M. K. Rathod, SVNIT, Surat
Dr. Arati Mulay, CoE, Pune
Dr. S. N. Pandya, SVNIT, Surat
Dr. A. S. Mulay, SVNIT, Surat
Dr. B. N. Sahoo, SVNIT, Surat
Dr. Pawan Sharma, SVNIT, Surat
Dr. Sudhanshu Kumar, MANIT, Bhopal
Dr. Prabhat Ranjan, BARC, Mumbai
Dr. Piyush Gohil, MSU, Baroda
Dr. V. J. Badheka, PDPU, Gandhinagar
Dr. K. P. Mehta, PDPU, Gandhinagar
Dr. Vishal John Mathai, AMCoE, Kerala
Dr. M. K. Chudasama, GEC, Dahod
Dr. Dilip B Jani, GEC, Dahod
Dr. Hiesh Panchal, GEC, Patan
Dr. M. P. Sutaria, CHARUSAT, Anand
Dr. J. D. Patel, MEC, Basna
Dr. Ravi Bhatt, CGPIT, Gujarat
Dr. Shakil Kagzi, SNPIT, Umrakh
Dr. K. D. Maniya, CKPCET, Surat
Dr. Abhay Utpat, SVERI, Pandharpur
Prof. J. M. Joshi, VGEC, Chandkheda

Preface

Since 2010, Department of Mechanical Engineering at Sardar Vallabhbhai National Institute of Technology, Surat, has been organizing a series of conferences on "Recent Advances in Manufacturing." In order to enable the sharing of knowledge in the areas of manufacturing technologies, we have organized six national conferences on "Recent Advances in Manufacturing" and now we have planned the International Conference on Recent Advances in Manufacturing (RAM-2020). The conference is organized to bring the academicians, researchers and practicing engineers for sharing their experiences in the field of advance manufacturing. RAM-2020 will provide the opportunity for networking among participant institutes/organizations/industries to systematically confront the challenges in mutual areas of interest to advance manufacturing technology in these areas.

The proceedings volumes are published in the Springer series Lecture Notes in Mechanical Engineering in two volumes, viz. Volume 1—Advance Manufacturing Processes and Volume 2—Advance Manufacturing Systems. We also acknowledge the academic support from Prof. Dumitru Nedelcu and Prof. K. P. Rajurkar while editing the volumes 1 and 2, respectively.

As the entire world is facing the threat from corona pandemic, the international as well as interstate travel is restricted. However, we have tried our best to carve out a comprehensive schedule, keynote speakers and oral presentations in both online and offline modes, all of which will facilitate stimulating insightful discussions within the research community. In spite of such a pandemic situation, 60 participants have presented their findings and exchanged ideas related to manufacturing domain.

We are thankful to the conference organizing committee members, the advisory committee members, the reviewers, the session chairs and the volunteers, without whose generous contributions this conference would not number of presentations and number of participants.

Most of all, we thank the participants for enriching the international conference with their active participation.

Surat, India

Dr. Shailendra Kumar
Dr. Harshit K. Dave
Organizing Secretary, RAM-2020

Contents

Proficiency of Electrical Discharge Machining in Fabrication of Microstructures 1
Promod Kumar Patowari, Siddhartha Kar, Tapas Debnath, and Amit Kumar Singh

Experimental Study of Effect of Machining Parameters on PMMA in Diamond Turning 19
Kuldeep A. Mahajan, Raju Pawade, and R. Balasubramaniam

A Comparative Study of Electro-discharge Drilling Process Using Solid and Tubular Electrodes 29
Sudhanshu Kumar and Harshit K. Dave

Optimization of Process Variables in Plasma Arc Machining of Inconel-718 Alloy Using Taguchi with Grey Relational Analysis 37
Khushboo Sharma, Jai Kishan Sambharia, and Alok Khatri

Optimization of WEDM Process Parameters for Aluminium Metal Matrix Material Al+SiC Using MCDM Methods 59
Jaksan D. Patel and Kalpesh D. Maniya

Multiple Parameter Optimization by Wire Electrochemical Discharge Machining Process on Quartz Glass 71
Rahul O. Vaishya, Ankit D. Oza, and Anurag Gupta

Effect of Process Parameters on Etch Depth of Aluminium Material in Photochemical Machining 87
Abhay Utpat, N. D. Misal, B. P. Ronge, and B. A. Kamble

Quartz Micro-machining Using Wire Electrochemical Spark Machining Process 95
Ankit D. Oza, Abhishek Kumar, Vishvesh Badheka, Aditya Nema, and Manoj Kumar

Stress Relaxation Study of Ultrafine-Grained AA 6061 Alloy Processed Through Combined Constrained Groove Pressing and Cold Rolling 111
K. Changela, K. Hariharan, and R. K. Digavalli

Effect of Friction Stir Welding Process Parameters on Tensile Strength and Forming Height of Tailor Welded Blanks 123
Sumit Patel, Shalin Marathe, Keyur Desai, and Harit Raval

A Brief Review on Formability, Wall Thickness Distribution and Surface Roughness of Formed Part in Incremental Sheet Forming 135
Kiran R. More, Vikas Sisodia, and Shailendra Kumar

Parametric Study of Non-axisymmetric Stretch Flanging Process on AA-6061-T6 Sheet 151
Sai Dhruthi Killi, Keyur Desai, and Harit Raval

Experimental Investigation and Effects of Process Parameters on Forming Time and Forming Accuracy in Incremental Sheet Forming 159
Bhavin Desai, Shalin Marathe, Keyur Desai, and Harit Raval

Effect of Operating Parameters on Forces During Backward Flow Forming Process for AA6061 173
Jaydeep B. Bhatt, Ravi J. Bhatt, Harit K. Raval, and Keyur P. Desai

Analysis of TIG-Welded Aluminum Alloys During Single Point Incremental Forming at Different Wall Angles 187
M. M. Ghadmode, Arpit R. Patil, B. U. Sonawane, and Amrut Mulay

Autogenous TIG Welding of Al-5083-H111 Butt Joint 205
Keyurkumar D. Tandel and Jyoti V. Menghani

Review on Friction Stir Welding of Polymer to Aluminium Alloys: Process and Properties 221
Rutvik Ghiya and Vishvesh J. Badheka

Effect of Process Parameters on Tensile Strength in FSW of Aluminium and Stainless Steel 239
Niraj Kumar, Dhrupal Kotadiya, Vishvesh J. Badheka, and Vijay S. Gadakh

Study of Friction Stir Welding on Aerospace Grade ZE41AMg Alloy and Its Comparison with Laser Beam Welding on ZE41AMg Alloy 249
Adithyan Annamalai, T. R. Kishore Babu, S. Karthikeyan, N. Siddharth, and S. Muralidharan

Influence of Layer Thickness, Infill Rate and Orientation on Thermal and Structural Loading of FDM Parts 263
Andrei-Danut Mazurchevici, Ramona-Iuliana Popa, Constantin Carausu, Simona-Nicoleta Mazurchevici, and Dumitru Nedelcu

Experimental Investigation on Influence of Process Parameters on Mechanical Properties of PETG Parts Made by Fused Deposition Modelling 283
Soham Teraiya, Swapnil Vyavahare, and Shailendra Kumar

Analysis of Compressive Strength of 3D Printed PLA Part 295
Ashish R. Prajapati, Shilpesh R. Rajpurohit, Naushil H. Patadiya, and Harshit K. Dave

An Experimental Study of Influence of Gradient Parameters on Compressive Strength, Stiffness, and Specific Energy Absorption (SEA) of Auxetic Structures Fabricated by FDM 305
Swapnil Vyavahare, Soham Teraiya, and Shailendra Kumar

Study on Compression Properties of ABS and FPU Parts Printed Using SLA 319
Ravi Teja Karumuri, Harshit K. Dave, Shilpesh R. Rajpurohit, and Ashish R. Prajapati

Experimental Investigation on Tensile Properties of Nylon Glass Fibre Material Made Using Fused Deposition Modelling Process 329
Mohit R. Chandravadia and M. K. Chudasama

Influence of Inter-Lamellar Spacing of Pearlite Phase on Spheroidized Annealed Structure 343
Partha Sarathi Sarkar, Bhavesh Rameshchandra Rana, and Pinkal D. Mistry

Review on Banana Fibre-Reinforced Composites 359
Chetan Prakash Chaudhari, Kiran Bhole, and Jayram Gholave

Experimental Evaluation of Mechanical Properties of Epoxy Based Composite Material Using Taguchi Method 381
Vishal Naranje, Ajay Rajan Sankar, Sachin Salunkhe, and Bhanudas D. Bachchhav

Experimental and Simulation Study on Permeability of Hybrid Composite 397
Prasanth Kottapalli, Rahul Narkhede, Harshit K. Dave, Himanshu V. Patel, and Rajesh kumar Verma

Impact and Flexural Testing of Jute and Flax Fiber Reinforced Composites Fabricated by VARTM Process 411
Chandresh B. Kumbhani, Harshit K. Dave, and Himanshu V. Patel

Influence of Fiber Orientation and Number of Layer on Tensile and Flexural Strength of Carbon Fiber-Reinforced Composites Fabricated by VARTM Process 421
Himanshu V. Patel, Shripal M. Patel, and Harshit K. Dave

About the Editors

Dr. Harshit K. Dave is currently Associate Professor at the Department of Mechanical Engineering, S. V. National Institute of Technology, Surat, India. His research interests include Additive Manufacturing Processes; 3D printing filaments & raw materials; Unconventional Machining processes; Micro machining processes; Modeling & optimization of machining processes; Robotics & Automation. He has published more than 100 papers in reputed international journals and conferences proceedings. He has successfully carried out several research projects funded by the DST, MHRD, GUJCOST, NPIU, etc.

Dr. Dumitru Nedelcu is a Professor at the "Gheorghe Asachi" Technical University of Iasi (TUIASI), Romania, Director of TUIASI Doctoral School. He is Manager of Fine Mechanics and Nanotechnology Laboratory, President of ModTech Professional Association, ModTech International Conference and Editor-in-Chief of the International Journal of Modern Manufacturing Technologies and Advanced Engineering Forum. He was a Visiting Professor at TAT, Institute of Engineering, Tokyo, Guest Professor at Osaka University, Japan and Grenoble Institute of Technology, France. He had Erasmus teaching internships in prestigious universities from Poland, Italy and Mexico. In October 2016 he was accepted as Visiting Professor at the Silesian University of Technology, Gliwice, Poland. As far as research is concerned, he coordinated 15 national and international projects as project manager/responsible. He has published more than 180 scientific papers on ISI and BDI journals and international conferences proceedings.

Proficiency of Electrical Discharge Machining in Fabrication of Microstructures

Promod Kumar Patowari, Siddhartha Kar, Tapas Debnath, and Amit Kumar Singh

1 Introduction

Electrical discharge machining (EDM) is used for precise machining of conductive materials irrespective of their strength and hardness. It removes material by melting and evaporation due to the generation of sparks between the tool and workpiece electrodes. In the EDM process, electrical energy is converted into thermal energy, which is visible in the form of a spark. The electrical sparks generate intense heat energy, which is responsible for the removal of material. A gap exists between the tool and the workpiece, which facilitates the occurrence of spark during the machining process. Thus, the process is contactless and free from the ill-effects of vibration and chatter-related problem which are quite relevant in conventional machining processes.

The traditional EDM is mainly used for the machining of dies and molds, where the tip of the tool acts as the negative impression of the machined cavity. Besides, it is also used for machining of parts, complex components, intricate shape components, drilling of micro- or nano-holes, etc. But over the years, the EDM process has been

P. K. Patowari (✉) · S. Kar · T. Debnath
Department of Mechanical Engineering, National Institute of Technology Silchar, P.O. Silchar, Assam 788010, India
e-mail: ppatowari@yahoo.com

S. Kar
e-mail: siddkar.nita@gmail.com

T. Debnath
e-mail: nit.tapas11@gmail.com

A. K. Singh
Department of Mechanical Engineering, National Institute of Technology Nagaland, Chumukedima, Dimapur 797103, India
e-mail: amit.kumar965@yahoo.com

H. K. Dave and D. Nedelcu (eds.), *Advances in Manufacturing Processes*, Lecture Notes in Mechanical Engineering, https://doi.org/10.1007/978-981-15-9117-4_1

successfully applied in different forms which can be termed as variants of EDM. The most commonly used EDM variants are die-sinking, wire-cut, block electrical discharge grinding (BEDG), milling and reverse. All these variants possess unique capabilities of fabricating a different kind of features and structures but with the same machining phenomenon. In this phenomenon, initially, the tool with specific geometry moves toward the workpiece, and resulting sparks are produced in between the tool and the workpiece in the presence of dielectric fluid. The first spark is produced at the location where the minimum distance between the electrodes is identified and breakdown occurs at the minimum resistance point. Subsequently, material is removed from both electrodes, and the discharge point shifts to a new location with the closest distance between the workpiece and tool electrode.

EDM is a non-contact machining process devoid of any vibration and chatter-related problems. This unique characteristic enables it to machine microfeatures from all electrically conductive material. But, the dimensional accuracy of machined features is affected by the tool wear, which is inevitable in any EDM variant. Proper selection of process parameters helps in minimizing the tool wear, although it cannot be eliminated. Application of suitable tool wear compensation techniques also augments the chances of achieving dimensional accuracy and precision. But, implementation of any compensation technique requires proper assessment and estimation of the tool wear and material removal. Thus, proper evaluation of the dimensional features of the machined components is paramount to the implementation of a compensation technique. Given all these factors, the present study demonstrates the capability of several variants of EDM in machining microfeatures and discloses their dimensional features to determine the quality of the machined product. Further, the challenges faced during the fabrication of various microfeatures and the measures taken to overcome those are also presented.

The organization of the present work is divided into five sections, and each of the section highlights briefly about the corresponding section. The first section introduces EDM with a focus on its background and importance. The second section discusses the differences between macro- and micromachining and their transition. The third section provides the detail information about microstructure fabrication techniques via different variants of EDM, in which the process mechanism and the capabilities of die-sinking EDM, wire-cut EDM, BEDG, reverse EDM (REDM) and milling EDM have been demonstrated. The fourth section delivers information about recent research trends and future scope. The last section provides a conclusion and summarizes the major findings of the study.

2 Macro- to Micromachining Using EDM: A Paradigm Shift

The material removal phenomenon in EDM is dependent on the energy discharged in the machining gap existing between tool and workpiece. In EDM, voltage is applied in the form of pulses at the alternate interval. During the application of voltage, the formation of spark takes place, which raises the temperature and leads to melting and vaporization of the material. The interruption in voltage supply leads to the formation of pressure shock waves in the machining zone, which ultimately leads to evacuation of material, leaving behind the impression of craters in the workpiece and tool. The full process takes place in the presence of a dielectric medium, which is also flushed in the machining zone to remove the debris. The material removal process takes place without any contact between the tool and the workpiece. This non-contact material removal mechanism of EDM enables the use of softer material such as copper and brass as tool electrode while machining hard and tough materials.

In EDM, the material is eroded in the form of craters. The dimension of the craters depends on the discharge energy of the process. At lower discharge energy, the effective unit removal of material is very less, and thus, shallow and tiny craters are formed, whereas the craters get wider and deeper with an increase in the discharge energy. This concept has been effectively utilized in micromachining of materials using the EDM process, where the supply of lower discharge energy is paramount to remove material in the micro-level. The ability of EDM in performing micromachining is termed as micro-electrical discharge machining (μEDM). The key differences present in the process mechanism of μEDM as compared to macro-EDM are:

Plasma channel diameter: The plasma channel radius is less in μEDM due to the supply of lower discharge energy. The plasma channel diameter expands with an increase in pulse duration, and unlike macro-EDM, it may equal or exceed the diameter of the tool. The expansion rate may vary once the plasma channel exceeds the tool diameter.

Electrode heating: In micromachining, the machining dimensions typically range from 1 to 999 μm, which necessitates the use of a micro-tool. The use of micro-tool restricts the quantity of energy supplied per discharge as the tool does not possess significant mass to carry the heat.

Viscous force of dielectric on the tool: In μEDM, the dielectric cannot be directly flushed in the machining zone. This is because the tool electrode is more sensitive to the viscous force of dielectric which may lead to deflection or bending of the micro-tool.

The absence of process forces (except the viscous force of dielectric) allows fabrication of microfeatures with high aspect ratio. A μEDM machine tool possesses the ability to fabricate two-dimensional (2D) and three-dimensional (3D) microfeatures with higher accuracy and precision. The minimization of discharge energy has been possible by the use of resistance capacitance pulse generator, where discharge energy is represented as Eq. (1):

$$\text{Discharge energy} = \frac{1}{2} \times \text{Capacitance} \times \text{Volatge}^2 \tag{1}$$

The capacitor stores the energy until discharge initiates, whereas voltage determines the breakdown limit at which the current flows through the inter-electrode gap. The increase in both capacitance and voltage leads to an increase in discharge energy, which expedites the process of material removal.

3 Microstructures Fabricated Using Different Variants of EDM

3.1 Die-Sinking EDM

Die-sinking is the most initial and basic form of EDM technology used mainly for manufacturing of dies/molds. It utilizes the concept of manufacturing of replica objects in large-scale production of micromechanical devices. The tool electrode acts as the positive impression of the desired shape of the product, where the machined cavity on the workpiece reflects the negative image of the tool electrode as depicted in Fig. 1. 3D cavities can be machined by this technique provided the desired pattern of tool electrode is fabricated beforehand. But the fabrication of complex patterned tool is a tedious task which requires some pre-processing techniques for their fabrication. The concept of sinking EDM is also utilized for drilling, which uses a simple cylindrical tool for machining micro-holes. To enhance the efficiency, most often rotation motion of the tool is applied during the drilling operation. Incorporation of rotation exhibits centrifugal force and agitation effect in the machining zone, which helps in better flushing of eroded materials. Circularity error (CE) also reduces with the incorporation tool rotation as compared to micro-holes drilled without rotation. But, overcut (OC) increases with the actuation of rotation motion in tool electrode due to the phenomenon of secondary sparking between trapped debris and sidewall of the workpiece. Another significant phenomenon affecting the taper angle (TA) of the drilled hole is the tool wear. The tool wears out in both frontal and corner

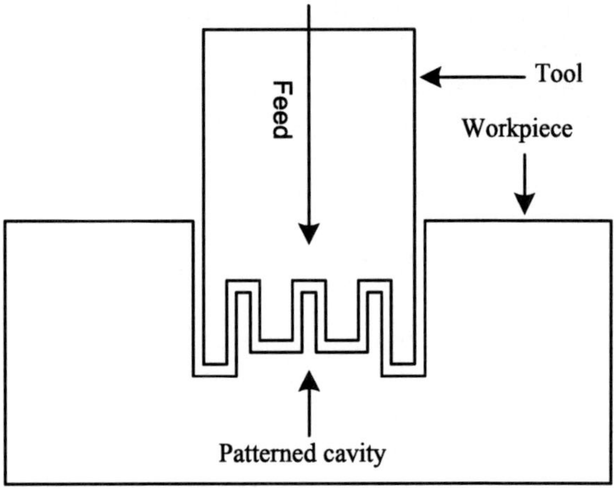

Fig. 1 Schematic diagram of die-sinking EDM

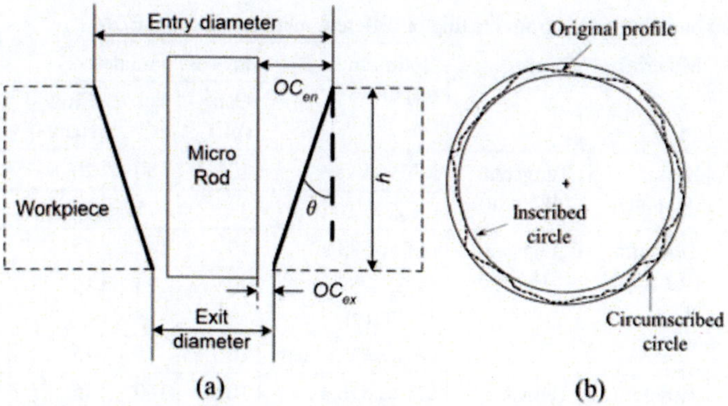

Fig. 2 Schematic diagram representing **a** overcut and taper **b** circularity error

portions, which instills dimensional inaccuracy in the final product. Frontal wear is easily compensated by feeding of tool for sufficiently higher length than the required depth of the micro-hole. But, compensation of corner wear is a very tedious task and cannot be eliminated. However, the process parameters can be chosen in such a way that the effect of corner wear is minimum to achieve higher accuracy and precision in the machined micro-holes. The most common dimensional features to determine the accuracy and preciseness of micro-holes are OC at entry (OC_{en}), OC at exit (OC_{ex}), TA and CE, which are schematically depicted in Fig. 2, and their corresponding equations are shown in Eqs. (2), (3), (4) and (5), respectively.

$$OC_{en} = \frac{\text{Hole entry diameter} - \text{Tool diameter}}{2} \tag{2}$$

$$OC_{ex} = \frac{\text{Hole exit diameter} - \text{Tool diameter}}{2} \tag{3}$$

$$TA(\theta) = \tan^{-1}\left(\frac{\text{Entry diameter} - \text{Exit diameter}}{2h}\right) \tag{4}$$

$$CE = \frac{\text{Circumscribed circle diameter} - \text{Inscribed circle diameter}}{2} \tag{5}$$

where h represents hole depth.

Over the years, μEDM has been successfully used for drilling of materials with various electrical and thermal conductivities. Table 1 shows a comparison of data extracted from the drilling of different materials using μEDM process. The range of parameters varied in each study is different and must not be considered the same for all the materials. EDM possesses the ability to machine materials with different thermal and electrical conductivity. From all the studies, it is perceived that minimum machining time (MT) is attained at higher parametric condition, especially at higher voltage and capacitance. On the contrast, the dimensional features such as OC, CE

Table 1 Comparison of data from drilling of different materials using μEDM

S. No.	Material	Tool	Optimum responses	Process parameters			
				Cap (pF)	Vol (V)	FR (μm/s)	TRS (rpm)
1 [1]	Brass (40 μm)	Tungsten (485 μm)	MT = 54 s	10^4	180	20	1500
			OC_{en} = 10 μm	33	75	10	0
2 [2]	Titanium (0.7 mm)	Tungsten (518 μm)	MT = 583 s	10^4	145	15	1000
			OC_{en} = 45.85 μm	10^4	75	15	1500
			OC_{ex} = 2.55 μm	10^4	75	15	1500
3 [3]	Titanium (0.7 mm)	Tungsten (518 μm)	MT = 510 s	10^4	180	15	1000
			OC_{en} = 14.7 μm	10^2	180	5	500
			OC_{ex} = 2.45 μm	10^2	180	10	1500
4 [4]	Silicon	Tungsten (518 μm)	MRR = 8.15 $μm^3$/min	10^4	200	10	0
			OC_{en} = 15 μm	10^2	140	10	0
			OC_{ex} = 13 μm	10^2	140	10	0
			TA = 1.145°	10^2	140	10	0
			CE_{en} = 7.5 μm	10^2	140	10	0
			CE_{en} = 5 μm	10^3	140	10	0
5 [4]	German silver	Tungsten (518 μm)	MRR = 124.8 $μm^3$/min	10^4	200	10	0
			OC_{en} = 28 μm	10^2	140	10	0
			OC_{ex} = 15 μm	10^2	140	10	0
			TA = 0.515°	10^2	140	10	0
			CE_{en} = 7 μm	10^2	140	10	0
			CE_{en} = 2 μm	10^2	140	10	0
6 [5]	Stainless steel (30 μm)	Tungsten (518 μm)	MT = 229 s	10^3	145	5	1500
			OC_{en} = 38.07 μm	33	75	5	500
			CE_{en} = 6.50 μm	33	75	5	1500
7 [6]	Metallic glass (20 μm)	Tungsten (513 μm)	MT = 2.14 min	10^4	100	10	1000
			OC_{en} = 7 μm	10^2	75	10	0
			OC_{ex} = 5.25 μm	10^2	75	10	0
			TA = 4.29°	33	150	10	0

Cap capacitance; *Vol* voltage; *MRR* material removal rate

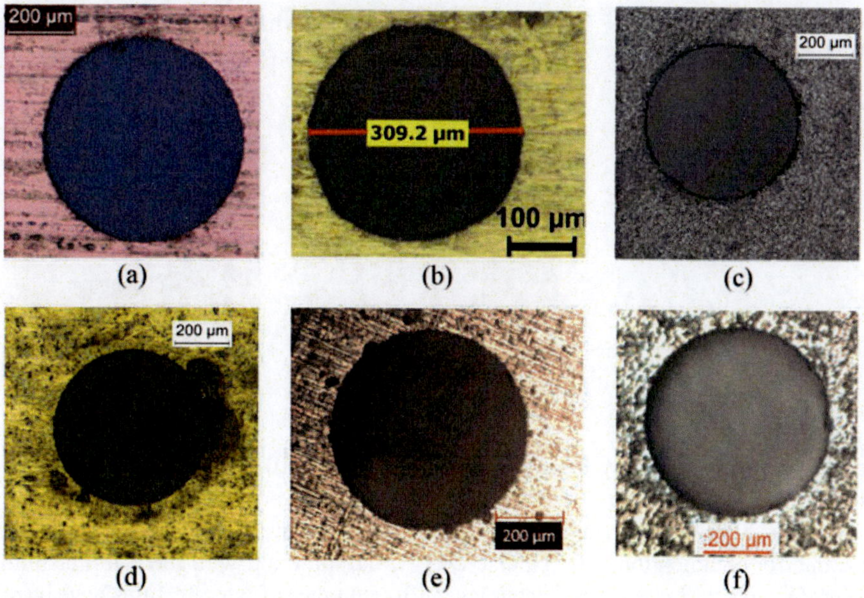

Fig. 3 Microscopic view of machined hole on different materials **a** Cu-based bulk metallic glass, **b** brass, **c** silicon, **d** German silver, **e** stainless steel, **f** titanium grade 2 alloy

and TA are minimum at lower capacitance and voltage. With an increase in voltage and capacitance, the discharge energy increases which consequently increases the spark intensity. High intense spark leads to the formation of bigger and deeper craters, which increases the unit material removal per spark but degrades the surface quality. This phenomenon is responsible for higher efficiency and lower surface quality at higher values of capacitance and voltage. Figure 3 shows microscopic images of micro-holes drilled on different materials such as Cu-based bulk metallic glass, brass, silicon, German silver, stainless steel (SS) and titanium (Ti) grade 2 alloy. The materials possess different physical properties, which further apprehend the usefulness of μEDM in micromachining a wide range of materials.

3.2 *Wire-Cut Electrical Discharge Machining*

Similar to die-sinking EDM, wire-cut EDM is another variant of EDM, which is mainly used to cut metal pieces. In this process, a thin brass or molybdenum wire is used to cut conductive materials irrespective of their other mechanical properties. Generally, two kinds of wire-cut EDM is used (a) consumable wire electrode type (where the wire is being discarded after a single use) and (b) non-consumable wire electrode type (where the wire is being used several times before it needs to be discarded). As wire-cut EDM is mainly used to cut a variety of metal pieces, the

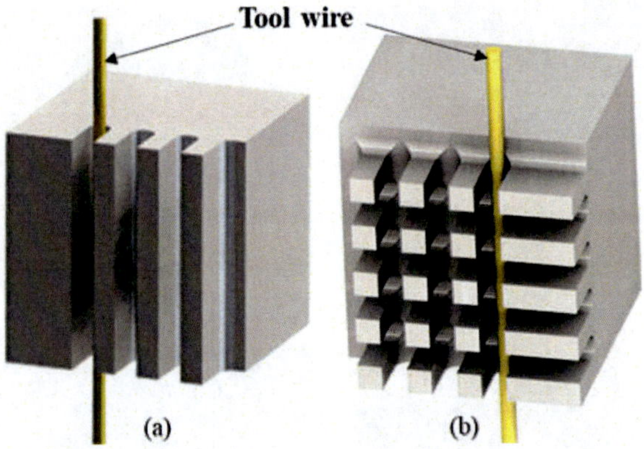

Fig. 4 Schematic of wire movement path to fabricate a square array of micro-fin **a** first cut, **b** second cut after 90° rotation of the workpiece (any direction)

pulse energy is generally high in comparison with μEDM but less than the die-sinking EDM. To create microfeatures using such wire-cut EDM, the parametric conditions have to be tuned properly. Debnath and Patowari [7] identified optimum parametric combination (lower pulse energy, duration and wire feed) to machine microfeatures with the desired accuracy. Different types of microfeatures have been successfully machined using the wire-cut EDM, which is elaborated in the subsequent sub-sections.

Micro-fins. Fins are attached to a hot body to increase the rate of heat transfer. Generally available fins are in macroscale, but micro-fins are very difficult to fabricate due to their delicate nature. Researches revealed micro-fins of different profiles have higher heat transfer rate than that of the macro-fins. The complete machining process has been accomplished utilizing multiple cuts (two or more) at different orientations of the workpiece. A computer numerical control code needs to be developed to define the wire path during machining. A schematic diagram for machining arrays of micro-fins is displayed in Fig. 4. After completing both the cuts, square extruded micro-fins have been fabricated. A typical image of the fabricated sample is shown in Fig. 5. The cross-sectional area of the chosen workpiece is 10 mm × 10 mm, and on that sample, total 25 × 25 (=625) micro-square fins have been fabricated. The average size of the generated fins is 165 μm × 165 μm × 800 μm. This type of micro-fins can also be used as a tool for drilling multiple holes of different shapes at a time. These tools help in faster production rate and also reduce the defects due to repeatability of the machining axis.

Similar to the machining procedure of square micro-fins, micro-fins of different profiles are also successfully fabricated using wire-cut EDM. Figure 6 shows the schematic wire path movement to fabricate 'Y' type of micro-fins, whereas Fig. 7 shows the actual image of the machined micro-fins by wire-cut EDM.

Micro-screws. Screws are one of the most important engineering elements used in assembling various components in the form of a temporary joint. Generally, machining is the prime fabrication process to make screws of different profiles and shapes, but the time consumed is very high. Nowadays, using a sliding die, smaller

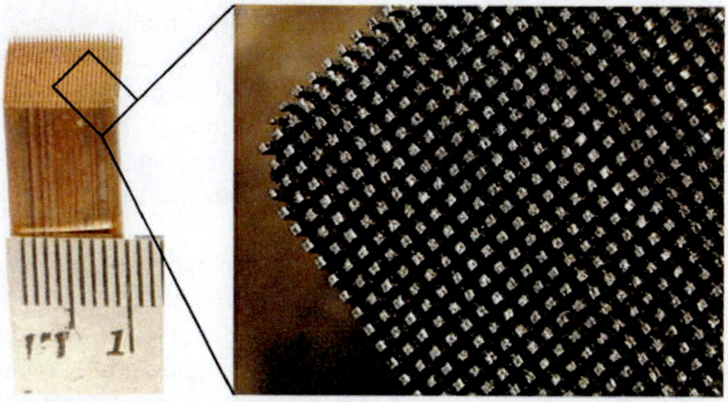

Fig. 5 Fabricated array of square micro-fin on a copper rod (10 mm × 10 mm)

Fig. 6 Schematic of wire path for fabricating 'Y' type micro-fins **a** initial cut **b** final cut

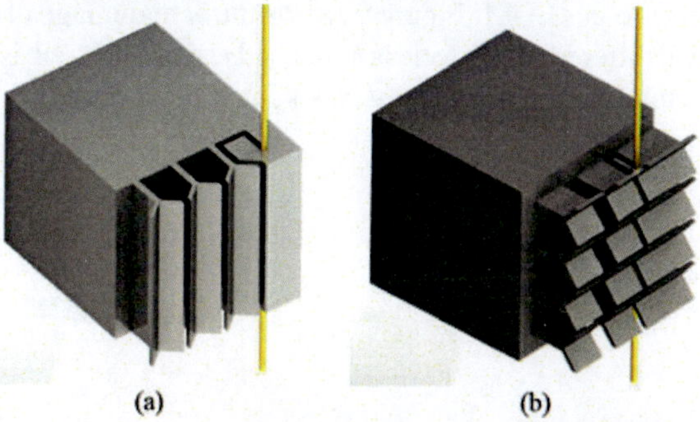

Fig. 7 Fabricated 'Y' type of micro-fins

rolled screws of limited shapes are also fabricated. The rolling process deforms the metal to form a screw shape, and for this reason, the metals which are easy to deform are only used. But to fabricate screws with tough metals, the rolling process faces difficulty, and thus, fabrication of micro-screws from such material is quite a challenging task. Wire-cut EDM can be used to machine such metals with greater accuracy, and to fabricate micro-screw, a separate attachment is required to rotate the workpiece with variable speed [8].

To make a micro-screw using wire-cut EDM, initially turning operation is performed to reduce the eccentricity of the rotating workpiece and also to get the near-net dimension of the desired micro-screw. Thereafter, the thread cutting operation yields the final shape of micro-screw. During turning, the workpiece rotational speed needs to be high keeping very low feed, whereas, in case of thread cutting, the workpiece rotational speed has to be minimum with higher feed. Figure 8 shows the concept for turning and thread cutting operation used for fabrication of micro-screw in wire-cut EDM. Figure 9 shows the actual image of the fabricated micro-screws of two different types (one is with head and another one is without head) with a dedicated notch to drive the micro-screws inwards or outwards based on the requirement.

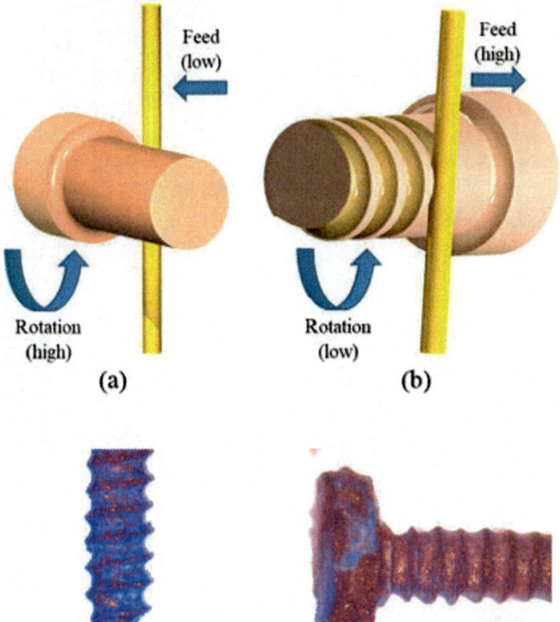

Fig. 8 Steps for fabricating micro-screw **a** turning operation **b** thread cutting operation

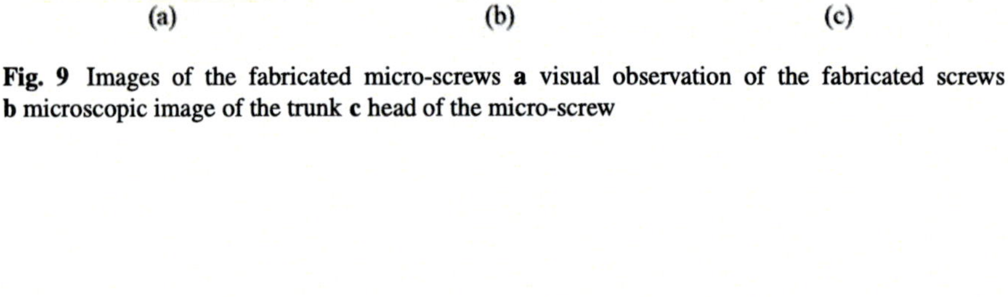

Fig. 9 Images of the fabricated micro-screws **a** visual observation of the fabricated screws **b** microscopic image of the trunk **c** head of the micro-screw

3.3 Block Electrical Discharge Grinding

BEDG is the process of using a sacrificial block in reducing the diameter of cylindrical rods. The sacrificial block acts as the tool electrode, whereas the rod is attached to the spindle, which represents the workpiece. The spindle is rotated at a certain speed and fed against the sacrificial block for sparks to occur. The schematic diagram of the BEDG process is shown in Fig. 10. BEDG is an effective method for fabrication of micro-rods, which besides a μEDM setup requires only a flat block of conductive material. The process suffers a setback in the form of block wear, which instills dimensional inaccuracy and imprecision in the fabricated micro-rods. Taper (positive and negative), position shift, overburn and breakage of tooltip are few of the errors and ill-effects encountered in BEDG process [9]. Holding of the collet, wobbling effect, block surface taper and higher energy are some of the causes of the errors. However, these errors can be eliminated to some extent by proper clamping of the rod in the collet, checking the block clamping and properly choosing the discharge energy to be used while machining.

Brass micro-rods have been successfully fabricated using the BEDG process, wherein a brass block is used as the sacrificial tool electrode [10]. It has been found that feed rate (FR) beyond 2.5 μm/s is detrimental to efficiency as well as the accuracy of the fabricated micro-rods. Higher TRS yields lower MT and standard deviation in diameter (SDD) but increases the average diameter (AD) due to increased block wear. Figure 11 shows a micro-rod fabricated by BEDG process at FR of 2.5 μm/s and TRS of 1250 rpm, where AD and SDD of 390.12 and 9.86 μm, respectively, are attained. Elsewhere, the effect of discharge energy on material removal rate (MRR) is a well-established fact, where MRR increases with an increase in discharge energy due to the increased volume of material removal with simultaneous decreased MT [9].

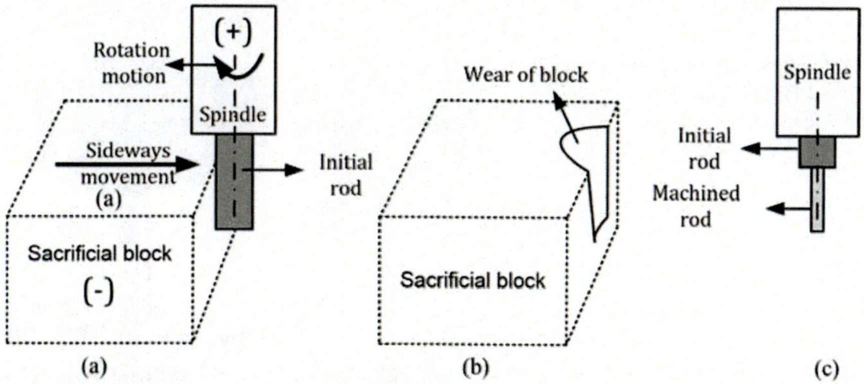

Fig. 10 Schematic diagram of BEDG process: **a** lateral feed of workpiece, **b** wear of sacrificial block and **c** machined rod

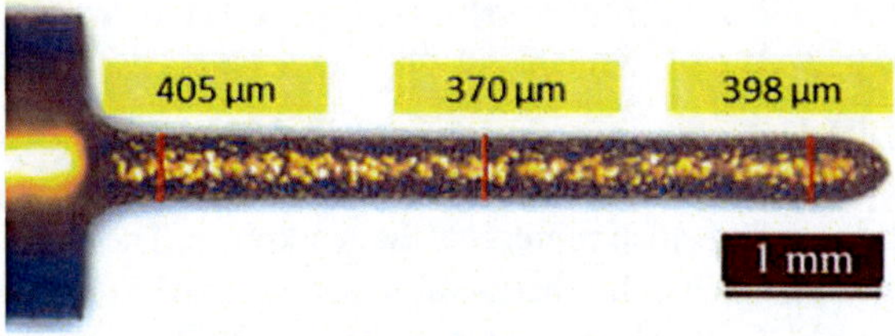

Fig. 11 Microscopic image of a micro-rod fabricated by BEDG

3.4 Reverse Electrical Discharge Machining

Reverse micro-electrical discharge machining (R-µEDM) is widely used for the fabrication of micro-rods and micro-fins which are in the form of circular and rectangular cross-sectional in shape. In this process, the workpiece in the form of a bulk rod is connected through spindle via the positive terminal, whereas the tool in the form of thin plate sheet with single or array of micro-holes is kept on the table and connected through the negative terminal. During the machining process, the workpiece is fed into the tool plate sheet up to a predefined length, which defines or indicates the fabricated length of the micro-rod as shown in Fig. 12.

Micro-tools in the form of micro-rods with circular cross-sectional area have been fabricated on tungsten bulk rod, where the diameter of bulk tungsten rod of diameter 800 µm has been reduced to a diameter of 200 µm using R-µEDM process [11]. An aspect ratio of five has been maintained during the experimentation. It has been observed that MT reduces with the increase in voltage, capacitance and FR. Voltage contributes more than the remaining process parameters toward MT and SDD. Surface roughness (Ra) of the machined micro-rods ranges from 2.12 to 5.94 µm, wherein lower capacitance yields lower Ra as compared to higher capacitance. Finally, a micro-tool with an aspect ratio of 18 (diameter 170 µm) has been successfully fabricated using the process.

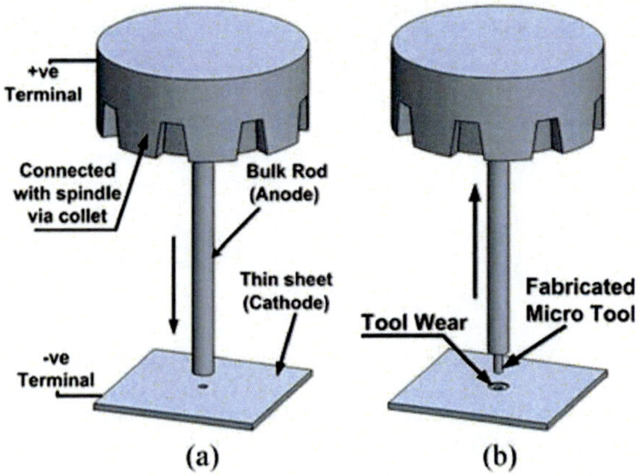

Fig. 12 Schematic diagram of R-µEDM **a** initial bulk rod **b** fabricated micro-rod

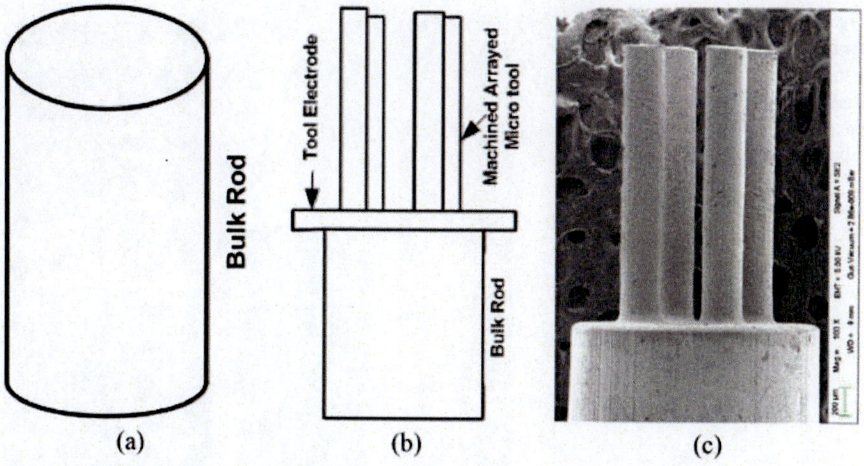

Fig. 13 Diagram of micro-rod **a** schematic of initial bulk rod **b** schematic of fabricated micro-rods **c** SEM image of fabricated micro-rods

Arrayed micro-rods have also been successfully fabricated on a brass bulk rod by R-μEDM process. Brass rod of diameter 1.8 mm has been considered as the initial workpiece. An array of 2 × 2 micro-tools has been fabricated through the machining of a bulk rod as shown in Fig. 13. It has been observed that the tool wear affects dimensions of the arrayed micro-tools during the machining process. An analytical model has been developed for evaluating the amount of tool wear and material removal from the workpiece. Tool wear and material removal increase with the increase in voltage and capacitance. It has been observed that straight micro-rod with longer length can be fabricated at lower parametric condition [12].

In another work, dimensional variation, MT and erosion rate have been analyzed while fabrication of micro-rods using R-μEDM. It is observed that as compared to the tip and root, the middle portion of the micro-rods undergoes lesser dimensional error [13]. Moreover, an array of 4 × 4 (=16) micro-rods has been fabricated successfully with each fabricated micro-rod having diameter 58 μm and length 830 μm as shown in Fig. 14.

3.5 *Milling Electrical Discharge Machining*

Milling EDM uses the lateral movement of tool and workpiece in the EDM environment to machine 3D cavities. The material removal mechanism is identical to drilling but with an additional feed in the lateral direction (x-y movement). Essentially, the tool feed takes place in two stages, i.e., initially, a rotating tool is fed to a certain depth in the z-direction (similar to drilling) followed by a horizontal feed of tool/workpiece to perform machining of the desired slot. It is capable of machining

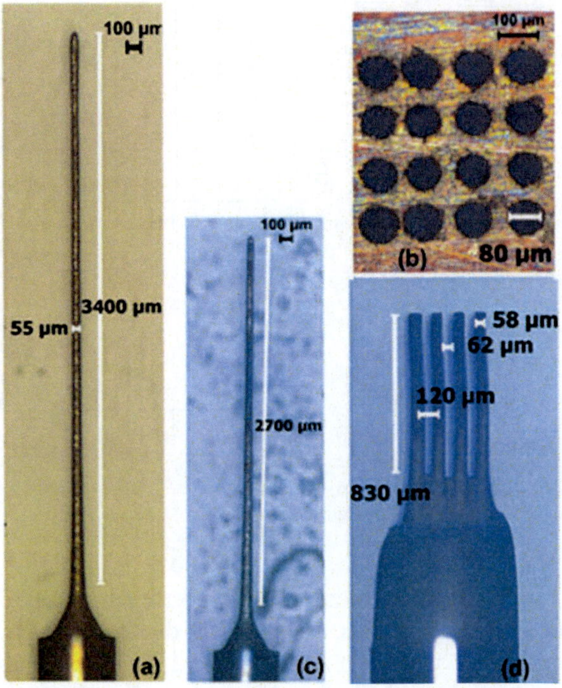

Fig. 14 Machining of arrayed micro-tools **a** micro-tool fabricated by BEDG process **b** array of micro-hole drilled using the fabricated micro-tool **c** condition of micro-tool after drilling micro-holes **d** arrayed micro-tools fabricated by R-µEDM on tungsten rod [13]

complex 3D cavities with a simple cylindrical rotating tool. Thus, the process eradicates the use of structured tool as in the case of die-sinking EDM in the fabrication of 3D cavities. But the effect of tool wear in milling EDM is more as compared to the die-sinking approach as it possesses horizontal (x-y) feed along with the z feed. A schematic diagram showing the effect of tool wear in the machined surface profile is displayed in Fig. 15. The surface profile inequality at the end side of the slot as

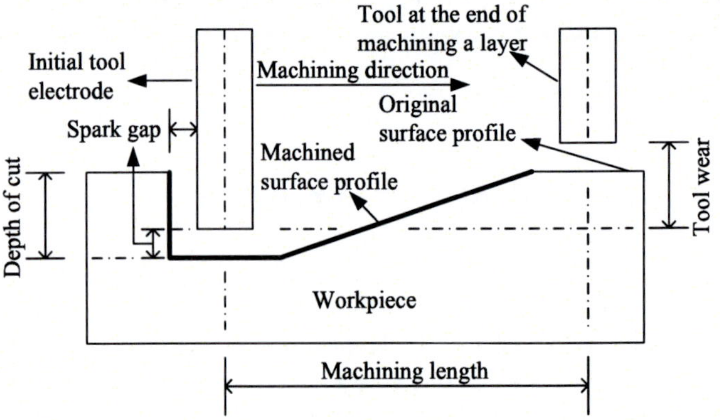

Fig. 15 Schematic of EDM milling showing effect of tool wear

compared to the start side is depicted. A simple technique to compensate this effect (tool wear) is to perform milling EDM in a to-and-fro layer approach, where z feed is given at the end of each to/fro layer to compensate the tool wear. Precise assessment of wear is necessary to give proper z feed at the end of each layer because an error in the initial assessment will lead to a compounding effect in the succeeding layers. Apart from the discharge energy, the depth of cut and horizontal FR also play a substantial part in determining the stability of EDM milling. A lower depth helps in stabilizing the process, where tool wear takes place only at the bottom and lateral wear is avoided.

EDM milling has been successfully applied for the fabrication of slots in different kinds of materials. Kar and Patowari [14] performed EDM milling on copper (Cu), SS304 and Ti grade 2 alloy, where they observed MRR and TWR in increasing order as $MRR_{SS} < MRR_{Ti} < MRR_{Cu}$ and $TWR_{Cu} < TWR_{Ti} < TWR_{SS}$, respectively. The higher electrical conductivity in Cu is responsible for the quick initiation of spark, which resulted in higher MRR. Elsewhere, the delay in spark formation in SS304 restricted the movement of tool electrode and allowed sufficient time for melting of tool material, which ultimately resulted in lower TWR. As compared to SS304 and Cu, Ti demonstrated improved surface morphology in respect to lower Ra, less redeposition, globule formation and stress concentration. The superior flushing of debris in Ti forms shallow craters and gives better surface morphology.

To compensate the effect of tool wear, 'to-and-fro' layer-by-layer machining strategy has been applied in EDM milling of Ti grade 2 alloy as shown in Fig. 16 [15]. The width of the micro-slot varies between 670 and 671 μm from one end to the other, whereas the standard deviation in the depth of the start, middle and end portions is 4.91 μm. Further, a micro-square pillar (side length = 269 μm) has been successfully machined (Fig. 17), where a standard deviation of 19.53 μm is attained in the depth of the four corners.

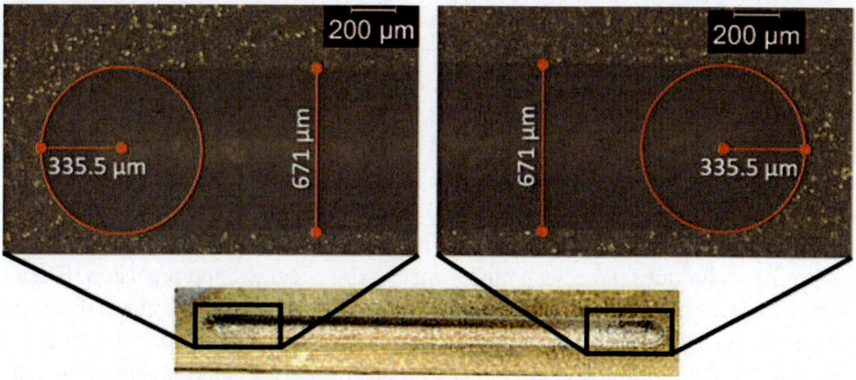

Fig. 16 Uniform slot fabricated by 'to-and-fro' machining approach

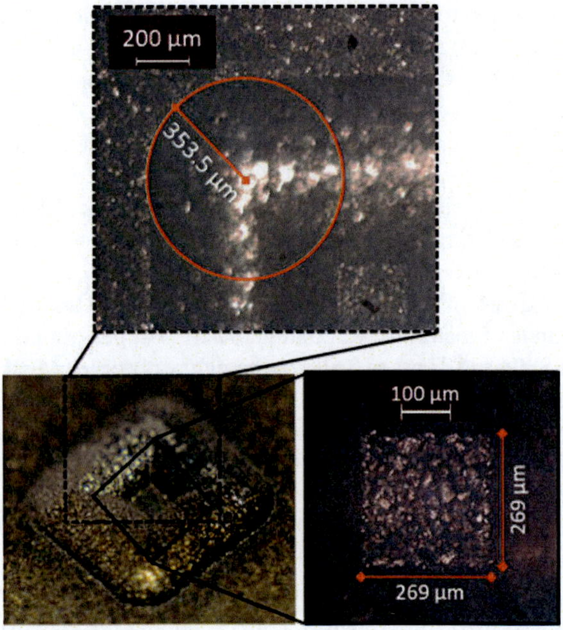

Fig. 17 Micro-square pillar fabricated by EDM milling

4 Research Trends and Future Scope

Nowadays, variety of advanced metals, alloys and metal matrix composites are being introduced to the in-demand market such as bio-medical, aerospace and defense sector, which are very difficult to machine using traditional machining methods. The EDM variants possess the unique ability to machine these hard and high strength materials provided they are electrically conductive or semi-conductive. Machinability study of such advanced materials needs to be investigated for better understanding of their appropriate machining condition. The non-contact mechanism of material removal also allows the fabrication of microfeatures from these materials.

Microfeatures, for instance, hole, tool, slot, fin, pin, screw, with various designs/shapes, are successfully fabricated by the different variants of EDM. μEDM has been successful in drilling micro-holes in various materials. However, achieving higher efficiency and dimensional accuracy requires a trade-off between these parameters. The effect of tool wear also needs to be compensated by some offline or online monitoring techniques to achieve higher machining accuracy and precision. BEDG process has been successful in fabricating micro-rods, where the customized diameter of micro-rods can be easily fabricated. However, the block wear instills dimensional inaccuracy in the micro-rods, which necessitates some form of compensation technique to be applied. Lower discharge energy helps in achieving higher dimensional accuracy but at the cost of lower efficiency. So, impetus should be also given to the miscellaneous factors such as block arrangement and tool fixation in the collet, which are normally ignored. Wire-cut EDM is the most versatile among all the variants of EDM due to its ability to machine a wide range of microfeatures such as pin, fin and screw. By proper controlling of wire path and positioning of the workpiece, complex 3D shapes can also be fabricated in wire-cut EDM. However, fluctuation of wire instills dimensional inaccuracy in the fabricated components, which requires further research to minimize or compensate its effect. R-μEDM allows machining of micro-rod and array of micro-rods with the use of predrilled hole or array of holes. The dimensional variation of the micro-rods can be minimized by proper selection of process parameters. EDM milling allows fabrication of 3D cavities with a simple rotating tool. However, the effect of tool wear diminishes dimensional accuracy and precision in the machined microfeatures. Several offline and online compensation techniques have been proposed and successfully applied for machining microfeatures with higher precision. But most of these techniques account for the lower depth of cut, which restricts in achieving higher efficiency. Thus, a tool wear compensation technique to oversee higher depth of cut to achieve higher efficiency, as well as higher dimensional accuracy, is yet to be flourished. Moreover, all these machining techniques require man, material, time and effort to perform these delicate tasks of fabricating microfeatures. Simulation or numerical modeling of these EDM variants would minimize the resources involved and help in checking the feasibility of fabricating microfeatures more concisely and profoundly before performing the actual machining.

5 Conclusions

The study presented an insight into the prospect of fabrication of microfeatures using EDM. The processing ability of different variants of EDM has been discussed in details. Different microfeatures such as micro-hole, micro-rod, micro-pin, micro-fin and micro-screw fabricated using various EDM variants have also been elaborated. From the effect of process parameters, it is ascertained that higher efficiency and dimensional accuracy is achieved at higher and lower discharge energy, respectively. The challenge of tool wear in EDM variants encountered during fabrication of the microfeatures and the steps undertaken to compensate it has also been demonstrated. Finally, it can be concluded that although EDM is capable of micromachining, still further research is required to expedite the capabilities of EDM variants for the fabrication of microfeatures with higher efficiency and dimensional accuracy.

References

1. Ranjan R, Kar S, Patowari PK (2018) Parametric optimization of micro drilling on brass in micro electrical discharge machining. AIP Conf Proc 1998:20005

2. Kar S, Sarmah P, Baroi BK, Patowari PK (2020) Drilling of micro-holes in titanium using micro-EDM: a parametric investigation. In: Biswal BB, Sarkar BK, Mahanta P (eds) Advances in mechanical engineering. Lecture notes in mechanical engineering. Springer, Singapore, pp 589–600
3. Kar S, Patowari PK (2019) Parametric optimization of micro-electrical discharge drilling on titanium. In: Shunmugam MS, Kanthababu M (eds) Advances in micro and nano manufacturing and surface engineering. Lecture notes on multidisciplinary industrial engineering. Springer, Singapore, pp 201–210
4. Deka S, Kar S, Patowari PK (2020) Machinability of silicon and German silver in micro electrical discharge machining: a comparative study. Silicon. https://doi.org/10.1007/s12633-020-00496-0
5. Bhattacharya T, Singh A, Singh NP, Kar S, Patowari PK (2019) Micro drilling of stainless steel by micro electrical discharge machining and its parametric optimization. Paper presented at 11th International Conference on Precision, Meso, Micro and Nano Engineering (COPEN 2019), IIT Indore, 12–14 December 2019
6. Kuriakose S, Patowari PK, Bhatt J (2019) Effect of micro-EDM machining parameters on the accuracy of micro hole drilling in Zr-based metallic glass. Eng Res Express 2(1):15001
7. Debnath T, Patowari PK (2019) Fabrication of an array of micro-fins using Wire-EDM and its parametric analysis. Mater Manuf Processes 34(5):580–589
8. Debnath T, Patowari PK (2019) Concept development for fabricating threaded micro-pin using wire-EDM. J Brazilian Soc Mech Sci Eng 41(10):402
9. Karthikeyan G, Ramkumar J, Dhamodaran S (2014) Block EDG: issues and applicability in multiple pass μED-milling. Mach Sci Technol 18(1):120–136
10. Kar S, Patowari PK (2019) Effect of non-electrical parameters in fabrication of micro rod using BEDG. Mater Manuf Processes 34(11):1262–1273
11. Singh AK, Patowari PK, Deshpande NV (2016) Experimental analysis of reverse micro-EDM for machining microtool. Mater Manuf Processes 31(4):530–540
12. Singh AK, Patowari PK, Deshpande NV (2017) Effect of tool wear on microrods fabrication using reverse μEDM. Mater Manuf Processes 32(3):286–293
13. Singh AK, Patowari PK, Deshpande NV (2019) Analysis of micro-rods machined using reverse micro-EDM. J Brazilian Soc Mech Sci Eng 41(15):1–12
14. Kar S, Patowari PK (2019) Experimental investigation of machinability and surface characteristics in microelectrical discharge milling of titanium, stainless steel and copper. Arab J Sci Eng 44(9):7843–7858
15. Kar S, Patowari PK (2017) Machining of micro slots in titanium using micro electrical discharge milling. Paper presented at 10th International Conference on Precision, Meso, Micro and Nano Engineering (COPEN 10), IIT Madras, 06–09 December 2017

Experimental Study of Effect of Machining Parameters on PMMA in Diamond Turning

Kuldeep A. Mahajan, Raju Pawade, and R. Balasubramaniam

1 Introduction

Single point diamond turning (SPDT) is used to produce ultra-precision parts with high levels of surface finish and form accuracy. The constructional features of SPDT offer flexibility in producing a variety of complex optical components. These components are having geometry like spheric, aspheric, and freeform and generally made up of metals and polymers. In the recent past, metals are replaced by polymers in most of the optical applications due to their favorable properties and lightweight. Among from all the polymer, two commonly used are polycarbonate (PC) and polymethylmethacrylate (PMMA) because these are clear and transparent plastic materials available for the optical applications [1]. PMMA is generally used in camera lenses for cell phones, intra-ocular lenses (IOLs), optical lenses, mirrors, sunscreens, and tail-lights [2]. The controllable machining parameters are having an important role in producing high-quality surface textured optics. These parameters include tool geometry, spindle speed, feed rate, and the depth of cut [3]. Most of the research is on the flat surface, and researchers found the optimum parameters combination to reduce the surface roughness and profile errors in the optical component. The optimized parameters employed for achieving surface finish may not give good profile accuracy [4–12]. In SPDT, the geometric surface finish is controlled by spindle speed, feed rate, and tool nose radius [13], whereas profile error is due to tool alignments like de-centering, material, and tool condition like wear [14]. In the case of spherical and aspheric optics to maintain the balance of surface finish and profile is difficult

K. A. Mahajan (✉) · R. Pawade
Mechanical Engineering Department, Dr. Babasaheb, Ambedkar Technical University, Lonere 402103, India
e-mail: kamahajan@gmail.com

R. Balasubramaniam
Bhabha Atomic Research Centre, Mumbai 400085, India

H. K. Dave and D. Nedelcu (eds.), *Advances in Manufacturing Processes*, Lecture Notes in Mechanical Engineering, https://doi.org/10.1007/978-981-15-9117-4_2

task as two, the slides move continuously. The movement of two machine slides to cut a contour is known as the interpolation. The SPDT is equipped with a superior (e.g., Diffsys) computer-aided manufacturing (CAM) system for generating optic surfaces [15]. The interpolation error is occurring due to the distance between the cutting points on the profile, more the distance cause increasing in error.

In this study, an attempt has been made to evaluate the surface finish (R_a) and profile error (P_t) of deep spherical PMMA disk produced using SPDT. The circular interpolation and point-to-point interpolation are considered to determine its effect on the surface along with the machining parameters like spindle speed (n), feed rate (f), and depth of cut (d). Further, by using statistical tools, the optimum parameter combination is investigated.

2 Materials and Methods

2.1 *Diamond Turning of a Spherical Surface*

The spherical surface cutting experiments are conducted on Precitech Nanoform 200 diamond turning machine. The workpiece material used for the experiment is polymethylmethacrylate (PMMA), having the size diameter (D) 25 mm with 8 mm thickness (t). The radius of curvature (R) of the component is given as 60 mm. The parameters used to conduct the experiments are feed rates, f (4, 20 μm/rev), speeds, n (2500, 4000 rpm), and depth of cuts, d (5, 10 μm). The cutting tool used for the experiment is contour to make a single point diamond tool having a tool nose radius 0.5 mm with 10° clearance angle and 0° rake angle. The cutting tool mounted on the tool post with a tool overhang of 14 mm. The coolant is not recommended due to the rainbow effect on the workpiece, so machining is carried out under dry conditions

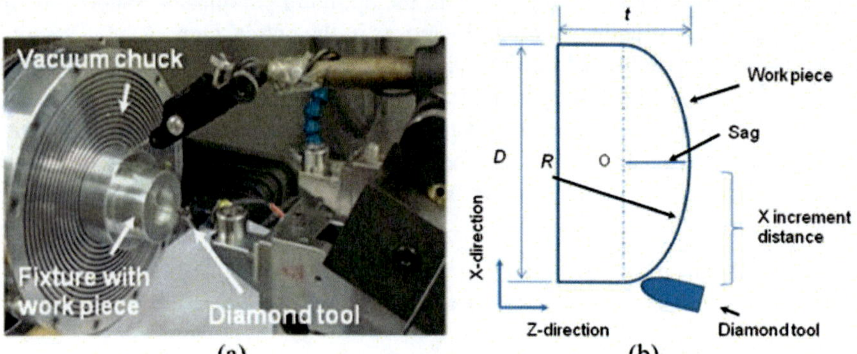

(a) (b)

Fig. 1 Diamond turn machining **a** experimental setup and b spherical cutting

[2]. The machining setup is as shown in Fig. 1a. The surface finish (R_a) and profile error (P_t) are measured with Taylor Hobson PGI talysurf series 2.

2.2 *Interpolation*

In the case of T-base two-axis SPDT machine, the X-axis slide carries the spindle and Z-axis slide carries tool holder. Both the axes have linear motion but oriented perpendicular to each other, as shown in Fig. 1a. While cutting the spherical surfaces, X- and Z-axes move simultaneously to divide the curve into several cutting points or locations so that interpolation taking place among those points. The interpolation is of two types: One is circular and another is a point-to-point. The circular interpolation requires a starting point, endpoint, and the radius of the curve. In this case, the tool moves from periphery to center of the workpiece in X-direction, whereas Z-direction moves front and back based on the concave or convex shape. According to the profile equation, the controller decides the Z-axis movement along with the feed rate and follows the profile for cutting. However, the point-to-point is carried out in the CAM system like Diffsys to decide machining points. In this case, the sag or depth of curve is divided into small equal parts for Z-axis movement based on the X-slide movement which is decided by the X-axis increment distance that ranges from 1 μm to 80 μm as shown in Fig. 1b.

2.3 *Experimental Work*

The Design of Experiment (DoE) is a scientific and systematic approach to conduct experimentation. The Taguchi L_{16} orthogonal array is considered for the experiments for studying the effect of parameters on the response. Spindle speed (*n*), feed rate (*f*), and depth of cut (*d*) are two-level input parameters, and surface finish (R_a) and profile error (P_t) are the responses as shown in Table 1. The effect of the tool nose radius is not considered in this study as the available tool is having a 0.5 mm size. The tool height and de-centering are done so that M- and W-shape errors are avoided. The single point diamond tool wear is not considered in this work.

3 Results and Discussion

3.1 *Interpolation of Error Analysis*

The sample surface roughness (R_a) plot of run order 1 and profile error (P_t) of run order 2 from Table 1 are as shown in Fig. 2a, b, respectively. It has been observed from

Table 1 L16 orthogonal array of experimental results with observed responses

Run	Spindle speed (rpm)	Feed rate (μm/rev)	Depth of cut (μm)	R_a (nm) for circular interpolation	R_a (nm) P-t-P for interpolation	P_t (μm) for circular interpolation	P_t (μm) for P-t-P for interpolation
1	2500	4	5	26	31	1.67	1.51
2	2500	4	5	28	32	1.65	1.23
3	2500	4	10	29	36	1.27	1.57
4	2500	4	10	31	35	1.34	1.38
5	4000	20	5	56	59	1.35	1.45
6	4000	20	5	55	59	1.44	1.41
7	4000	20	10	54	57	1.37	1.42
8	4000	20	10	53	57	1.35	1.31
9	2500	20	5	46	48	1.45	1.50
10	2500	20	5	46	49	1.55	1.33
11	2500	20	10	51	52	1.43	1.36
12	2500	20	10	51	52	1.44	1.45
13	4000	4	5	33	34	1.56	1.68
14	4000	4	5	33	34	1.46	1.62
15	4000	4	10	33	35	1.28	1.45
16	4000	4	10	33	35	1.34	1.44

Table 1 that the surface roughness for the circular interpolation is comparatively less than the point-to-point interpolation. In contrast, profile error is marginally less for point-to-point interpolation. The comparison of surface roughness (R_a) and profile error (P_t) for the given run order is as shown in Fig. 3. The surface roughness (R_a) is observed higher between the run order 5 and order 12, and this is due to higher feed rate. The tool nose radius used is having size 0.5 mm, and during successive feed, it leaves some material uncut, which is commonly known as a scallop.

These scallops make the peaks influencing the surface roughness (R_a), whereas low spindle speed and the low feed rate produce better surface roughness (R_a) . There is a marginal difference between the surface roughness (R_a) of two interpolations. The profile error (P_t) is quite difficult to maintain with optimized surface roughness, and sometimes with low surface roughness, it may be observed high and vice versa. The profile error (P_t) is dependent on the various parameters like tool nose radius, the coolant used, and clamping of the workpiece along with machining parameters. Here, the profile error (P_t) is observed comparatively low for point-to-point interpolation as it involves more cutting points on the profile due to the X-incremental distance. The lowest profile error (P_t) is observed with low spindle speed, low feed rate, and low depth of cut in point-to-point interpolation. As the tool nose radius is small, the profile error (P_t) is normally observed on the higher side, which is similar to reported in the literature.

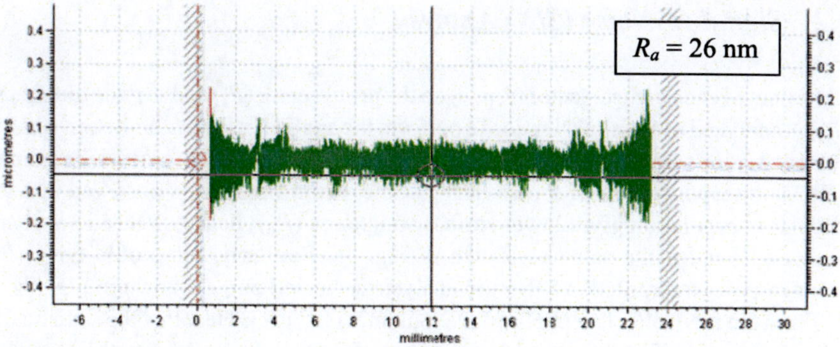

(a)

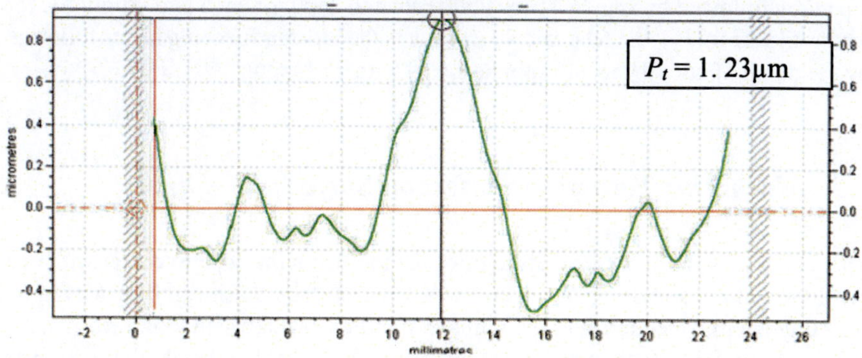

(b)

Fig. 2 Surface quality **a** lowest surface roughness (R_a) and **b** lowest profile error (P_t)

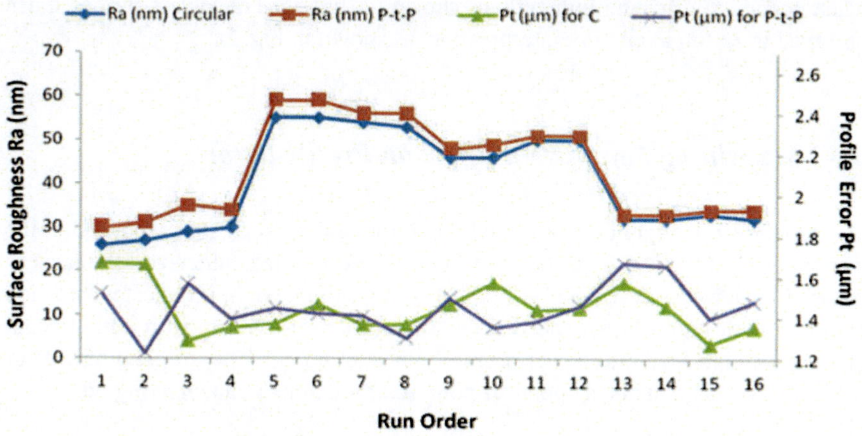

Fig. 3 Comparison of surface roughness (R_a) and profile error (P_t)

3.2 Signal-to-Noise (S/N) Analysis

The effect of each control factor (f, n, d) on the profile error (P_t) and surface roughness (R_a) is analyzed by using S/N ratio. To perform the analysis, 'lower-the-better' quality characteristic is used for the calculation of the S/N ratio. The process parameter effects on responses are as shown in Fig. 4a, b. From Fig. 3, it is observed that the circular interpolation gives lower surface roughness (R_a); hence, S/N ratio analysis is carried out for the same as shown in Fig. 4a. Similarly, for profile error (P_t) point-to-point interpolation S/N ratio analysis is carried out, as shown in Fig. 4b.

From the main effects plot of surface roughness (R_a), it is clear that the statistically most influencing factor in minimizing the surface roughness (R_a) is the feed rate 4 μm/rev followed by spindle speed 2500 rpm and depth of cut 5 μm. Similarly, for the profile error (P_t), the feed rate 20 μm/rev is the most influencing factor followed by the depth of cut 10 μm and spindle speed with 2500 rpm.

3.3 Machining Parameters Effect on Surface Roughness

The feed rate is the successive distance of the tool from the edge to center with one revolution of the spindle in X-direction. A better surface finish can be achieved by selecting the low feed rate and low spindle. It indicates that when tool travel toward the center, with low feed, it removes material with less scallop height that improves the surface roughness. The surface roughness is deteriorating when spindle speed and feed rate both are increasing, as shown in Fig. 5a. Depth of cut is not that significant on the surface roughness, but it should maintain low, as it generates heat, vibration when goes high affecting indirectly as shown in Fig. 5b. The lowest surface finish observed is 26 nm in circular interpolation, as shown in Fig. 2a.

3.4 Machining Parameters Effect on Profile Error

The profile error might be due to tool condition and residual stress induced in the workpiece. In SPDT, a small layer of the workpiece is machined, and the layer is stressed, which causes residual stress. As shown in Fig. 6a, b, it indicates that high feed rate with high spindle speed gives low profile error. Also, with a high depth of cut with high feed rate, lower profile error is observed. The lowest profile error observed is 1. 51 μm during point-to-point interpolation as shown in Fig. 2b.

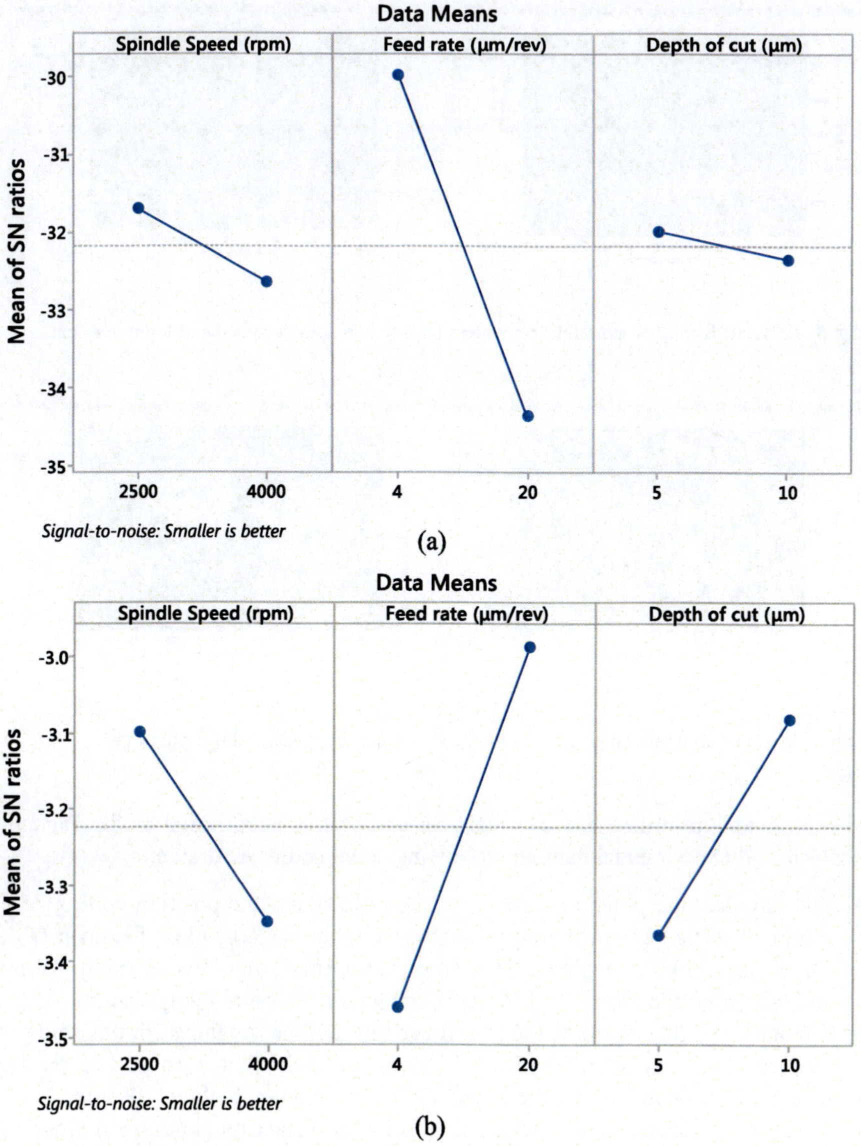

Fig. 4 Process parameter effect on S/N ratio **a** surface roughness (R_a) and **b** profile error (P_t)

4 Conclusions

In this work, an experimental investigation of the effect of machining parameters on PMMA in diamond turning has been carried out. The experiments are carried out

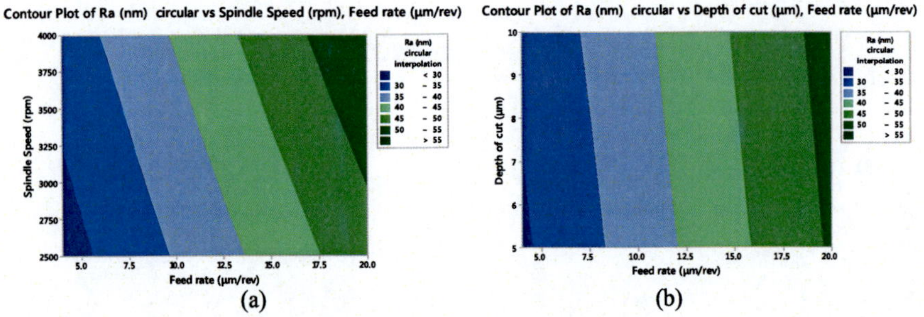

Fig. 5 Effect of feed rate on surface roughness (R_a) with **a** spindle speed and **b** depth of cut

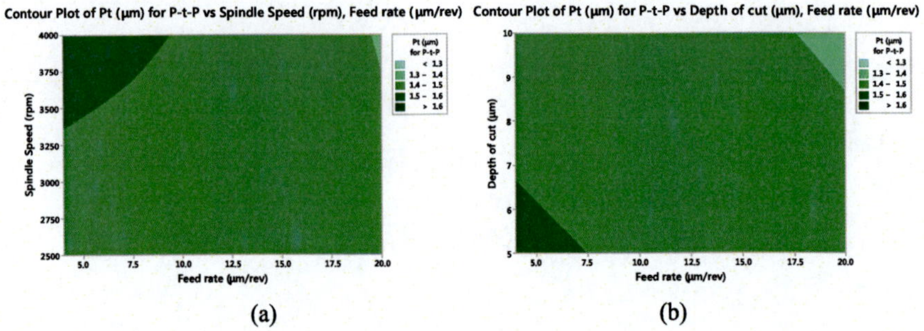

Fig. 6 Effect of feed rate on profile error (P_t) with **a** spindle speed and **b** depth of cut

to explore and minimize surface roughness and profile error. Based on the Taguchi analysis and experimental data, the following conclusions are drawn:

- The circular interpolation cutting and point-to-point interpolation cutting of a spherical surface give variation in surface roughness (R_a) and profile error (P_t). The lowest surface roughness (R_a) noted is 26 nm during circular interpolation and lowest profile error (P_t) 1.23 µm during point-to-point interpolation.
- The most suitable combination to achieve low surface roughness in this study is feed rate 4 µm/rev, spindle speed 2500 rpm and depth of cut 5 µm and for profile is feed rate 20 µm/rev, spindle speed 2500 rpm, and depth of cut 10 µm.
- The low feed rate is the most influencing factor for surface roughness (R_a), whereas high feed rate is a dominating factor for profile error (P_t) along with the low spindle speed.

Acknowledgements The authors are grateful to and would like to express sincere thanks to Mr. Vinod Mishra, Dr. Neha Khatri, Dr. Rohit Sharma, and Dr. Vinod Karar from CSIR-CSIO, Chandigarh, for collaboration work.

References

1. Mishra V, Sabui D, Burada DR, Karar V, Jha S, Khan GS (2020) Experimental investigations on slow tool servo process parameters for freeform optics machining. J Mater Manuf Process 35(7): 797–810. https://doi.org/10.1080/10426914.2020.1743849
2. Goel B, Singh S, Sarepaka RV (2016) Precision deterministic machining of polymethyl methacrylate by single point diamond turning. J Mater Manuf Process 31(14):1917–1926. https://doi.org/10.1080/10426914.2016.1140186
3. Mishra V, Sharma R, Khatri N, Garg H, Khan GS (2018) Processing of polycarbonate by ultra-precision machining for optical application. Mater Today: Proc 5(3(11)Part):25130–25138. https://doi.org/10.1016/j.matpr.2018.10.314
4. Singh H, Vaishya RO, Sing K, Mishra V, Sarepaka RV (2013) Analysis of surface roughness and waviness during diamond turning of polycarbonate. Int J Sci Res 2(6):2277–8179. https://doi.org/10.15373/22778179/JUNE2013/85
5. Khatri N, Mishra V, Sarepaka RGV (2012) Optimization of process parameters to achieve nano level surface quality on polycarbonate. Int J Comput Appl 48:39–44. https://doi.org/10.5120/7412-0507
6. Gubbels GPH (2004) Diamond tool wear when cutting amorphous polymers. Ann CIRP 53(1):25130–25138. https://doi.org/10.1016/j.matpr.2018.10.314
7. Kobayashi A, Hirakawa K (1984) Ultraprecision machining of plastics. Part 1. Polymethyl Methacrylate, Polym-Plastics Technol Eng 22(1):15–25. https://doi.org/10.1080/03602558408070029
8. Carr JW, Feger C (1993) Ultraprecision machining of polymers. Precis Eng 15(4):221–237. https://doi.org/10.1016/0141-6359(93)90105-J
9. Jagtap K, Pawade R (2014) Experimental Investigation on the Influence of Cutting Parameters on Surface Quality obtained in SPDT of PMMA. Int J Adv Des Manuf Technol 7(2):53–58
10. Olufayo OA, Hossein K A (2013) Preliminary investigation of surface finish of a contact lens polymer in ultra-high precision diamond turning. In: Proceedings—2013 6th robotics and mechatronics conference, RobMech 2013. 117–122. https://doi.org/10.1109/RoboMech.2013.6685502
11. Mahajan KA, Sadaiah M, Gawande SH (2010) Experimental investigations of surface roughness on OFHC copper by diamond turning machine. Int J Eng Sci Technol 2(10):5215–5220
12. Mahajan KA, Brahmankar PK, Balasubramanium R, Dhakad MR (2017) Experimental investigation of tool vibration effect on the surface roughness in facing operation of PMMA material. Int Rev Mech Eng (IREME) 11(10):718–724. https://doi.org/10.15866/ireme.v11i10.12847
13. Zhang SJ, To S, Wang SJ, Zhu ZW (2015) A review of surface roughness generation in ultra-precision machining. Int J Mach Tools Manuf 91:1–16. https://doi.org/10.1016/j.ijmachtools.2015.02.001
14. Lee WB, Cheung CF, Chiu WM, Leung TP (2000) An investigation of residual form error compensation in the ultra-precision machining of aspheric surfaces. J Mater Process Technol 99(1–3):129–134. https://doi.org/10.1016/S0924-0136(99)00403-3
15. Zhu Z, To S (2015) Adaptive tool servo diamond turning for enhancing machining efficiency and surface quality of freeform optics. Opt Express 23(16):20234–20248. https://doi.org/10.1364/OE.23.020234

A Comparative Study of Electro-discharge Drilling Process Using Solid and Tubular Electrodes

Sudhanshu Kumar and Harshit K. Dave

1 Introduction

Electro-discharge machining process extensively applied for machining of difficult to machine electrically conducting materials. Material is removed with the application of spark generated on tool by either melting and/or vaporization of workpiece. A very small gap is maintained by the dielectric medium between workpiece and tool electrode [1]. Since it is a noncontact kind of removal process, this machining process is widely used for thin or fragile workpiece. EDM process has been successfully used for drilling of holes in stainless steels (AISI 304) with high dimensional accuracy. Drilling of through holes is often used in food processing industries, automobile parts and plastic molding industries. The working of EDM process depends on several governing factors like peak current, pulse-on and pulse-off time, duty cycle, etc.

Literature review indicates that a good quantum of works has been published on observing of material removal rate (MRR), tool wear rate (TWR) and surface roughness. Majority of the works are focused on obtaining the appropriate parametric combinations for maximum MRR with minimum TWR and surface roughness. However, very limited works have been reported on EDM process with solid as well as tubular electrodes except in a few. In previous work, Dave et al. [2] observed that flushing pressure and gap voltage are insignificant parameters in determining the MRR, TWR and surface roughness using solid and tubular tool electrodes. In another work, Dave et al. [3] performed machining with bundled tool electrode considering lower values of peak current. Authors observed that MRR with solid tool electrode is higher than bundled tool electrode. This may be because of orbital tool movement

S. Kumar (✉)
Department of Mechanical Engineering, MANIT, Bhopal, India
e-mail: sudhanshuk27@gmail.com

H. K. Dave
Department of Mechanical Engineering, SVNIT, Surat, India

H. K. Dave and D. Nedelcu (eds.), *Advances in Manufacturing Processes*, Lecture Notes in Mechanical Engineering, https://doi.org/10.1007/978-981-15-9117-4_3

of bundled electrode. Orbital movement was required to remove the center material with hollow tool shape in case of blind hole. Orbital movement of tool electrode increases the total machining time which reduces the MRR. In EDM process, the tool shape also affects the production cost. Gu et al. [4] performed electro-discharge drilling using solid and tubular tool electrode at very high peak current ranging from 40 to 127 A. It was observed that tubular tool sustained the higher current than solid tool electrode. Tubular tool yielded much higher MRR with relatively lower TWR. In EDM process, it can be observed that the rate of material erosion has increased with the improvement in debris removal rate from the machining zone [5]. In order to improve the flushing between the tool and workpiece, Murugesan et al. [5] tried a tool electrode with number of small holes. Authors observed that machining time was reduced in case of multi-hole tool electrode because of better flushing condition. Flushing conditions also depend upon the tool design. Ojha et al. [6] checked the performance of three different tool electrodes with varying edge angles keeping constant cross-sectional area. It was obtained that MRR increases with increase in edge angle of tool due to improved flushing. Yilmaz et al. [7] obtained the tool electrode with single channel that produces more MRR than multichannel tool electrode. Mohan et al. [8] compared the MRR observed using solid and tubular tool rotating during EDM process. It was reported that rotating tubular tool produced more MRR than that of rotating tubular tool electrode. Recently, Selvarajan et al. [9] used hollow brass tool for drilling of through holes in conductive ceramics. Sultan et al. [10] reported that hollow tool experience less tool wear during drilling of holes in EDM process.

The majority of research works are found about drilling of blind hole at smaller depth. With increase in depth of cavity, flushing of dielectric fluid becomes difficult that may affect the machining performance. In the present research work, deep through holes (20 mm) have been drilled on AISI 304 steel using tubular and solid tool electrodes. The input parameters, namely peak current and pulse-on time, are most important input parameters in EDM process [11, 12]; hence, pulse-on time and peak current are picked for the present investigation. Results have been analyzed and compared for solid and tubular tool electrodes.

2 Experimentations

Experiments have been conducted keeping one factor constant and other factors varied at a time. Each factor has been varied at three different levels. During the experimentations, pulse-off time has been selected keeping duty factor constant at 0.7. Each experiment is repeated twice, and average of both results has been used to analyze and compare the responses.

Fig. 1 Flushing of dielectric fluid **a** jet flushing **b** through flushing

2.1 Experimental Setup

Experimentations have been done using EDM machine of Joemars make. Electrolytic copper has been used as tool material and AISI 304 as workpiece material. Two types of tools, i.e., solid (9.5 mm diameter) and tubular (9.5 outer diameter and 6.5 inner diameter), have been used to drill through holes of 20 mm deep. In case of machining with solid tool electrode jet flushing and with tubular tool electrode through flushing of dielectric fluid has been used, as illustrated in Fig. 1.

2.2 Input and Response Parameters

Peak current and pulse-on time are input parameters considered at three different values, as shown in Table 1.

Material removal rate (MRR), tool wear rate (TWR) and surface roughness (Ra) are response parameters. MRR and TWR have been obtained using weight difference methods that are mathematically expressed as follows.

$$\text{MRR or TWR} = \frac{\text{weight before machining} - \text{weight after machining}}{\text{machining time} \times \text{density of material(workpiece or tool)}} \quad (1)$$

Surface roughness has been evaluated using Mitutoyo Surftest instrument and is expressed as mean surface roughness (Ra) value. Ra value has been evaluated at six

Table 1 Input parameters with values

Parameter	Unit	Values
Peak current (Ip)	A	13, 21, 28
Pulse-on time (ton)	μs	100, 210, 315

different sections of machined cavity, and average of all values has been considered as surface roughness of cavity.

3 Results of Experiments

The main aim of present work is to compare the electro-discharge drilling process performed using solid and tubular tool electrodes with respect to MRR, TWR and Ra. Table 2 shows the measured value of MRR, TWR and Ra.

3.1 *Analysis of MRR*

Figure 2a represents the deviation of MRR (mm^3/min) with respect to peak current (A) at different pulse-on time machined using solid and tubular tool electrode. It can be noted that MRR rises with rise in peak current. At lower peak current, i.e., at 13 A, MRR is lowest for all selected pulse-on time, and MRR is highest at higher peak current, i.e., at 28 A. The increment of MRR with peak current may be because of increased spark energy in working gap during machining. Higher discharge energy erodes more material per cycle from workpiece.

The experimental results plotted in Fig. 2a indicate that MRR calculated for the hole machined with tubular tool electrode is greater than that with solid tool electrode. Highest MRR has been observed with tubular tool electrode at 28 A peak current and 100 μs pulse-on time which is 70% more than the solid tool electrode at same parameter settings. Tubular tool electrode has also produced more MRR at lower peak current and pulse-on time (13 A and 100 μs) setting than the solid tool electrode. Tubular tool electrode yields 60% more MRR than solid tool electrode at lower setting

Table 2 Experimental results

S. No.	Parameters		Tubular tool electrode			Solid tool electrode		
	Ip	ton	MRR	TWR	Ra	MRR	TWR	Ra
1	13	100	27.014	1.6211	7.885	10.642	0.17	6.430
2	13	210	18.084	0.3669	10.295	14.713	0.08	9.130
3	13	315	11.787	0.2882	11.695	18.271	0.05	10.160
4	21	100	59.737	11.5147	9.905	14.557	0.55	7.025
5	21	210	30.452	4.6905	13.225	26.359	0.26	11.155
6	21	315	35.190	0.5126	13.555	21.682	0.13	12.340
7	28	100	70.083	12.3496	13.285	20.718	1.76	8.505
8	28	210	50.168	6.1874	12.930	23.757	0.41	12.545
9	28	315	60.786	16.8474	10.515	29.397	0.31	14.750

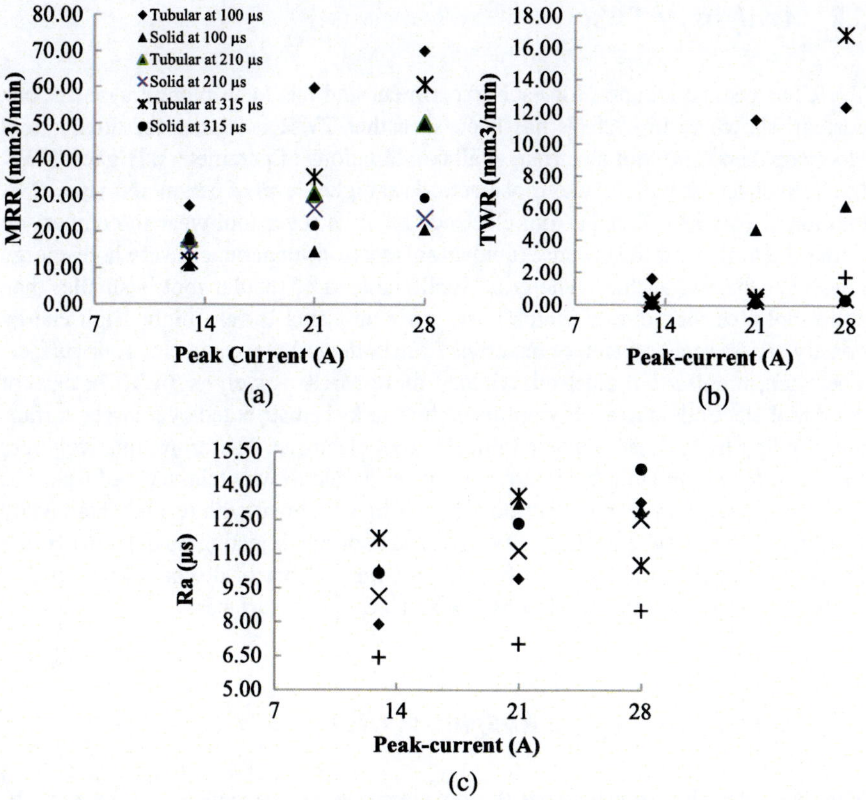

Fig. 2 Effect of peak current on **a** MRR **b** TWR and **c** Ra at different pulse-on time

of parameters. Further, it has been noted that tubular tool electrode has consumed less machining time for drilling of through hole in EDM process than solid tool electrode. The reason of greater MRR with tubular tool electrode may be because of smaller area of erosion than solid tool electrode. In case of solid tool electrode, surface area or area to be machined is larger than tubular tool electrode. Therefore, tubular tool electrode has to remove relatively less material whereas solid tool electrode has to remove all materials equal to cross-sectional area of tool. Better flushing of dielectric fluid may be the other reason for higher MRR in case of machining with tubular tool electrode. As presented in Fig. 1, dielectric fluid is more focused (with tubular electrode) at machining point because of flushing. Better flushing plays a crucial role in EDM process because it helps to remove eroded particles effectively. Flushing becomes more critical in case of deep hole drilling. The obtained results are in line with the observation reported by the researchers Gu et al. [4] in which MRR was observed to be higher with tubular tool. MRR is also found to be increasing with increasing peak current. Similar trend is reported by researchers Dave et al. [13]. MRR has been found minimum at higher value of pulse-on time.

3.2 Analysis of TWR

TWR has been calculated for each experiment and has been plotted against peak current, shown in Fig. 2b. It has been seen that TWR is higher for tubular tool electrode than solid tool electrode at all combinations of parameters. Highest TWR has been observed with tubular tool electrode at higher level of parameters (Ip = 28A and ton = 315 μs). Tubular tool electrode exhibits sever tool wear in comparison with solid tool electrode because tubular tool electrode experiences very high energy density at the end surface. The cross-sectional area of tubular tool is smaller than solid tool electrode due to which energy per unit area is very high. High energy density causes rapid erosion of materials from both workpiece and electrode surface. Therefore, tubular tool electrode yielded more MRR as well as TWR. In case of solid tool electrode, the energy produced by sparks is distributed over larger surface area, and partially, it is conducted through tool electrode. Therefore, relatively less spark energy is available for removal of material due to which lesser tool wear has been observed on solid tool electrode. This is in agreement with results obtained by Ozgedik et al. [14] where higher TWR has found during injection flushing than other flushing because in the injection flushing a large number of discharge-type pulses occur which increase the burning of tool and workpiece surfaces.

3.3 Analysis of Surface Roughness (Ra)

The effect of peak current between the tool electrode and workpiece on surface roughness has been depicted in Fig. 2c. It can be observed from Fig. 2c that surface roughness of machined cavity is significantly affected by peak current in EDM process. Ra increases with increase in peak current. Lower value of peak current produces sparks with lesser discharge energy which leaves smaller crater on machined surface. Due to smaller and sallower craters, lower surface roughness has been observed. Effect of pulse-on time on surface roughness can also be observed in Fig. 2c. Surface roughness of machined cavity is increasing with increase in pulse-on time with both types of tool (solid and tubular) electrode. Longer pulse-on time implied spark will sustain for longer time that makes deeper craters. Further, it has been noted that surface roughness obtained is relatively higher with tubular tool electrode.

4 Conclusions

An experimental comparative study has been done on electro-discharge drilling process of through holes using two different types of tool, namely solid and hollow. It has been found that MRR, TWR and Ra are higher for tubular tool electrode. Tubular tool electrode has produced 60–70% more MRR than solid tool electrode. The increase in MRR is observed with peak current for both of the tool electrodes, solid and tubular. TWR and surface roughness have also been observed higher with tubular tool electrode. Surface roughness of machined cavity has been found increasing with increase in peak current. From this research work, it can be noted that machining rate can be increased in electro-discharge drilling process with tubular tool electrode but at the cost of higher tool wear and more surface roughness. Therefore, tubular tool electrode may be suitable for rough machining.

References

1. Ghosh A, Mallik A (1986) Manufacturing science. Ellis Horwood
2. Dave HK, Kumar S, Raval HK (2012) Electro discharge drilling of through holes under different conditions. Paper presented at 4th International & 25th All India Manufacturing Technology, Design and Research Conference (AIMTDR12) Jadavpur University, Kolkata, 14–16 December 2012
3. Dave HK, Kumar S, Rana NC, Raval HK (2014) Electro discharge machining of AISI 304 using solid and bundled electrodes. Paper presented at 5th International & 26th All India Manufacturing Technology, Design and Research Conference (AIMTDR 14), IIT Guwahati, 14–16 December 2014
4. Gu L, Li L, Zhao W, Rajurkar KP (2012) Electrical discharge machining of Ti6Al4V with a bundled electrode. Int J Mach Tools Manuf 53(1):100–106
5. Murugesan S, Balamurugan K, Narayanan CS, Venkatakrishnan PG (2012) Study on EDM of Al-15% Sic MMC using solid and multihole Electrodes—a Taguchi approach. Eur J Sci Res 68(2):161–171
6. Ojha K, Garg RK, Singh KK (2011) Parametric optimization of PMEDM process using chromium powder mixed dielectric and triangular shape electrodes. J Minerals Mater Character Eng 10(11):1087–1102
7. Yilmaz O, Okka MA (2010) Effect of single and multi-channel electrodes application on EDM fast hole drilling performance. Int J Adv Manuf Technol 51(1–4):185–194
8. Mohan B, Rajadurai A, Satyanarayana KG (2004) Electric discharge machining of Al–SiC metal matrix composites using rotary tube electrode. J Mater Process Technol 153:978–985
9. Selvarajan L, Mouri P, Ramesh RR (2018) Experimental investigation of EDM parameters on machining Si_3N_4-TiN conductive ceramic composite using hallow tube electrode for improving geometrical accuracy. Mater Today. Proc 5(2):8080–8088
10. Sultan T, Kumar A, Gupta RD (2014) Material removal rate, electrode wear rate, and surface roughness evaluation in die sinking EDM with hollow tool through response surface methodology. Int J Manuf Eng. https://doi.org/10.1155/2014/259129
11. Li L, Gu L, Xi X, Zhao W (2012) Influence of flushing on performance of EDM with bunched electrode. Int J Adv Manuf Technol 58(1–4):187–194
12. Wong YS, Lim LC, Lee LC (1995) Effect of flushing on electro-discharge machined surfaces. J Mater Process Technol 48 (1–4):299–305
13. Dave HK, Desai KP, Raval HK (2011) Effect of the orbital tool movement on the material removal rate during electro discharge machining. Paper presented at international conference on advances and trends in engineering materials and their Applications, Montrel, Canada, 1–5 August 2011
14. Ozedik A, Cogun C (2008) An experimental investigation of tool wear in electric discharge machining. Int J Adv Manuf Technol 27(5–6):488–500

Optimization of Process Variables in Plasma Arc Machining of Inconel-718 Alloy Using Taguchi with Grey Relational Analysis

Khushboo Sharma, Jai Kishan Sambharia, and Alok Khatri

Abbreviations

PAM	Plasma arc machining
PAC	Plasma arc cutting
MRR	Material removal rate
SR	Surface roughness
HAZ	Heat-affected zone
GRA	Grey relational analysis
GRG	Grey relational grade
OA	Orthogonal array
ANOVA	Analysis of variance

1 Introduction

Inconel-718 is a nickel-based precipitation-solidifying amalgam intended for extremely high return quality, rigidity and creep crack properties at temperature till 13000 F (705 °C). Inconel-718 materials are oxidation-consumption safe materials suitable for extended conditions exposed to weight and warmth. Inconel 718 contains

K. Sharma (✉) · J. K. Sambharia (✉) · A. Khatri
Mechanical Engineering Department, Engineering College Ajmer (an Autonomous Institution of Govt. of Rajasthan), Ajmer, Rajasthan 305025, India
e-mail: sharmak9027@gmail.com

J. K. Sambharia
e-mail: jsambharia@gmail.com

A. Khatri
e-mail: alok.khatri@ecajmer.ac.in

H. K. Dave and D. Nedelcu (eds.), *Advances in Manufacturing Processes*, Lecture Notes in Mechanical Engineering, https://doi.org/10.1007/978-981-15-9117-4_4

nickel components 55–55%, chromium 17–21% and the base material tantalum 0.005% [1]. Inconel-718 utilized in the GE airplane motors with application in the airfoils, the critical turning parts, the supporting structures and the weight vessels. Inconel-718 is also used in fabrication of cryogenic tank. Inconel-718 is regularly utilized for jet motor, gas turbine tasks, rocket engine, manufacturing aircraft engine (blades sheet dices), and the aerospace applications [2]. Due to high hardness, high rigidity and creep crack properties at 705 °C temperature Inconel-718 materials, it is difficult to process with traditional machining process. So, thermal energy-based cutting processes are widely used, and mainly plasma arc machining (PAM) process is used for cutting such materials. Plasma arc machining is widely used in industries to overcome the cutting cost and increase the productivity.

A schematic sketch of PAC method is shown in Fig. 1, arc is made among electrode and workpiece, and then it is confined by means of a fine-bore, copper nozzle. This will increase the temperature and speed of the plasma originating from the nozzle. The plasma temperature is more than 20,000 °C, and also the speed will increase till the speed of sound. In metal cutting activities, the plasma flow is expanded, so the profoundly infiltrating plasma stream slices in the material and liquid material is expelled in the efflux plasma. In the nozzle, principally a pilot arc is generated amid the negative anode (cathode) and the workpiece. The energy of arc results in ionization of the cutting gas. At that point, the gas enthalpy increases suddenly, and it transforms into plasma state. The enthalpy of gas then increases suddenly, and it becomes a plasma state. A mixture of plasma and hot gas is used to flow through the narrow passage of torch and high velocity jet of high temperature ionized gas is produced. Outside the nozzle, the supersonic stream crest strikes to the workpiece, rapidly liquefying the metal, and the liquid material flows through the kerf. Shield gas assists in the cutting zone, is helping the cooling of the torch and is contracting the plasma jet. Experiments are conducted using four input variables and two output variables which are significantly affecting the cutting quality.

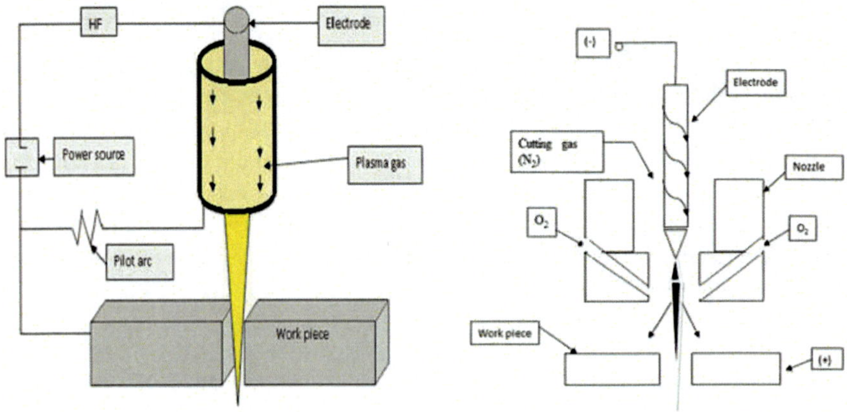

Fig. 1 Schematic diagram of plasma arc machining

Das et al. [3] optimized process variables for PAC of EN31 steel using GRA. Author also observed material removal rate and surface roughness. Aldazabal et al. [4] observed HAZ by using mini-tensile test for observed micro-hardness. Author also performed the measurement of residual and examined X-ray result. Maity and Bagal [5] studied the MRR, avg. surface roughness, chamfer, dross and kerf of AISI 316 stainless steel using plasma arc machining by GRA and response surface methodology (RSM). Ozek et al. [6] investigated surface roughness (SR) of AISI 4140 steel using plasma arc cutting (PAC). Also fuzzy rule modelling was used for estimation of surface roughness. Ilii et al. [7] investigated the SR of an austenitic stainless steel sheet, and (AISI 304) was cut using different conditions and also determines the most influencing parameters on R_a. Ramakrishan et al. [8] investigated the SR, kerf width and HAZ of SS321using PAC by Taguchi and genetic algorithm (GA). The author observed best value of SR. Hutchinson et al. [9] observed HAZ, toughness and SEM-EBSD method. Author also compared the result with hardness value. Lazarevic [10] investigated the kerf, surface roughness (SR) and bevel angle of stainless steel X10CrNiMn-16-10-2 using PAC. Hatal et al. [11] contemplated the plan and assessment of plasma cutting mechanical procedure. Impact of innovative factors on SR parameters avg. surface roughness of the steel surface ISO Fe510 had been assessed utilizing arranged analyses. Bhuvenesh et al. [12] were used to cut standard AISI 1017 steel based on the selected based parameters value. The capability and condition of the plasma cutting process air pressure (bar), cutting current (A), cutting speed (mm/min) and gap (mm) four parameters were used. Shiva Teja et al. [13] contemplated the different parameters which were on plasma circular segment cutting procedure while machining mellow steel. The trials were led utilizing Taguchi L16 symmetrical cluster with current, voltage, speed, plate thickness as the input parameters, surface unpleasantness and kerfs as the reaction factors. Ananth kumar et al. [14] contemplated the measurement exactness and surface nature of PAC with basic components, for example MMR, kerf at top, base surface, HAZ and execution of plasma cut surface. The impact of PAC parameters such bend current, cutting pace, SOD and gas pressure assessed the part quality attributes, for example MRR, KT and HAZ of Monel-400 super composite (TOPSIS). Naik and Maity [15] considered the impact of procedure parameter and dimensional exactness of PAC process. The info procedure parameters were chosen as bend voltage, stand-off separation and cutting pace. A rectangular plate of 304L hardened steel of 10 mm thickness was taken for the investigation as a workpiece. Adakarsan et al. [16] examined the quality attributes of the cut surface which were evaluated by estimating the surface unpleasantness and kerf width while cutting the 304L hardened steel. The exploratory preliminaries were planned by Taguchi L18 orthogonal cluster dissimilar to the focal composite structure utilized with customary reaction surface technique (RSM). Pitayachaval and Sato [17] utilized 588GrA, 242 Type I metal 10 mm thickness. Explored were three elements such as cutting current, cutting pace and gas compel that effect to wear of plasma spout. Computerized magnifying lens used to quantify distances across of spout. Sovan et al. [18] used the plasma for cutting round portion of AISI 304 steel. The investigation had been performed by taking three system parameters and three factor levels for each parameter. Peko et al. [19] used kerf width in PAC process developed artificial neural network (ANN) model. Used aluminium was sheet of 3 mm thickness. The cuts were performed to use of compressed air as plasma gas. Outlet diameter of nozzle was 1.2 mm and plasma gas pressure 6 bar.

2 Material and Methodology

On the bases of literature review, most of the researchers have worked on stainless steel and other grades of stainless steel and mild steel. From literature review, there are very less studies which are observed on Inconel-718 material with composition in Table 1, so Inconel-718 material is selected for experimental study. Experimental studies are conducted on a nickel-based alloy Inconel-718 material of size 200 mm × 100 mm × 8 mm (Fig. 2) of rectangular shape using PAC. As shown in Table 1, nickel percentage is highest, i.e. 55–55%, which will be responsible for the hardness of materials. [1] Plasma arc machine of Model Hypertherm Powermax 125 is used for plasma metal cutting and gouging framework for handheld cutting 38 mm and motorized puncturing 25 mm (Fig. 4).

For measurement of surface roughness, the surface roughness measuring instrument (SURFTEST SJ-210) is used before and after the experiment. Digital microscope is used to capture high resolution images of workpiece surface to identify the heat-affected zone and changes in microstructure on cutting surface. The high precision weight machine is used to measure weight after and before cutting process of the specimens. Material removal rate is defined as the proportion of weight variance of the workpiece before and after machining to the machining, and it is denoted by the unit gm/sec. The below expression shown in Equation 1 shows the formula for the MRR.

$$\text{MRR} = \frac{\text{Weight difference(After and before machining)}}{\text{Toal time taken for cutting}} \tag{1}$$

2.1 Experimental Setup

During the experiment, the CNC plasma arc machine is used as shown in Fig. 4 that is fully programmable at different cutting conditions and parameters. The movable drawers attached with the coordinate table which deliver fast and effective removal of waste material. Cutting speed range depends on thickness of workpiece and the machine cuts 1.5 times of thickness. Plasma arc machining process is mostly used to cut stainless steel (SS) and mild steel (MS) material and the optimum value of variables are cutting speed 1800 mm/s, cutting current 105 A, voltage 152 V, and SOD 0.5 mm. The machine workplace was integrated with a ventilation system with exhaust fan to exhale the smoke and provide fresh air to machining zone.

Table 1 Chemical composition of Inconel-718

Element	C	Mg	P	S	Cr	Ni	Mo	Nb	Ti	Al	Co	B	Ta	Cu	Fe
Percentage	0.08 max	0.35 max	0.015 max	0.05 max	17–18	50–55	2.80–3.30	4.75–5.50	0.65–1.15	0.20–0.80	1.00 max	0.006 max	0.05 max	0.30 max	Bal.

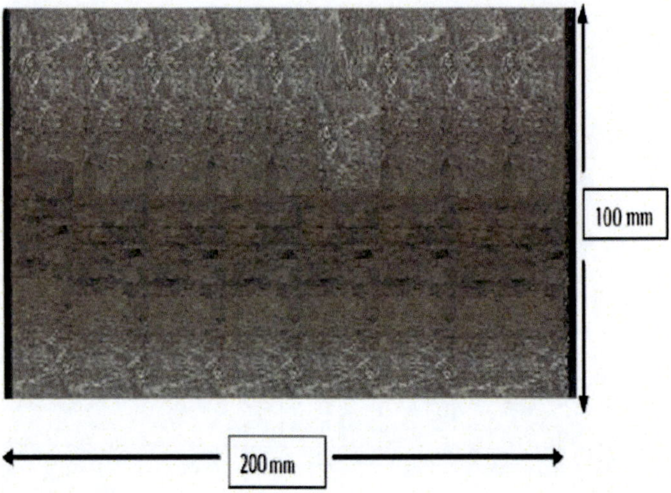

Fig. 2 Inconel-718 workpiece

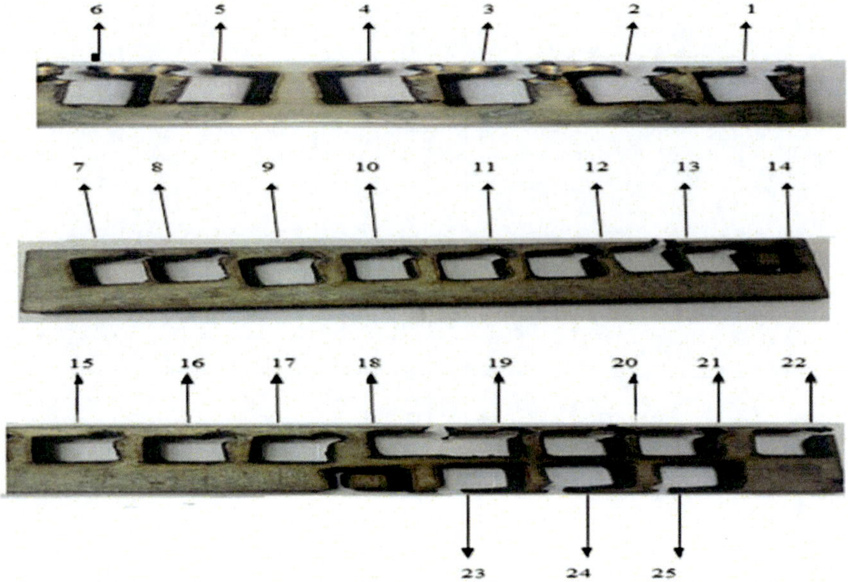

Fig. 3 Run 25 experiments on Inconel-718

2.2 Design of Experiment (DOE)

The optimization started by conducting an investigation on the condition that is essential for a better cut quality. For this purpose, many experiments were conducted.

Fig. 4 View of the PAC experimental setup

The four variables or factors and the five levels are potentially affecting the quality of the parts. Minitab-16 software was used in this research for planning the L25 experimental design. Five levels were considered for four of the factors (1 = low and 4 = high). The method of fitting the regression model related the response to the factor levels are shown in Table 3). The orthogonal array L25 was used to diminish the number of runs and still obtains the maximum information which allows easy interpretation of results. Among the orthogonal array approaches, an L-25 orthogonal array is selected augmented. There were four factors that investigated this test: current ampere, cutting speed, SOD and voltage as shown in Table 2. The current ampere (102, 104, 105, 107, 110 A), cutting speed (1200 1400, 1600, 1800, 2000 mm/min), SOD (0.2, 0.3, 0.4, 0.5, 0.6 mm) and voltage (152, 153, 154, 155, 156 V) were applied to cut metal (Inconel-718) 8 mm thickness. The specimen size is 100 mm × 200 mm × 8 mm. The cutting parameter establishes as presented in Table 3.

Table 2 Input and output parameters

Input parameters	Output parameters
1. Cutting speed—1200 to 2000 (mm/min)	1. Materials removal rate (MRR)
2. Stand-off distance (SOD)—0.2 to 0.6 (mm)	
3. Voltage—152 to 157 (V)	2. Materials removal rate (MRR)
4. Current—102 to 110 (A)	

Table 3 Parameters and their respective levels

Parameter	Units	Level-1	Level-2	Level-3	Level-4	Level-5
Speed	mm/min	1200	1400	1600	1800	2000
SOD	mm	0.2	0.3	0.4	0.5	0.6
Current	A	102	104	105	107	110
Voltage	V	152	153	154	155	156

2.3 Examination of Grey Relational Grade

In current study, MRR and surface roughness are optimized as a multi-response optimization problem. Grey relation analysis with Taguchi method is used to multi-response optimization. The experimental results are obtained from roughness of surface and rate of material removal. The results are analysed with Taguchi investigation with grey relation grade which is acquired by following arrangement of calculations.

Step-I Normalization (grey relational generation)

The initial step is the grey relational generation in which the consequences of the analyses are standardized in scope of 0–1. Equation (2) stands for normalizing of MRR values, higher-the-best (HB), and Eq. (3) stands for normalizing of R_a values, lower-the-best (LB).

$$A = \frac{\text{max} - \text{actual}}{\text{max} - \text{min}} \tag{2}$$

$$B = \frac{\text{actual} - \text{min}}{\text{max} - \text{min}} \tag{3}$$

Step-II Grey relational coefficients

Grey relation coefficients are determined and express connection between the perfect and the genuine trial results. The grey relational coefficient can be determined as:

$$\text{GRC} = 1 - \text{actual} \tag{4}$$

Step-III Grey relational grade

The multi-response MRR and surface roughness are converted into overall response of the grey relational grade. The $\Delta = 0.5$ is used to calculate grey relational coefficients.

$$\text{GRG} = \frac{\text{min} + \Delta}{\text{actual} + \Delta} \tag{5}$$

Step-IV Overall grey relational grade

The average of grey relational coefficient is presented in the overall grey relational grade. In this whole analysis, multi-response optimization problem is transformed into single response optimization problem.

Step-V Rank

The higher value of grey relational grade is stronger relational degree between the ideal and given sequences. The optimum value is obtained at cutting speed 1400 mm/s, SOD 0.3 mm, cutting current 105A and voltage 155 V. The maximum value of MRR is 5.4 gm/s, and the minimum value of R_a is 13.55 μm. Superior value of grey relational grade impels stronger relational degree. Table 5 shows the experimental result for the grey relational grade and their ranks. As a result, multi-response optimization problem is transformed into single response problem.

3 Result and Discussion

3.1 Machining on Inconel-718 Workpiece

The Taguchi L_{25} orthogonal approach (OA) is used to design 25 set of experiments as shown in Table 4. All 25 experiments are run on specimen material as shown in Fig. 3. Experiments are directed on the CNC plasma arc cutting with computer-controlled cutting machine. Stand-off distance maintained between the torch tip and workpiece for best cut quality. The response variables are material removal rate (MRR) and surface roughness. The investigational results for MRR and surface roughness are including in Table 5.

The result may be expressed in terms of either the S/N ratio or the mean. The response table for the mean of grey relation grade is shown in Table 6, the relative magnitude of the effects which contains ranks based on delta statistics. The delta statistic is the maximum average for each factor minus lowest average for the same. Ranks are given on the basis of delta values.

3.2 Regression Analyses

Regression is a well-known statistical method to investigate between some independent variables and one dependent variable. It is very useful and probably the most used statistical technique for developing empirical models. The best-fitting curve has to obtain for a dependent variable in a multi-dimensional space, with all autonomous variables being a dimension. This is also used to recognize the autonomous variables that are correlated to the dependent variable and to investigate the forms of these relationships. The linear regression equations for MRR and surface roughness are specified below.

Table 4 Experimental design and measured response

Expt. No.	Speed (mm/sec.)	SOD (mm)	Current (A)	Voltage (V)	MRR (gm/sec.)	R_a (μm)
1	1200	0.2	102	152	4.0000	21.670
2	1200	0.3	104	153	4.2000	18.320
3	1200	0.4	105	154	4.2000	17.400
4	1200	0.5	107	155	4.2000	16.300
5	1200	0.6	110	156	3.8000	17.890
6	1400	0.2	104	154	4.2000	19.270
7	1400	0.3	105	155	4.2000	17.840
8	1400	0.4	107	156	4.2440	17.040
9	1400	0.5	110	152	4.2000	20.600
10	1400	0.6	102	153	4.6190	15.270
11	1600	0.2	105	156	4.2000	21.828
12	1600	0.3	107	152	4.2000	18.080
13	1600	0.4	110	153	4.2634	20.870
14	1600	0.5	102	154	4.4000	18.830
15	1600	0.6	104	155	4.2000	16.880
16	1800	0.2	107	153	4.1570	21.230
17	1800	0.3	110	154	4.2000	21.880
18	1800	0.4	102	155	4.0000	22.283
19	1800	0.5	104	156	4.2000	19.910
20	1800	0.6	105	152	4.1632	20.640
21	2000	0.2	110	155	4.4000	21.660
22	2000	0.3	102	156	4.0000	27.218
23	2000	0.4	104	152	4.0394	23.590
24	2000	0.5	105	153	4.0000	24.010
25	2000	0.6	107	154	3.9519	26.800

$$\begin{aligned}\text{MRR} = &-110.4 - 0.000354\text{speed} + 69.5\text{SOD} + 0.990\text{current} \\ &+ 0.646\text{voltage} - 0.3790\text{SOD} * \text{current} - 0.1915\text{SOD} * \text{voltage} \\ &- 0.00546\text{current} * \text{voltage}\end{aligned} \tag{6}$$

$$\begin{aligned}\text{SR} = &-608 - 0.1154\text{speed} - 372\text{SOD} + 4.76\text{current} \\ &+ 6.95\text{voltage} + 0.001193\text{speed} * \text{current} + 6.047\text{SOD} * \text{current} \\ &- 1.74\text{SOD} * \text{voltage} - 0.0592\text{current} * \text{voltage}\end{aligned} \tag{7}$$

These equations are depended on input parameters speed, SOD, current and voltage. These equations are single response multivariable equations.

Table 5 Grey relational grades for response factors

Ex. No	Experimental		Normalized		Grey relational coefficient		Grey relational grade		Overall GRG	Rank
	MRR	R_a	MRR	R_a	MRR	R_a	MRR	R_a		
1	4	21.67	0.9	0.502786	0.1	0.497214	0.833333	0.501397	0.667365	10
2	4.2	23.03	0.8	0.586997	0.2	0.413003	0.714286	0.547643	0.630964	12
3	4.2	25.6	0.8	0.74613	0.2	0.25387	0.714286	0.663244	0.688765	9
4	4.2	20.33	0.8	0.419814	0.2	0.580186	0.714286	0.462883	0.588585	13
5	3.8	17.89	1	0.268731	0	0.731269	1	0.406085	0.703042	8
6	4.2	19.27	0.8	0.35418	0.2	0.64582	0.714286	0.436369	0.575327	15
7	4.2	17.84	0.8	0.265635	0.2	0.734365	0.714286	0.405066	0.559676	18
8	5.4	13.55	0.2	0	0.8	1	0.384615	0.333333	0.358974	25
9	4.2	16.88	0.8	0.206192	0.2	0.793808	0.714286	0.386456	0.550371	19
10	5	15.27	0.4	0.106502	0.6	0.893498	0.454545	0.358809	0.406677	24
11	4.2	14.18	0.8	0.039009	0.2	0.960991	0.714286	0.342234	0.52826	21
12	4.2	18.08	0.8	0.280495	0.2	0.719505	0.714286	0.410003	0.562144	17
13	5.8	24.23	0	0.6613	1	0.3387	0.333333	0.596161	0.464747	23
14	4.4	18.83	0.7	0.326935	0.3	0.673065	0.625	0.426234	0.525617	22
15	4.2	16.88	0.8	0.206192	0.2	0.793808	0.714286	0.386456	0.550371	19
16	3.8	21.23	1	0.475542	0	0.524458	1	0.488063	0.744031	5
17	4.2	28.84	0.8	0.946749	0.2	0.053251	0.714286	0.903749	0.809018	2
18	4	24.49	0.9	0.677399	0.1	0.322601	0.833333	0.607828	0.720581	7
19	4.2	19.91	0.8	0.393808	0.2	0.606192	0.714286	0.452001	0.583143	14
20	4	25.46	0.9	0.737461	0.1	0.262539	0.833333	0.655704	0.744519	4

(continued)

Table 5 (continued)

Ex. No	Experimental		Normalized		Grey relational coefficient		Grey relational grade		Overall GRG	Rank
	MRR	R_a	MRR	R_a	MRR	R_a	MRR	R_a		
21	4.4	21.66	0.7	0.502167	0.3	0.497833	0.625	0.501086	0.563043	16
22	4	29.7	0.9	1	0.1	0	0.833333	1	0.916667	1
23	4.2	23.59	0.8	0.621672	0.2	0.378328	0.714286	0.569263	0.641775	11
24	4	27.11	0.9	0.839628	0.1	0.160372	0.833333	0.75715	0.795241	3
25	4.2	26.8	0.8	0.820433	0.2	0.179567	0.714286	0.735763	0.725024	6

Table 6 Grey relation grade for response levels

Level	Speed	SOD	Current	Voltage
1	0.655744	2.883283	3.017981	3.166174
2	0.490205	3.478469	3.236907	3.041662
3	0.526228	3.143138	3.316461	3.323751
4	0.720258	2.978759	2.978759	2.982255
5	0.72835	3.129634	3.252202	3.090087
Delta	0.238145	0.595186	0.273444	0.341496
Rank	4	1	3	2

In the main effects plot for MRR as shown in Fig. 5, if the line for a specific parameter is near horizontal, then the parameter has no substantial effect. On the other hand, the parameter for which the line has the highest inclination will have the most substantial effect. The main effects plot show that almost all parameters have significant. The effect of cutting parameters on the performance characteristics are observed that higher the value represent the desirable performance characteristics. The importance of each factor can be interpreted from the minimum and maximum values of grey relational grade. Therefore, the order of importance of cutting parameters in plasma cutting is current, speed, voltage and SOD.

Main effects plot for surface roughness is revealed in Fig. 6. Roughness value of surface is a constant trend. The R_a value increases with increase of speed and thickness due to the fact that if traverse speed increases, the flame travels fast and without correct cutting of the surface. So the roughness value of surface increases. The effect of cutting parameters on the performance characteristics are observed that higher the value represent the desirable performance characteristics. The importance of each factor can be interpreted from the minimum and maximum values of grey

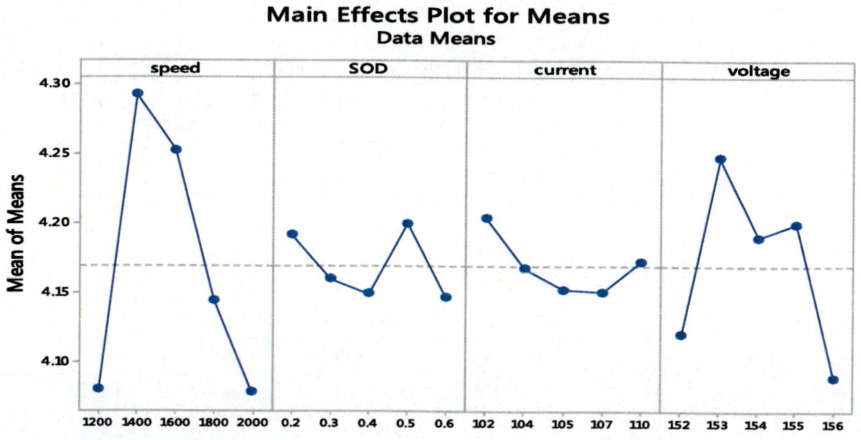

Fig. 5 Main effects plot for MRR

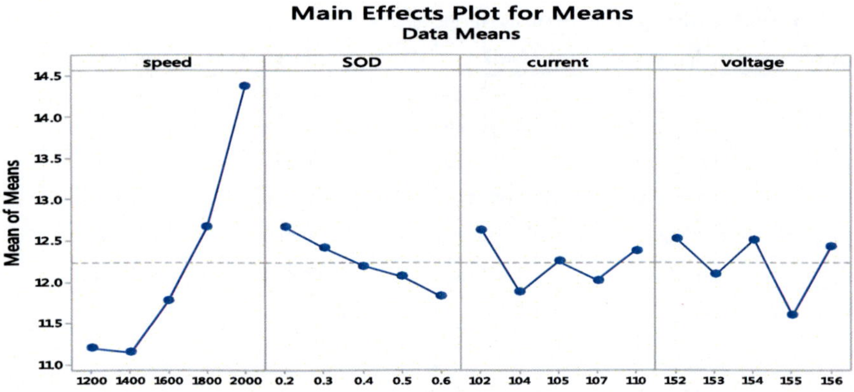

Fig. 6 Main effects plot for surface roughness

relational grade. Therefore, the order of importance of cutting variables in plasma cutting is current, speed, voltage and SOD.

3.3 *Analysis of Variance (ANOVA) for MRR*

Analysis of variance for MRR is shown in Table 7 shows the maximum contribution of voltage than current and speed. Fitted model capability is measured by F-test in ANOVA. In this model, R^2 (adj.) has the value of 81.82% which mean that output is near the approximate value. Model fitness quality is represented by R^2 value. S is the standard deviation of residuals, and its value is produced as 0.07092033. As the P-value is less than 0.05 and F-value is greater than 5% of critical value, so most of these values are significant. Variation of values above the mean is represented

Table 7 Analysis of variance for MRR, using adjusted SS for tests

Source	DF	Adj. SS	Adj. MS	F-value	P-value
Regression	7	0.56333	0.08048	16	0
Speed	1	0.18589	0.18589	36.96	0
SOD	1	0.1641	0.1641	32.63	0
Current	1	0.01269	0.01269	2.52	0.131
Voltage	1	0.01204	0.01204	2.39	0.14
SOD * current	1	0.4146	0.4146	82.43	0
SOD * voltage	1	0.02789	0.02789	5.54	0.031
current * voltage	1	0.00924	0.00924	1.84	0.193
Error	17	0.08551	0.00503		
Total	24	0.64883			

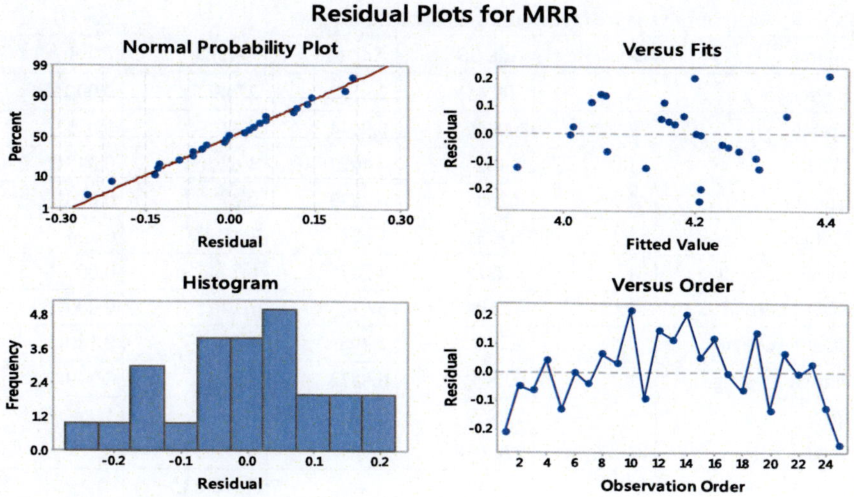

Fig. 7 Residual plots for MRR

by F-value. Mean square ratio is calculation of F-value. These residual plots for MRR show the normal probability plot, versus fits, histogram and observation order. Normal probability plot shown in Fig. 7 the straight line and the residual are normally distribution.

3.4 *Analysis of Variance for SR*

Analysis of variance for SR is shown in Table 8 shows the maximum contribution as current than SOD, voltage and speed. R denotes an observation with a large standardized residual.

Fitted model capability is measured by F-test in ANOVA. In this model, R^2 (adj.) has the value of 89.83% which mean that output is near the approximate value. Model fitness quality is represented by R^2 value. S is the standard deviation of residuals, and its value is produced as 0.975922. As the P-value is less than 0.05 and F-value is greater than 5% of critical value, so most of these values are significant. Variation of values above the mean is represented by F-value. Mean square ratio is calculation of F-value. These residual plots for surface roughness show the normal probability plot, versus fits, histogram and observation order. Normal probability plot shown in Fig. 8, the straight line and the residual are normally distribution.

Table 8 Analysis of variance for SR, using adjusted SS for tests

Source	DF	Adj. SS	Adj. MS	*F*-value	*P*-value
Regression	8	209.459	26.1824	27.49	0.000
Speed	1	6.994	6.9944	7.34	0.015
SOD	1	4.402	4.4023	4.62	0.047
Current	1	0.258	0.2579	0.27	0.610
Voltage	1	1.123	1.123	1.18	0.294
speed * current	1	8.287	8.2875	8.7	0.009
SOD * current	1	67.104	67.1037	70.46	0.000
SOD * voltage	1	2.298	2.298	2.41	0.140
current * voltage	1	0.887	0.8872	0.93	0.349
Error	16	15.239	0.9524		
Total	24	224.698			
S	R^2	R^2 (Adj.)		R^2 (Pred.)	
0.975922	93.22%	89.83%		84.64%	

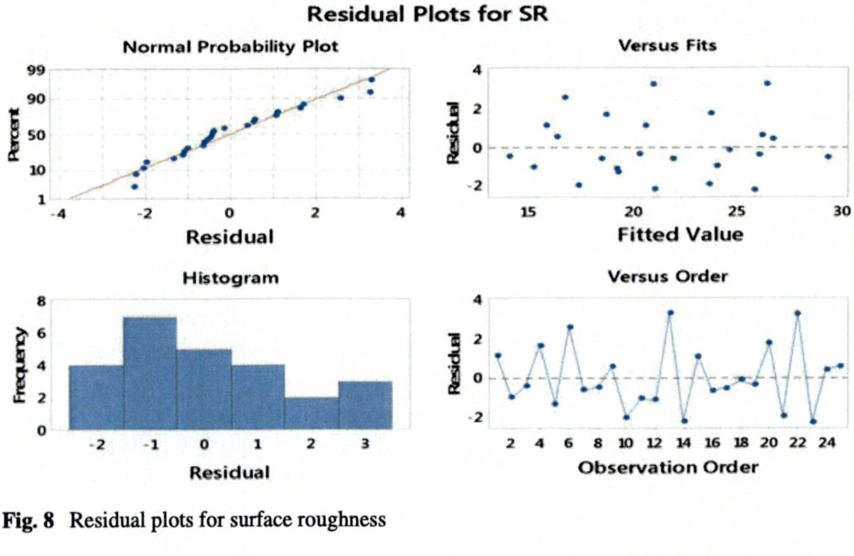

Fig. 8 Residual plots for surface roughness

3.5 Verification Test

The verification test has been performed to validate the applied optimization technique. Verification test are calculated by taking averages of reading pf all the experiments. The actual value of MRR is 4.288 g/sec, and actual value of surface roughness is 21.2928 µm. The predicted value of MRR is 3.633, and SR is 1.1968 obtained by regression equations. After the complete experiments, the parametric study have

Table 9 Result of verification test

Best combination				MRR (mm/sec.)		Surface roughness (μm)	
Speed	SOD	Current	Voltage	Avg. actual (MRR)	Predicted (MRR)	Avg. actual (R_a)	Predicted (R_a)
1600	0.4	105	154	4.288	3.633	21.2928	20.09
Error %				0.66		1.1968	
S			R^2	R^2 (Adj.)		R^2 (Pred.)	
0.07092033			86.82%	81.40%		72.11%	

been performed and then confirmation test is done to check the accuracy of experimental outcomes. The minimum 0.66% of error for MRR and 1.1968% for SR are shown in Table 9 that the experimental model is closest to the true values.

3.6 MRR Effect on Input Parameters

In this experiment, four input parameters are used and find out the effect on output parameters. The effect of output variables is obtained like cutting speed, cutting current, SOD and voltage on MRR. Plot various graph with the help of Origin-8 software.

In Figure 9a, the graph indicates that if speed increases, then MRR decreases. As the speed increases, the contact time between workpiece and arc are decrease results in energy coming down to arc also decrease. So, the melting and vaporization of material will decreases, that impact the overall MRR. During high cutting speed, the interaction time is exceptionally ostensible which impact deficient plasma energy moves to the workpiece bringing about abatement in MRR.

In Fig. 9b, the graph indicates that if current increases, then MRR also increases. With increase in current results in material increment and little amount of material liquefy and further in the base of mixing zone causes higher MRR. As circular segment current speeds up, high grouping of plasma vitality is moved to the workpiece that prompts fast softening and vaporization of metal and therefore expands MRR. As, current increases to a specific level (107–110 A), the high volatility of material increases results in softening of little material. Further, development of vaporization causes increment of MRR.

Figure 9c shows the graph between SOD and MRR. With increase in SOD causes decrease in kinetic energy of plasma causes resulting in decreases in MRR. The maximum MRR value is obtained at SOD 0.3 mm. Higher stand-off distance causes arc coherence leading to swerving of plasma arc which peripheral drag from the adjacent backgrounds. Therefore, increase in stand-off distance causes decrease of kinetic energy of plasma; as a result, MRR decreases.

In Figure 9d, the graph shows that if voltage value increases in first four reading, then small variation is observed in MRR during cutting process. Overall observation

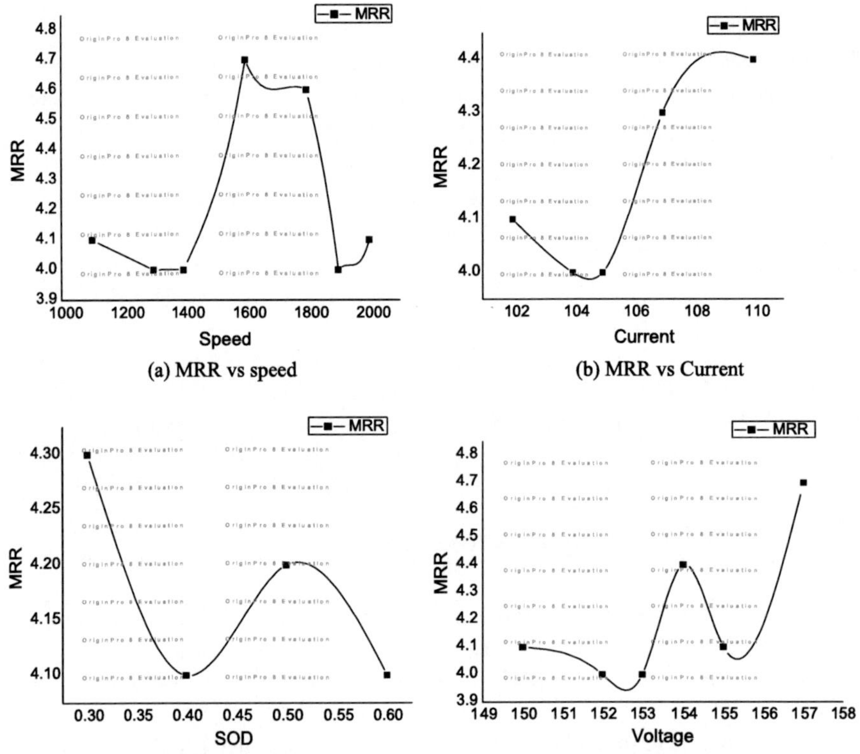

(a) MRR vs speed

(b) MRR vs Current

(c) MRR vs SOD

(d) MRR vs voltage

Fig. 9 MRR effect on input parameters

shows that if voltage value increases, then MRR also increases. With increase in voltage, the increase plasma concentration and kinetic energy of molecules increases, and results in increase in MRR. The graph indicates that effect of applied voltage on MRR and voltage increases, then MRR also increase. Maximum MRR value obtains at voltage 157 V.

3.7 Surface Roughness Effect on Input Parameters

In this experiment, the roughness value is obtained from Mitutoyo Surftest SJ-210. Before conducting machining process, roughness value of material was 19.02 micron, and after machining, roughness value was increased. The figure also shows some value after cutting. The genuine profile is the profile coming about because of the crossing point of the workpiece surface and a plane common to that surface and towards a way that increases the surface repulsiveness regard at right edges to the machining marks.

The roughness profile is the profile resulted in separating of the essential profile with a cut-off wavelength λ c.

In Figure 10a, the graph shows that if SOD increases, then R_a value also increases at certain point and then starts decreasing. This is because of decrease in kinetic energy. The minimum value of R_a was observed obtains at 0.3 mm of SOD.

In Fig. 10b, the graph indicates that if current increases, then R_a also increases. With increase in current, results in material increment and large amount metal liquefy and further increase in vaporization at the base of machining zone causes increment R_a at the minimum value of 107 A current.

In Fig. 10c, the graph indicates that if voltage increases, then R_a decreases. Value of roughness varies with voltage. When voltage is increase roughness value increase first two reading than value or roughness decrease and further reading also show variation with voltage. Overall observation shows that if voltage value increases, then roughness decreases. Minimum R_a obtains at 157 V voltages.

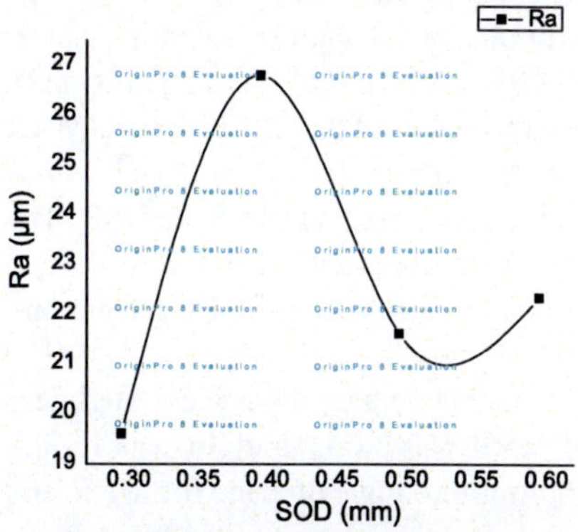

(a) Surface roughness vs SOD

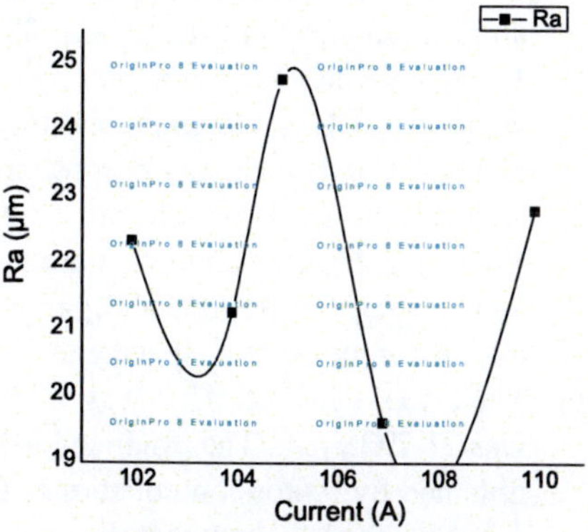

(b) surface roughness vs current

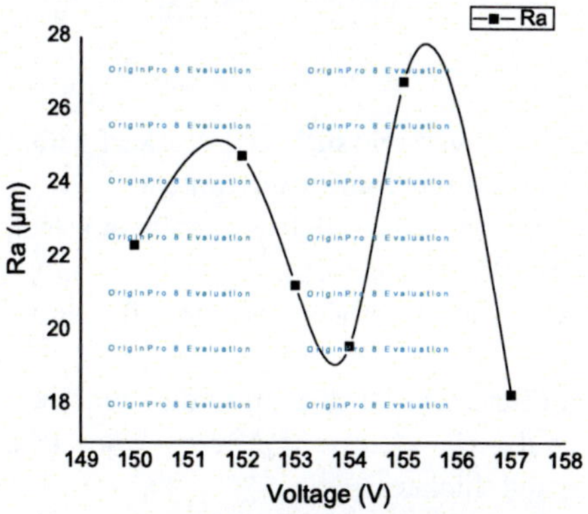

(c) surface roughness vs voltage

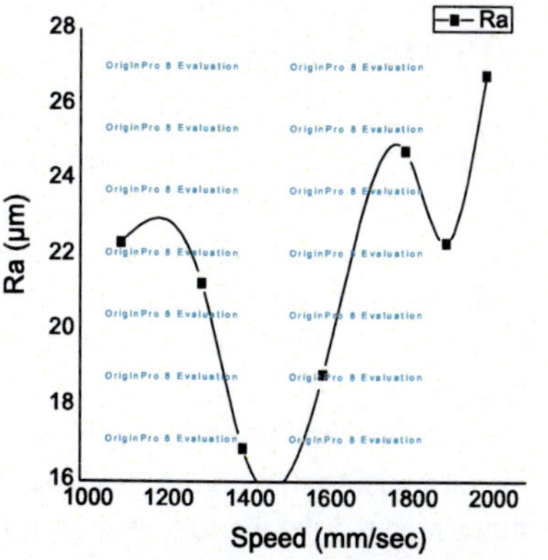

(d) surface roughness vs speed

Fig. 10 Surface roughness effect on input parameters

In Fig. 10d, when cutting speed increases, the value of R_a also increases. The R_a value decrease with increase in cutting speed till 1400 mm/s, but above 1400 mm/sec. cutting speed affects the increased now R_a value for higher side. At higher cutting speed the heat affected zone has been observed on the cutting surfaces, which is hard to estimate on cutting surface.

4 Conclusion and Future Action

4.1 Conclusion

Conclusions of plasma arc cutting on the basis of research work are as follow:

(a) The combination of Taguchi and GRA approach has been effectively utilized to decide the most excellent parameter blends for plasma cutting of Inconel-718 under multi-response approach. The optimum value obtains at cutting speed 1400 mm/s, SOD 0.3 mm, cutting current 105 A, voltage 155 V. The maximum value of MRR is 5.4 gm/s, and the minimum value of R_a is 13.55 μm. MRR and R_a value is inversely proportional to each other.
(b) ANOVA study shows the most influencing parameter as MRR and SR. For MRR, significant parameters are speed, SOD, current and interaction of SOD and current. For SR, most significant parameters are speed, SOD, current and interaction of current, speed and SOD.
(c) The actual value of MRR was 4.288 g/s, and actual value of surface roughness was 21.2928 μm. The predicted value of MRR was 3.633, and SR was 1.1968 obtained by regression equations. The minimum 0.66% of error for MRR and 1.1968% for SR show that the experimental model is closest to the true values.

4.2 Future Action

(a) Further study can be performed for responses variable such as dross, kerf width, and bevel angle used new using high precision measuring equipment.
(b) Another study can also be made using other materials such as aluminized steel, titanium alloy, beryllium copper, Monel-400 and Ti6Al.14A.
(c) Study on nozzle wear, nozzle diameters and material of nozzle still can be explored.
(d) Another system can be utilized in a similar material to analyse the outcomes. Surface honesty attributes, for example subsurface harm and different workpiece materials, can be evaluated with fluctuated thickness.

Acknowledgements The author is thankful to "Laxmi Engineering and Work, Ajmer" industry for providing the facilities to perform experimental studies on PAM of Inconel-718 material.

References

1. Special material INCONEL alloy 718, www.specialmetals.com
2. INCONEL alloy 718: Properties and applications. https://www.corrotherm.co.uk/blog/inconel-alloy-718properties-and-applications
3. Das MK, Kumar K, Barman TK, Sahoo P (2014) Optimization of process parameters in plasma arc cutting of EN 31 steel based on MRR and multiple roughness characteristics using grey relational analysis. Proc Mater Sci 5:1550–1559. https://doi.org/10.1016/j.mspro.2014.07.342
4. Aldazabal J, Meizoso AM, Klimpel A, Bannister A, Cicero S (2018) Mechanical and microstructural features of plasma cut edges in a 15 mm thick S460M steel plate. Metals-Open Access Metallurgy J 8(447):1–13. https://doi.org/10.3390/met8060447
5. Maity KP, Bagal DK (2015) Effect of process parameters on cut quality of stainless steel of plasma arc cutting using hybrid approach. Int Adv Manuf Technol 78(1–4):161–175. https://doi.org/10.1007/s00170-014-6552-6
6. Cebeli Ö, Ulaş Ç, Engin Ü (2011) A fuzzy model for predicting roughness in plasma arc cutting 62 of ASIS 4140 steel. Mater Manuf Process 95–102. https://doi.org/10.1080/10426914.2011.551952
7. Sanda B, Margareta C, Munteanu A (2010) Experimental results concerning the variation of surface roughness parameter (R_a) at plasma arc cutting of a stainless steel workpiece. Int J Modern Manuf Technol 2(1):31–36. ISSN 2067-3604
8. Ramakrishnan H, Balasundaram H, Ganesh RN, Karthikeyan N (2018) Experimental investigation of cut quality characteristics on SS321 using plasma arc cutting. J Brazilian Soc Mech Sci Eng 40:1–11. https://doi.org/10.1007/s40430-018-0997-8
9. Bevis H, Komedna J, Rohrer GS, Beladi H (2015) Heat affected zone microstructures and their influence on toughness in two micro alloyed HSLA steels. Acta Mater 17:380–391. https://doi.org/10.1016/j.actamat.2015.05.055
10. Lazarevic A (2014) Experimental research of the plasma arc cutting process. J Appl Eng Sci 12(4):304: 291–296. https://doi.org/10.5937/jaes12-6778
11. Hatala M, Zajac J, Čep R, Orlovský I (2011) Research of the technological parameters importance for plasma arc thermal cutting. Appl Mech Mater 110–116, 3742–3749. https://doi.org/10.4028/www.scientific.net/amm
12. Bhuvenesh R, Norizaman MH, Abdul Manan MS (2012) Surface roughness and MRR effect on manual plasma arc cutting machining. Int J Industr Manuf Eng 6(2). https://doi.org/10.5281/zenodo.1062516
13. Siva Teja S, Karthik G, Sampath S, Shaj Md., (2015) Experimental investigations to study the impact of machining parameters on mild steel using plasma arc cutting. Int J Eng Res Appl 5(8):83–88. ISSN: 2248-9622
14. Ananthakumar K, Rajamani D, Balasubramanian E, Paulo Davim J (2018) Measurement and optimization of multi-response characteristics in plasma arc cutting of Monel 400 using RSM and TOPSIS. Measurement 135:725–737. https://doi.org/10.1016/j.measurement.2018.12.010
15. Naik DK, Maity KP (2018) Optimization of dimensional accuracy in plasma arc cutting process employing parametric modeling approach. Mater Sci Eng 338:012039. https://doi.org/10.1088/1757-899X/338/1/012039
16. Adalarasan R, Santhanakumar M, Rajmohan M (2015) Application of grey Taguchi-based response surface methodology for optimizing the plasma arc cutting parameters of 304L stainless steel. Int J Adv Manuf Technol 78:1161–1170. https://doi.org/10.1007/s00170-014-6744-0
17. Pitayachaval P, Sato M (2018) Investigation parameters that effect to wear of plasma nozzle. In: 6th Asia conference on mechanical and materials engineering (ACMME2018), vol 213. https://doi.org/10.1051/matecconf/201821301010
18. Sovan B, Jisnu B, Gautam M, Asish B (2018) Experimental study of plasma arc cutting of AISI 304 stainless steel. Materials Today: Proc 5(2):4541–4550. https://doi.org/10.1016/j.matpr.2017.12.024

19. Peko I, Nedic B, Djordjevic A, Veza I (2018) Modelling of Kerf width in plasma jet metal cutting process using ANN approach. Tehnicki Vjesnik 25(2):401–406. https://doi.org/10.17559/tv-20161024093323

Optimization of WEDM Process Parameters for Aluminium Metal Matrix Material Al+SiC Using MCDM Methods

Jaksan D. Patel and Kalpesh D. Maniya

1 Introduction

WEDM considered as unique adoption of the conventional EDM process, which comprises of a continuous wire as electrode. Wire electrode usually made of brass copper, molybdenum of diameter 0.05–0.30 mm which converts the electrical energy into the thermal energy. The wire is wound on a wire drum. The wire is fed continuously from wire drum through the work piece, which is supported by a pair of wire guides. In WEDM, the eroded particles are flushed away by the continuously pumped dielectric fluid. Kuriakose and Shunmugam et al. [1] has been carried out the experiment on WEDM machine considering the Ti6Al4V as a work material and zinc coated wire with 0.25 mm diameter as a tool material. Bamberg and Rakwal [2] investigates the electrical discharge machining of gallium-doped p-type germanium with a relaxation type pulse generator. A series of experiments were performed to establish the slicing rate for different types and sizes of electrode wires. It was also investigates the use of small wire diameters (50–200 μm) to enhance slicing rate and surface characteristics. It was found that increase in slicing rate with decrease in wire diameter. Maher et al. [3] investigation the evolution of EDM wire electrode technologies from using copper to the widely employed brass wire electrodes and from brass wire electrodes to the latest coated wire electrodes for maximum productivity and quantity. It was also found that increasing the wire diameter increasing the cutting velocity and material removal rate. Also it was found that different wire shapes patented to enhance material removal rate. Mitsuyasu and Tasaki et al. [4]

J. D. Patel (✉)
Department of Mechanical Engineering, Merchant Engineering College, Mehasana, India
e-mail: jaksanpatel_mech1986@yahoo.com

K. D. Maniya
Department of Mechanical Engineering, C.K. Pithawala College of Engineering & Technology, Surat, India

H. K. Dave and D. Nedelcu (eds.), *Advances in Manufacturing Processes*, Lecture Notes in Mechanical Engineering, https://doi.org/10.1007/978-981-15-9117-4_5

studied the effect of coating the cutting surface with brass wire in this context, as well. Accordingly, we found that brass could be coated on a part 1 mm from its upper face and could hold the cores. Patil and Brahmnakar [5] study the effect of wire electrode material type and volume fraction of reinforcement on wire electrical discharge machining performance of A359/SiCp composite it was found that the cutting rate was greater in the case of coated wire as compared to the plain brass wire. The improvement in the cutting rate was found to be in greater than 58% for the coated wire. Goyal et al. [6] focuses the effect of process parameters on material removal rate and surface roughness in wire electric discharge machining of Inconel 625. Machining was done by using a normal zinc coated wire and cryogenic treated zinc coated wire. Analysis of variance is employed to optimize the material removal rate and surface roughness. Based on analysis it is found that pulse on time, tool electrode and current intensity are the significant parameters that affect the material removal rate and surface roughness. Ishfaq et al. [7] evaluated parametric effects WEDM process considering work piece orientation, layer thickness of individual layer, wire diameter, and pressure ratio of dielectric fluid on the cutting speed. Taguchi's L18 orthogonal array has been used for experimental design. It has been observed that the individual layer thickness of the cladded material plays a vital role in controlling the cutting speed of WEDM. The contribution of stainless steel layer thickness on cutting speed is found to be two times as compared to the contribution of mild steel layer. In this study effects type of wire electrode material, wire diameter and electrical parameter like pulse on time and pulse off time on machining performance of aluminium matrix composites are reported. Many experimental studies have been undertaken on WEDM of aluminium matrix composites but studied effects of wire electrode material and wire diameter less work have been reported. The objective of this study was to find out optimum parameter set of wire electrode material and wire diameter of the commercially employed wire electrodes. WEDM process performance efficiency closely depends on how wire electrode tool relative the machining parameter such as pulse on time, pulse off time, flushing pressure and gap voltage. So wire electrode tool selection based on following objectives: avoid short circuit, high electrical conductivity without breakage, high straightness between two guides, avoid bridging of the gap by severe contamination etc.

2 Experimental Setup

The following section highlights the material and its properties, methods of composite preparation, designing for the experimental studies [8].

Table 1 Chemical composition of Al6061 by wt%

Element	Si	Fe	Cu	Mn	Ni	Pb	Zn	Ti	Sn	Mg	Cr	Al
%Wt	0.43	0.7	0.24	0.139	0.05	0.24	0.25	0.15	0.001	0.802	0.25	Balance

Table 2 Chemical composition of SiC by wt%

Element	SiC	Si	SiO_2	Fe	Al	C
%Wt	98.5	0.3	0.5	0.08	0.1	0.3

2.1 *Material*

Aluminium matrix material for present study is Al+SiC. Table 1 gives the chemical composition of Al6061 and Table 2 represents chemical composition of SiC which is used as reinforcing material of particle sizes 220 mesh.

2.2 *Composite Fabrication Process*

Aluminium 6061 grade melted in coal furnace at temperature of 680 °C then the molten metal cooled to its semi solid temperature. Aluminium at semi solid temperature de-moisturized powder of 220 mesh size mixed manually with the aluminium. Then mixer again heated to melting temperature of aluminium and mechanical stirring of 3 min carried out so the particle of de-moisturized powder distributed well. Then the molten mixer poured in the sand mould prepared previously. After solidification of work piece sand removed and the plate of 150 mm * 150 mm * 17 mm cleaned. Then plate cut so get 25 mm * 25 mm * 17 mm piece each from it for experiment. Figure 1 show some process of work piece preparation.

Fig. 1 Work piece sample

2.3 Designing for the Experimental Studies

To understanding the behaviour of the WEDM process with different wire electrodes having different diameters a systematic plan was made for conducting the experiments. ELPULS-40 DLX WEDM machine used for experiment as shown in Fig. 2. Experiments were conducted by using statistically designed Taguchi's L_{16} orthogonal array. The machining process parameters chosen for this study are shown in Table 3 with their levels and ranges of variation of these factors are decided on the basis of some preliminary experiments as well as operator experience and literature survey.

The responses to measure the performance of the process were material removal rate, surface roughness and cutting velocity. Material removal rate can be found by using Eq. 1 where weight of work piece measure by weight measurement machine as shown in Fig. 3. Surface roughness was measured by using mitutoyo surface roughness tester as shown in Fig. 4 in terms of CLA values (Ra) having stroke length 0.8 cm and speed of probe of tester fixed 0.5 mm/s. Cutting velocity was found by using Eq. 2 where as cutting length is measured by vernier caliper. Table 4 represents the process parameters and result of output parameter of WEDM machine for aluminium metal matrix Al+SiC.

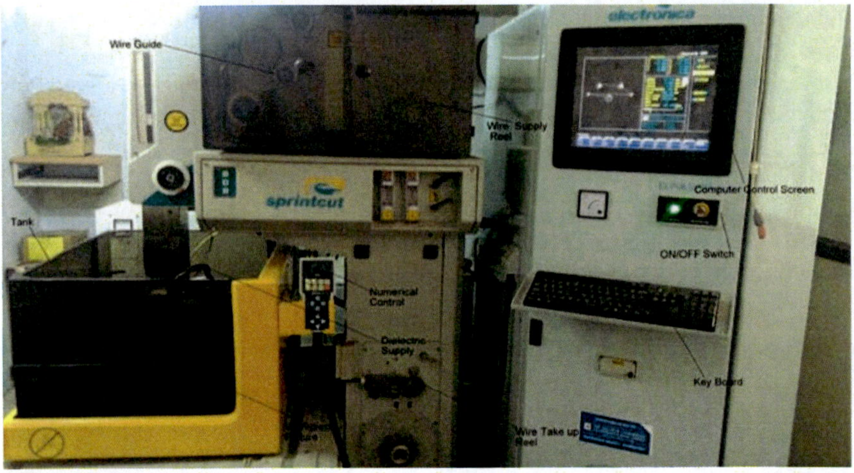

Fig. 2 WEDM machine setup

Table 3 Process parameter and their level

Factor	Level 1	Level 2	Level 3	Level 4
Pulse on time (T_{on}, μs)	108	115	123	130
Pulse off time (T_{off}, μs)	50	54	58	62
Wire diameter (d_w, mm)	0.25	0.3	–	–

Fig. 3 Weight measurement machine setup

Fig. 4 Surface roughness tester setup

$$\text{MRR} = \frac{W_b - W_a}{\rho \times T} \tag{1}$$

where

W_b = Weight before experiment
W_a = Weight after experiment
T = Time taken to cut
ρ = Density of material

$$\text{Cutting Velocity} = \frac{\text{Cutting Length of Workpiece}}{\text{Time of Cut}} \tag{2}$$

Table 4 Process parameter and result of Al + SiC

Exp. No.	T_{on}	T_{off}	d_w	Uncoated brass wire			Coated brass wire		
				MRR mm^3/min	SR μm	CV mm/min	MRR mm^3/min	SR μm	CV mm/min
1	108	50	0.25	13.5282	3.275	2.9453	15.0263	3.33	3.2715
2	108	54	0.25	10.7841	2.921	2.3574	12.3071	3.207	2.7514
3	108	58	0.3	9.6277	3.03	1.8590	10.4997	3.431	1.9800
4	108	62	0.3	7.7604	2.917	1.6655	9.4022	3.375	1.7559
5	115	50	0.25	26.8732	3.787	4.9702	35.8388	3.721	6.9944
6	115	54	0.25	23.3531	3.247	4.5896	28.4956	4.024	5.5385
7	115	58	0.3	17.1950	3.718	3.3468	23.8378	4.221	4.1215
8	115	62	0.3	14.9425	3.885	2.9405	21.6176	3.805	3.7156
9	123	50	0.3	30.1724	4.302	5.4386	48.5394	4.115	7.7707
10	123	54	0.3	26.4009	4.423	4.7461	47.5189	4.372	7.5000
11	123	58	0.25	32.3732	4.092	5.7412	39.1618	4.586	7.1976
12	123	62	0.25	26.7668	4.032	4.9703	33.3786	4.818	6.0345
13	130	50	0.3	30.4305	4.161	5.5654	56.1890	4.322	9.0837
14	130	54	0.3	25.7126	3.853	4.7013	46.5211	4.980	7.3964
15	130	58	0.25	30.5662	4.558	5.7531	39.3785	4.800	7.3457
16	130	62	0.25	27.2591	4.516	4.9802	33.3973	4.328	6.0775

3 MCDM Methods

Multi criteria decision making methods is best tool for the decision maker when selection of best alternative or value among of the value or alternative as per the decided goal and objective. Figure 5 shows the step of multi criteria decision making methods.

MCDM refers creating decisions in the presence of various, typically conflicting criteria. Depending on whether the problem is a selection problem or a design problem, the problems of MCDM can be generally categorized into two (1) Multiple Attribute Decision Making (MADM) and (2) Multiple Objective Decisions Making

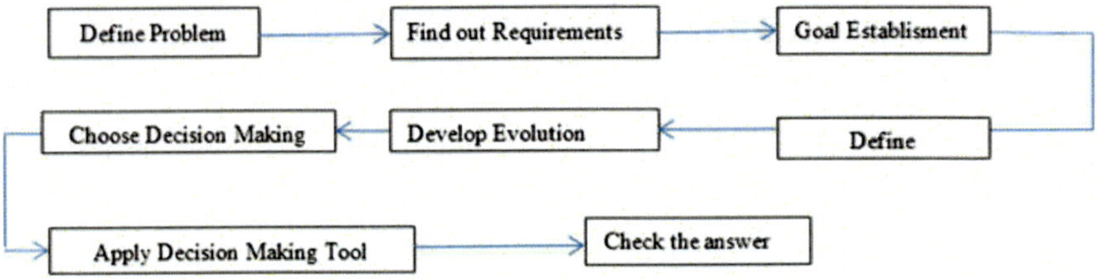

Fig. 5 Surface roughness tester setup

(MODM). When infinitive or a large number of choices are present among which solution is suitable for the decision maker for that MODM method is used. Genetic algorithm, particle swarm optimization, ant colony optimization, teaching learning based optimization, ant bee colony etc. are MODM methods. When finite number or limited number of alternatives available amongst which solution is suitable for decision maker for that MADM methods use. AHP, PROMOTHEE, TOPSIS, MOORA, ARAS, OCRA, GRA etc. are MADM methods.

3.1 Genetic Algorithm as MODM Methods

GA is general optimization methods based on the principles of natural selection and evolutionary theory. The algorithm was provided with a set of possible solutions (represented by chromosomes) called as a population. The solutions of a population are taken and used to for a new population. This was motivated by new population will achieve better than its predecessors. The solutions chosen to form new solutions (excluding springs) are selected according to their reliability the better they are, the better their chances of being reproduced. This selection process is repeated until a predetermined condition. The procedure for solving the discrete optimization problem mentioned using GA is shown in Fig. 6.

3.2 Operational Competitiveness Rating Analysis (OCRA) Method as MADM Methods

Operational competitiveness rating analysis method developed by Parkan et al. [9] uses for finding out the important performance set for response of production unit. OCRA uses an intuitive method for deciding the important parameters set as per the decision maker criteria and preference ratings of the alternatives in OCRA method reflect the decision maker's preferences for the criteria. Benefits of OCRA method is it can deal with different weight criteria assign to the alternative or some weight criteria not applicable to alternative same as the MCDM situations. Steps and illustration of WEDM process parameter using OCRA methods listed below.

1. Constructed decision matrix.
2. Construct normalized matrix using Eq. 3.

$$R_{ij} = \frac{X_{ij}}{\sum_{i=1}^{m} X_{ij}} \quad (3)$$

3. Find out non beneficial and beneficial criteria using Eqs. (4) and (5), respectively.

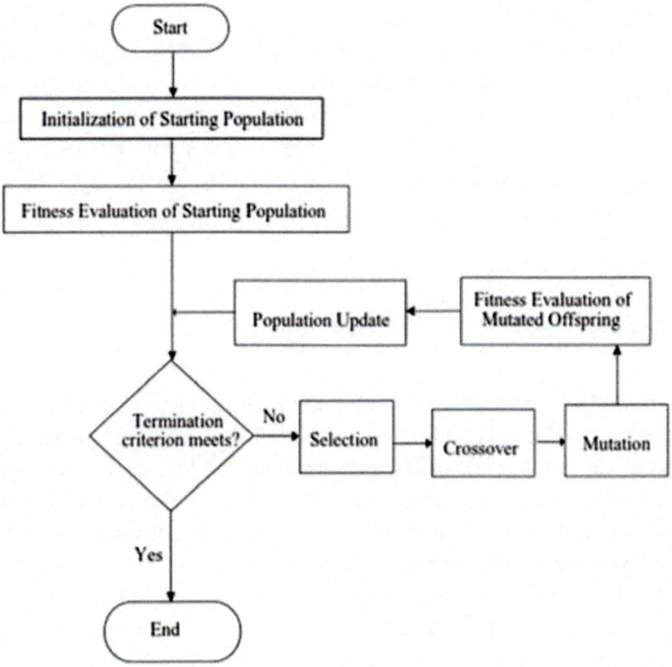

Fig. 6 Surface roughness tester setup

$$\overline{I}_i = \sum_{j=1}^{n} W_j \frac{\max\left(X_j^m\right) - X_j^i}{\min\left(X_j^m\right)} \tag{4}$$

$$\overline{O}_i = \sum_{h=1}^{H} W_h \frac{X_h^i - \min\left(X_h^m\right)}{\min\left(X_h^m\right)}, \tag{5}$$

4. Find out linear preference rating of input and output criteria using Eqs. (6) and (7), respectively.

$$\overline{\overline{I}}_i = \overline{I}_i - \min\left(\overline{I}_i\right) \tag{6}$$

$$\overline{\overline{O}}_{\mathrm{i}} = \overline{O}_i - \min\left(\overline{O}_{\mathrm{i}}\right) \tag{7}$$

5. Find out overall preference ratings by using Eq. (8).

$$P_i = \left(\overline{\overline{I}}_i + \overline{\overline{O}}_i\right) - \min\left(\overline{\overline{I}}_m + \overline{\overline{O}}_m\right) P_i = \left(\overline{\overline{I}}_i + \overline{\overline{O}}_i\right) - \min\left(\overline{\overline{I}}_m + \overline{\overline{O}}_m\right) \tag{8}$$

Optimization of result is carried out by using MCDM methods like genetic algorithm (GA) and OCRA methods by considering seven different weight age criteria for different output parameter like C_1: $W_1 = 0.3333$, $W_2 = 0.3333$, $W_3 = 0.3333$; C_2: $W_1 = 0.5$, $W_2 = 0.2$, $W_3 = 0.3$; C_3: $W_1 = 0.3$, $W_2 = 0.5$, $W_3 = 0.2$; C_4: $W_1 = 0.3$, $W_2 = 0.5$, $W_3 = 0.2$; C_5: $W_1 = 0.2$, $W_2 = 0.5$, $W_3 = 0.3$; C_6: $W_1 = 0.3$, $W_2 = 0.2$, $W_3 = 0.5$; C_6: $W_1 = 0.2$, $W_2 = 0.3$, $W_3 = 0.5$.

3.3 Objective Function for Optimization Using GA Algorithm

Function prepared for multi objective optimization problem by normalizing and combining the function. Function is normalized by weighted average normalizing method. Three objective functions MRR, SR and CV conflict each other because MRR and CV is maximizing function and SR is minimizing function. So MRR, SR and CV was normalized and combined as per weighted average normalizing method as per below.

$$\text{Normalized function } Y = W_1\frac{\text{MRR}}{\text{min}.MRR} + W_2\frac{\text{max}.SR}{SR} + W_3\frac{CV}{\text{min}.CV} \tag{9}$$

Normalized function Y is maximizing problem for MATLAB it must be converted to minimization problem. For convert maximization problem to minimization problem the function was inverted.

$$\text{Combined objective function} y = \frac{1}{\text{combined normalized function} Y} \tag{10}$$

For function MRR, CV, and SR is generated by general regression in MINITAB 16.

Al+SiC_Uncoated_MRR = $-64.1377 + 1.16766\,T_{\text{on}} - 0.800811\,T_{\text{off}} - 30.9084\,W_d$ Al+SiC_Coated_MRR = $-53.6477 + 1.16766\,T_{\text{on}} - 0.800811\,T_{\text{off}} - 30.9084\,W_d$ Al+SiC_Uncoated_SR = $-3.04627 + 0.0578984\,T_{\text{on}} + 0.00724479\,T_{\text{off}} - 1.48875\,W_d$

Al + SiC_Coated_SR = $-2.84166 + 0.0578984\,T_{\text{on}} + 0.00724479\,T_{\text{off}} - 1.48875\,W_d$ Al + SiC_Uncoated_CV = $-5.91675 + 0.182121\,T_{\text{on}} - 0.125554\,T_{\text{off}} - 16.6906\,W_d$ Al+SiC_Coated_CV = $-4.49206 + 0.182121\,T_{\text{on}} - 0.125554\,T_{\text{off}} - 16.6906\,W_d$

$$y_{\text{Al - SiC\_UC}} = 1./\left(\begin{array}{l}(1./3).*\left(\left(\begin{array}{l}-64.1377\ +\ (1.16766.*\ x_1)\\ -\ (0.800811.*\ x_2)\ -\ (30.9084.*x_3)\end{array}\right)./7.7604\right)\\ +(1./3).*\left(4.558./\left(\begin{array}{l}-3.04627+(0.0578984.*x_1)+(0.00724479.*x_2)\\ -(1.48875.*x_3)\end{array}\right)\right)\\ +(1./3).*\left(\left(\begin{array}{l}-5.961675+(0.182121.*x_1)-(0.125554.*x_2)\\ -(16.6906.*x_3)\end{array}\right)./1.6655\right)\end{array}\right)$$

$$y_{Al-SiC\_UC} = 1./\left(\begin{array}{l} (1./3).*\left(\left(\begin{array}{l} -53.6477 + (1.16766.*x_1) - (0.800811.*x_2) \\ -(30.9084.*x_3) \end{array}\right)./9.4022\right) \\ +(1./3).*(4.818./\left(\begin{array}{l} -2.84166 + (0.0578984.*x_1) \\ +(0.00724479.*x_2) - (1.48875.*x_3) \end{array}\right).+(1./3). \\ *\left(\left(\begin{array}{l} -4.49201 + (0.182121.*x_1) - (0.125554.*x_3) \\ -(16.6906.*x_3) \end{array}\right)./1.7559\right) \end{array}\right)$$

Here constraints are

0.25	$\leq$	X_1	$\leq$	0.30	Where $X_1 = D_W$ in mm
108	$\leq$	X_2	$\leq$	130	Where $X_2 = T_{on}$ in μs
050	$\leq$	X_3	$\leq$	062	Where $X_3 = T_{off}$ in μs

4 Coefficient of Variation (COV)

The standard deviation is an absolute measure of dispersion. The coefficient of variation is a relative measure of dispersion and is denoted by COV.

$$\text{COV} = \frac{\sigma}{x} \chi 100 \quad (11)$$

where σ is SD and x is mean of given series. The coefficient of variation has great practical significance and is the best measure of comparing the variability of two series. The series or groups which the coefficient of variation is greater is said to be more variable or less consistent. On the other hand the series for which the variation is lesser is said to be less variable or most consistent.

5 Result and Discussion

MCDM methods like MODM (GA) and MADM (OCRA) methods applied to experimental results considering different weight age criteria for responses like MRR, CV and SR, respectively. After optimization process optimum parameter listed in Table 5 for GA and Table 6 for OCRA methods. Also coefficient of variation analysis can be applied to output result and results listed in Table 7.

From the Tables 5 and 6 optimum condition is pulse on time: 123 μs, pulse off time: 50 μs, wire diameter: 0.25 mm when experiments perform using uncoated brass wire use as electrode and pulse on time: 123 μs and 130 μs pulse off time: 50 μs, wire diameter: 0.25 and 0.3 mm when experiments perform using coated brass wire use as electrode. Also from the Table 7 coated wire experiment is more

Table 5 Optimization result by genetic algorithm

Sr. No.	Weightage			Optimum input parameter					
				Uncoated			Coated		
	W_1	W_2	W_3	T_{on}	T_{off}	d_w	T_{on}	T_{off}	d_w
1	0.333	0.333	0.333	123	50	0.25	123	50	0.25
2	0.5	0.3	0.2	123	50	0.25	123	50	0.25
3	0.5	0.2	0.3	123	50	0.25	123	50	0.25
4	0.3	0.5	0.2	123	50	0.25	122	50	0.25
5	0.2	0.5	0.3	123	50	0.25	123	50	0.25
6	0.3	0.2	0.5	123	50	0.25	123	50	0.25
7	0.2	0.3	0.5	123	50	0.25	123	50	0.25

Table 6 Optimization result by OCRA

Sr. No.	Weightage			Optimum input parameter					
				Uncoated			Coated		
	W_1	W_2	W_3	T_{on}	T_{off}	d_w	T_{on}	T_{off}	d_w
1	0.333	0.333	0.333	123	58	0.25	130	50	0.3
2	0.5	0.3	0.2	123	58	0.25	130	50	0.3
3	0.5	0.2	0.3	123	58	0.25	130	50	0.3
4	0.3	0.5	0.2	123	58	0.25	130	50	0.3
5	0.2	0.5	0.3	123	58	0.25	130	50	0.3
6	0.3	0.2	0.5	123	58	0.25	130	50	0.3
7	0.2	0.3	0.5	123	58	0.25	130	50	0.3

Table 7 Summary of coefficient of variation

Material type	Wire type	Output parameter		
		MRR (mm^3/min)	SR (μm)	CV (mm/min)
Al+SiC	Uncoated	36.71	14.34	32.82
	Coated	45.59	13.39	40.08

consistent than the uncoated wire experiment for material removal rate and cutting velocity whereas for surface roughness uncoated wire experiment is more consistent than coated wire.

6 Conclusion

Material removal rate and cutting velocity was improved by performing the experiment by zinc coated brass wire whereas surface roughness increase by performing the experiment uncoated brass wire for aluminium silicon carbide work piece material. Also lower the wire diameter improves the material removal rate and cutting velocity whereas higher the wire diameter increases the surface roughness for aluminium silicon carbide material.

References

1. Kuriakose S, Shunmugam MS (2005) Multi-objective optimization of wire-electro discharge machining process by non-dominated sorting genetic algorithm. J Mater Process Technol 170:133–141
2. Bamberg E, Rakwal D (2008) Experimental investigation of wire electrical discharge machining of gallium-doped germanium. J Mater Process Technol 197:419–427
3. Maher I, Sarhan AD, Hamdi M (2015) Review of improvements in wire electrode properties for longer working time and utilization in wire EDM machining. Int J Adv Manuf Technol 76:329–351
4. Mitsuyasu T, Tasaki K, Kawano M (2016) The coating technology with the wire electrode. Proc CIRP 42:226–230
5. Patil N, Brahmankar P (2016) On the effects of wire electrode and ceramic volume fraction in wire electrical discharge machining of ceramic particulate reinforced aluminum matrix composites. Proc CIRP 42:286–291
6. Goyal A (2017) Investigation of material removal rate and surface roughness during wire electrical discharge machining (WEDM) of Inconel 625 super alloy by cryogenic treated tool electrode. J King Saud Univ Sci 29:528–535
7. Ishfaq K, Mufti NA, Mughal MS, Saleem MQ (2018) Investigation of wire electric discharge machining of stainless-clad steel for optimization of cutting speed. Int J Adv Manuf Technol 96:1429–1443
8. Patel JD, Maniya KD (2019) WEDM process parameter selection using preference ranking method: a comparative study. Int J Manuf Res 14(2):118–144
9. Parkan C, Lam K, Hang G (1997) Operational competitiveness analysis on software develop. J Oper Res Soc 48(2):892–905

Multiple Parameter Optimization by Wire Electrochemical Discharge Machining Process on Quartz Glass

Rahul O. Vaishya, Ankit D. Oza, and Anurag Gupta

1 Introduction

Quartz glass is a pure form of SiO_2 which has extraordinary mechanical, thermal, and optical properties like transmissibility in UV and IR wavelengths, thermal shock resistance, low thermal expansion coefficient, good resistance to chemicals, and outstanding electrical insulator courtesy of which it finds itself in various applications in optical, water purification, photovoltaic, chemical, pharmaceutical, and semiconductor industries. Quartz being hard, brittle, and non-conductive in nature proves a little difficult to machine through conventional machining processes because of crack propagation, fracture, poor surface texture, cutting through grain boundaries, etc. [1]. Also, developing microfeatures seems quite must be cumbersome through conventional processes. Entering into picture, non-conventional machining processes also have some inherent disadvantages. For example, some non-conventional machining processes like EDM, ECM, W-EDM cannot machine non-conductive materials like quartz used in this case. Other processes like ultrasonic machining, abrasive jet machining, laser beam machining, and ion beam machining can machine electrically non-conductive materials but not without giving poor surface finish, inferior dimensional accuracy, inadequate material removal rate, heat-effected zone, etc. Laser beam

R. O. Vaishya
Department of Production and Industrial Engineering, Punjab Engineering College (Deemed to Be University), Chandigarh 160012, India

A. D. Oza (✉)
Department of Industrial Engineering, Pandit Deendayal Petroleum University, Gandhinagar 382007, India
e-mail: ankit.ophd15@sot.pdpu.ac.in

A. Gupta
Department of Mechanical Engineering, KIET—Group of Institutions, Delhi-NCR, Ghaziabad 201206, India

H. K. Dave and D. Nedelcu (eds.), *Advances in Manufacturing Processes*, Lecture Notes in Mechanical Engineering, https://doi.org/10.1007/978-981-15-9117-4_6

can machine hard, brittle, and non-conducting materials but it is expensive to do it with precision [2]. Abrasive water jet machining can be used for the same purpose but a taper cut is unavoidable [3]. Utilizing the benefits of two or more non-conventional machining processes we can develop a hybrid machining procedure which can assist in machining of electrically non-conductive materials. One such process is ECDM and the method of which this content would be is WECDM [4–7].

The basic block diagram of WECDM setup is shown in Fig. 1.

Wang et al. [8] made a vertically operating WECDM setup and machined quartz glass with aid of oil film insulation with diamond wire sawing using vertical WECDM setup. Yadav et al. [9] performed WECSM on alumina epoxy nanocomposites (AENC) and investigated effect of applied voltage, electrolytic concentration, pulse on-time, and pulse off-time on quartz. Rattan et al. [10] used a magneto hydro-dynamic (MHD) convection process to increase the electrolyte flow in TW-ECSM process. Paul and Hiremath [11] performed WECDM process on semiconductive silicon wafers with mixed electrolyte of NaOH and KOH solution with equal weight

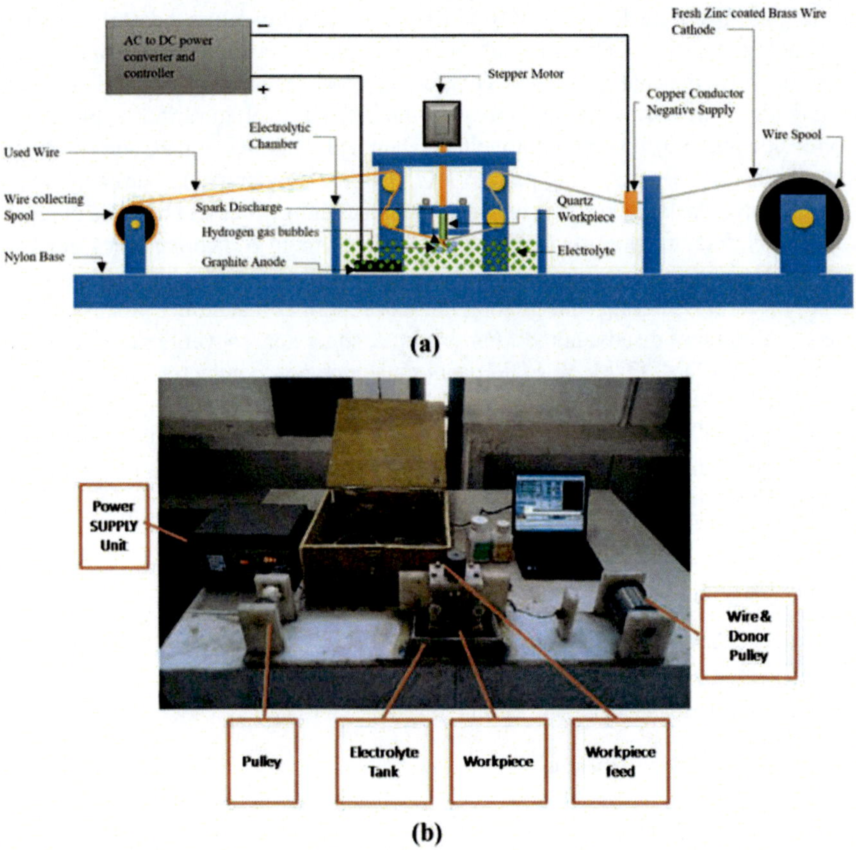

Fig. 1 Schematic diagram of WECDM setup

fraction. Kumar et al. [12] experimented with NaOH, KOH, and NaCl solutions used as electrolytes in ECDM and studied their effects on MRR and Overcut. Singh et al. [13] conducted experiments to reduce wire breakage problem while machining fine slots through optimum parametric setting of voltage and electrolytic concentration. Kuo et al. [14] performed WECDM under titrated electrolyte flow with the addition of SiC powder in KOH electrolyte using brass wire electrode of 150 μm diameter. Jain et al. [15] studied effects of voltage and electrolyte concentration on MRR, diametral overcut, tool wear rate on glass-epoxy and Kevlar-epoxy composites using NaOH as electrolyte on a TW-ECSM process. WECDM is also known as high-temperature etching process [16–18].

In this paper, WECDM setup is developed and slots are machined on quartz glass using NaOH electrolyte solution. Material removal rate and width of cut are optimized using Minitab software employing Taguchi and ANOVA techniques for input parameter combination of DC voltage, electrolyte concentration, wire speed, and interelectrode gap. Surface texture of the slots is obtained and compared using optical microscopic images. Also, process is optimized through multiattribute utility function.

2 Experimentation

2.1 Basic Principle

This process utilizes the energy of discharge to vaporize the workpiece thus causing micromachining. WECDM has a wire as cathode and an auxiliary electrode extraordinarily larger in size than the wire as anode. Anode is dipped and cathode just touches the electrolyte placed in an electrolytic cell. The electrolyte is used as a working fluid here to complete the electrical circuit between them and also for other purposes like cooling. DC power is connected to the electrodes with supplied voltage varying with electrolyte concentration to be and the workpiece used as shown in Fig. 1. The workpiece used in this process is essentially non-conducting. A DC voltage is applied between the electrodes which cause electrolysis and the formation of H_2 gas bubbles at cathode (wire) and O_2 gas bubbles at anode. When the number of H_2 gas bubbles formed at wire becomes substantial, these coalesce to form an insulating layer around the wire which causes huge resistance between wire electrode and electrolyte interface. This heightened resistance causes more heat, which vaporizes the electrolyte causing in turn an electrically non-conducting layer of H_2 and vapor mix layer around the wire. When the applied voltage overcomes this resistance, that is, when it crosses a critical level, discharge of electrons occurs in form of spark at the wire electrolyte interface creating radiation powerful enough to vaporize the workpiece. Energy of that spark varies proportionally to the applied voltage between the two electrodes. In order to machine the workpiece, it has to be placed in a zone close to the discharge so that spark energy can be utilized to melt, vaporize, and erode it [19–21].

2.2 Experimental Procedure

The experiments were performed on quartz glass slices with dimensions of 75 mm × 25 mm × 1 mm using NaOH electrolyte solutions, using Taguchi methodology L_9 OA generated from Minitab software. Graphite rod was taken as anode and zinc-coated brass wire with 0.1 mm diameter used as cathode. The properties of quartz glass, electrolytes, and wire are mentioned as the hard zinc-coated brass wire of base material CuZn37% having tensile strength 900–1050 N/mm^2 and diameter of 0.1 mm is used. The NaOH electrolyte is used having boiling point temperature of 1388 °C and having softening point 1665 °C. The input parameters were voltage, electrolyte concentration, wire speed, and interelectrode gap. The responses optimized were material removal rate (MRR) and width of cut (WOC). Each experiment was performed for 20 min. Based on trials that were performed on WECDM setup for various combinations, three levels were taken for each factor to perform the experiments as shown in Table 1. L_9 table generated from Minitab software. The width of cut was observed under metallurgical optical microscope and the dimensions of the cut were obtained taking the snaps of the images and processing them in Digimizer software. Some of the images with dimensions are shown in Fig. 2a–c.

It is observed from Fig. 2a, i.e., at 25 V and 10% concentration slot geometry is quite better as compared to Fig. 2b, c. Material removal rate was measured by weighing the slice, before and after machining and dividing the difference in weight to the time, which calculates machining rate. In order to obtain a single optimized combination of input parameters to justify both the output response, utility-based multiattribute optimization is conducted as given by Verma et al. [21]. Using this technique, a proper setting of input factors would be achieved simultaneously to obtain a higher material removal rate (MRR) and lower width of cut (WOC).

It involves giving a utility value to each response variable. This is done by giving a preference number (P_i) as in Eq. 1.

$$P_i = A \times \log\left(\frac{X_i}{X_i'}\right) \tag{1}$$

Table 1 Levels of input parameters

Level	Input parameter			
	V (A)	E.C. (B) (% wt/v)	Wire speed (C) (%)	Interelectrode gap (D) (cm)
1	25	10	100	3
2	33	20	120	8
3	40	30	140	13

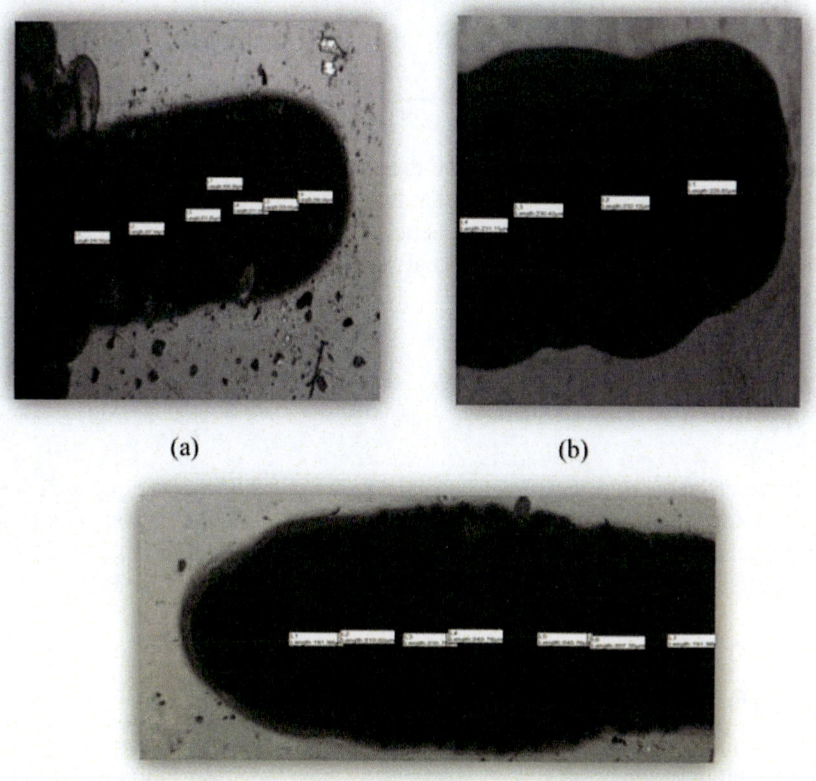

(a) (b)

(c)

Fig. 2 Machined slot. Where **a** slot cut at 25 V and 10% wt/v, **b** slot cut at 40 V and 10% wt/v, and **c** slot cut at 40 V and 30% wt/v

where

- A constant
- X_i value of the performance measure i
- X_i' minimum admissible value of performance measure in case of higher the better type output and maximum admissible value of performance measure in case of smaller the better type output. The value of A is given by Eq. 2 according to Verma et al. [22]

$$A = \frac{9}{\log \frac{X^*}{X_i'}} \tag{2}$$

where $X^* =$ Predicted optimal value of output response

The overall utility value can be computed from Eq. 3:

$$U = \sum_{i=1}^{n} W_i P_i \tag{3}$$

where W_i is priority weight allocated to the characteristics.

It has been advocated by Taguchi that the utility function computed from these characteristics would be of larger the better type only irrespective of the nature of individual output characteristic. The goal would be to optimize the overall utility function for larger the better nature

3 Results and Discussion

All the experiments were performed as per the L_9 OA and the sequence of experiments was randomized. Two trials for each combination of input parameters were performed to obtain the value of material removal rate (MRR) and width of cut (WOC). The data is presented in tabular form representing the results as shown in Table 2.

3.1 *MRR as an Output Response*

Using Minitab software, graphs are obtained to show the effect of input parameters on material removal rate (MRR). From S/N ratio graph (Fig. 3), the ranking of the input factors can be observed and it can be seen that voltage has a tremendous effect on the machining rate and it can be observed that to obtain maximum MRR by using

Table 2 Output responses readings

Exp. no.	MRR (mg/min)		WOC (mm)	
	R_1	R_2	R_1	R_2
1	0.0054	0.0036	0.146	0.118
2	0.0185	0.0131	0.188	0.158
3	0.0143	0.0112	0.175	0.137
4	0.1872	0.1287	0.232	0.198
5	0.2132	0.1765	0.287	0.251
6	0.1730	0.1579	0.256	0.204
7	0.2326	0.2075	0.297	0.265
8	0.3261	0.2281	0.345	0.289
9	0.2742	0.1917	0.354	0.302
Mean	0.1423		0.334	

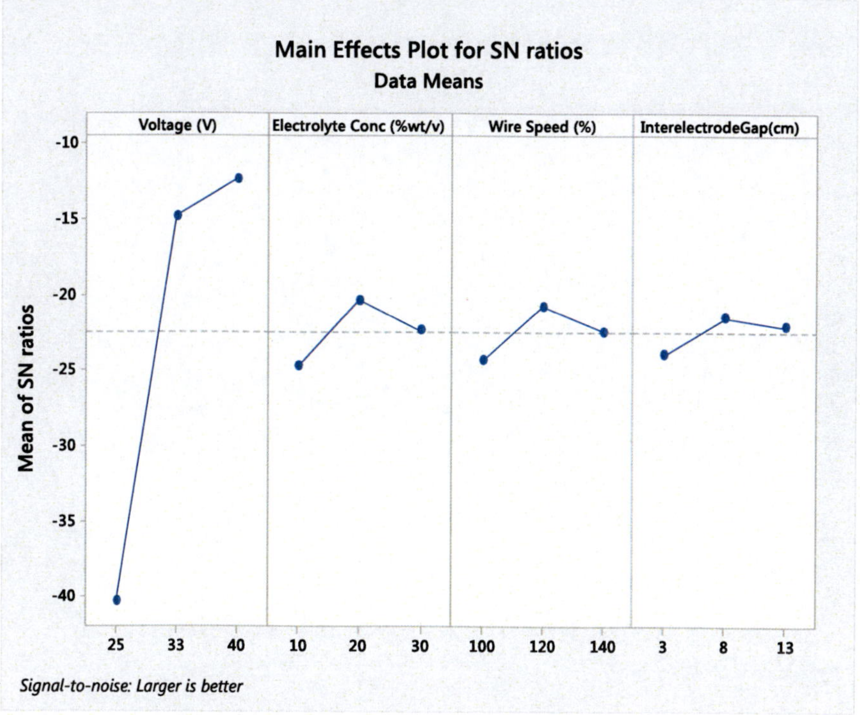

Fig. 3 S/N ratio graph for MRR

NaOH as electrolyte, the optimized combination is A3, B2, C2 and D1 or 40 V DC supply, 20% wt/v concentration, 120% wire speed, and 30 mm interelectrode gap. As from Fig. 4 it can be seen that higher voltage causes an increment in MRR. At 40 V, maximum MRR is observed. The contact area of the cathode wire and the free surface of electrolyte are small, if there is an increase in voltage, the current density in that area becomes higher which ultimately means more supply of electrons. The high supply of electrons makes electrolysis more aggressive which leads to evolution of extremely high quantity of hydrogen gas bubbles and fumes which in turn provide stable and frequent gas blanket insulation. Thus, discharge energy increases resulting in more material removal.

In case of electrolyte concentration, it can be observed that maximum material removal rate is obtained when the concentration of the electrolyte is 20% wt/v. The maximum conductivity possible for NaOH at 25 °C is achieved at 20% where the conductivity is 414,000 microcentimeter. After that the conductivity begins to decrease. It means that around 20% wt/v concentration, NaOH electrolyte gives supreme conduction through itself allowing more current to pass through and causing electrolysis at best possible speed making more bubbles and more discharge. Mate-

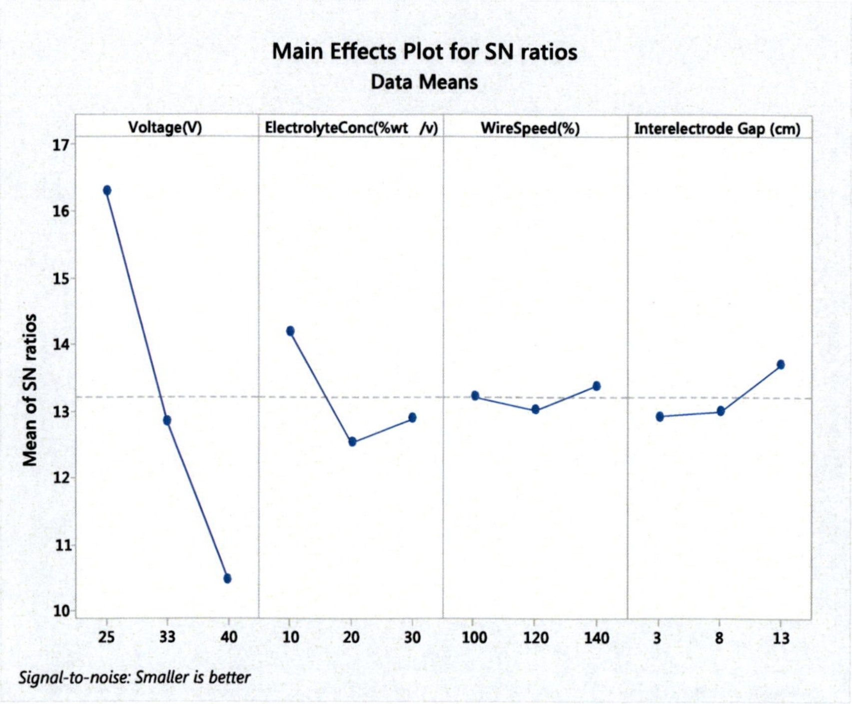

Fig. 4 S/N ratio graph for WOC

rial removal rate is best attained at 120% wire traveling speed. Although it is not significant as per the ANOVA Table 3, low wire speed resembles a stationary tool which gets heated up quickly, melts and breaks down easily. Also, low speed does not help in flushing the slot easily. High speed on the other hand carries away the heat and helps in flushing the slot but sweeps the hydrogen bubbles with it reducing the insulation film causing low discharge. An optimal speed is best to carry the heat and remove the debris from the slots, allowing maximum possible gas insulation. Keeping the anode closest to the cathode wire in the electrolyte gives maximum material removal rate. This is because less the ions travel in the electrolytic medium to complete the circuit, more readily and quick the circuit will get completed or in other words less resistance will be faced and electrolysis will occur at high rate. Here, maximum MRR is obtained at 30 mm gap.

Results for MRR were analyzed using ANOVA to find significant factors which are affecting response factor is given in Table 3. The confidence interval of confirmation experiments (CICE) is calculated by using Eq. 4.

$$\text{CICE} = \sqrt{F_{\alpha}(1, f_e)V_e\left[\frac{1}{n_{\text{eff}}} + \frac{1}{\text{R}}\right]} \tag{4}$$

Table 3 ANOVA for MRR

Source	DF	Seq SS	Contribution (%)	Adj SS	Adj MS	*F*-value	*P*-value
Voltage (V)	2	0.1702	91.27	0.1702	0.0851	69.40	0.000
Electrolyte conc (% wt/v)	2	0.0039	2.12	0.0039	0.0019	1.61	0.253
Wire speed (%)	2	0.0005	0.29	0.0005	0.0002	0.22	0.806
Interelectrode gap (cm)	2	0.0007	0.40	0.0007	0.0003	0.30	0.745
Error	9	0.0110	5.92	0.0110	0.0012		
Total	17	0.1865	100.00				

where

$F_\alpha\ (1, f_e)$ The *F*-ratio at the confidence level of $(1-\alpha)$ against DOF 1 and error degree of freedom f_e
X_i' error DOF
N Total number of result
R Sample size for confirmation experiments
V_e Error variance

$$n_{\text{eff}} = \frac{N}{1 + [\text{DOF associated in the estimate of mean response}]} \tag{5}$$

Here,

$\alpha = \text{risk} = 0.05, f_e = 9$, $V_e = 0.001227$, $N = 18$, DOF associated in the estimate of mean response = 8, $R = 2$, F0.05 (1, 9) from table = 5.1174, Thus, $n_{\text{eff}} = 18/(1 + 8) = 2$, and, CICE = ± 0.0792, the estimated mean $\mu\rho$ is given by Eq. 6

$$\mu_\rho = T_\rho + (A - T_\rho) + (B - T_\rho) + (C - T_\rho) + \ldots \tag{6}$$

where

T_ρ mean of response
$A, B, C\ldots$ extreme value of significant factors obtained from response table of means

Here, $T_\rho = 0.1423$, as voltage is the only significant factor, its highest value from response table of means is 0.2420. Thus, $\mu_\rho = 0.2420$ mg/min.

The predicted optimal range for 95% confidence interval is given by Eq. 7

$$[\mu_\rho - \text{CI}] < \mu_\rho < [\mu_\rho + \text{CI}] \tag{7}$$

The predicted optimal range for 95% confidence interval for the confirmation of experiments is shown in Eq. 8:

$$0.1628 < \mu_\rho < 0.3212 \quad (8)$$

Two possibilities arise, one is the prescribed combination of parameters level is identical to one of those in the experimental design and second is the prescribed combination of parameters level is not included in the experiment. In the present investigation, second situation exists and optimization of results has been carried out through the prescribed combination. The experiment performed on these input factors give a value of 0.2842 mg/min which comes well in this range.

3.2 *WOC as an Output Response*

Using Minitab software, graphs are obtained to show the effect of input factors on width of cut (WOC) when NaOH is used as an electrolyte. From S/N ratio graph (Fig. 5) and from mean graph (Fig. 6) and ANOVA (Table 4), it can be observed that to obtain minimum width of cut (WOC) by using NaOH as electrolyte solution, the optimized combination is A1, B1, C3, and D3, that is, 25 V DC supply, 10% wt/v concentration, 140% wire speed, and 13 cm interelectrode gap.

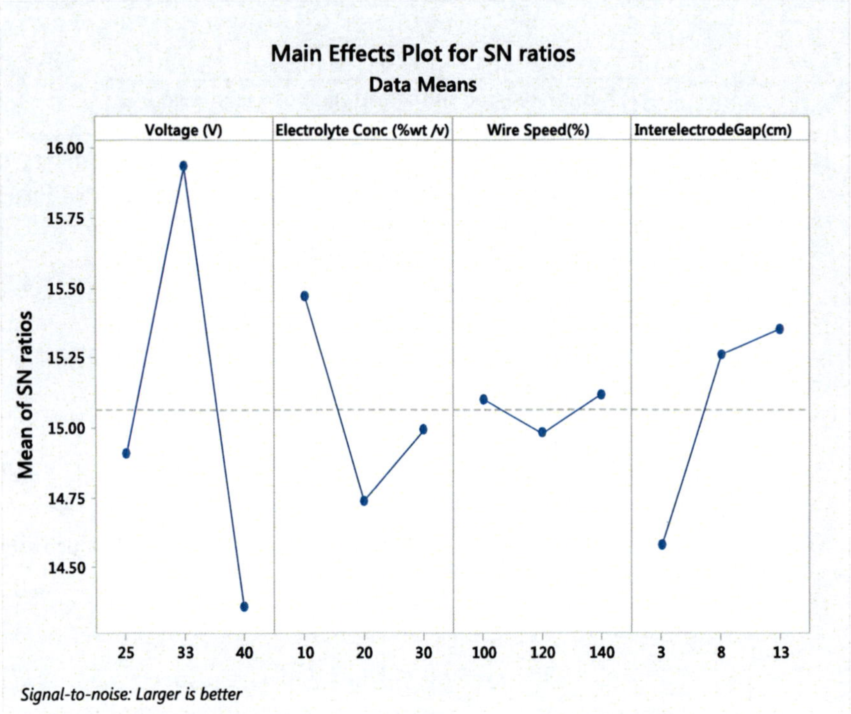

Fig. 5 S/N ratio graph for utility function

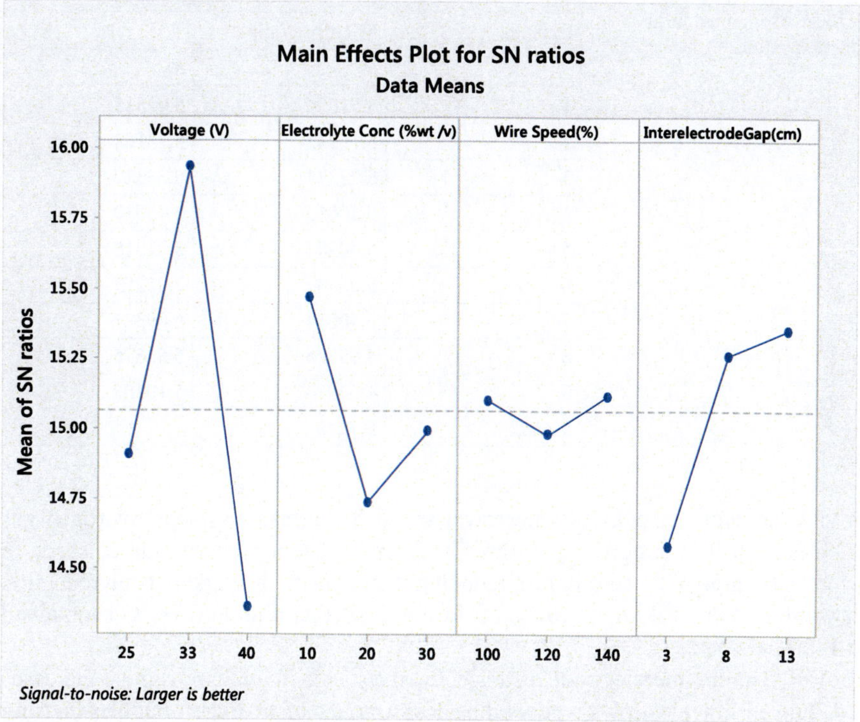

Fig. 6 Means graph for utility function

Table 4 ANOVA for WOC

Source	DF	Seq SS	Contribution (%)	Adj SS	Adj MS	*F*-value	*P*-value
Voltage (V)	2	0.07226	83.00	0.07226	0.036131	42.93	0.000
Electrolyte conc (%wt/v)	2	0.00590	6.79	0.00590	0.002954	3.51	0.075
Wire speed (%)	2	0.00048	0.56	0.00048	0.000244	0.29	0.755
Interelectrode gap (cm)	2	0.00082	0.95	0.00082	0.000414	0.49	0.627
Error	9	0.00757	8.70	0.00757	0.000842		
Total	17	0.08705	100.00				

Low voltage reduces electrolysis process and spark intensity leading in low-intensity spark discharges. The spark annular area around the wire remains small. So, a narrow slot is obtained. Here, at 25 V, WOC obtained is minimum. At 10% wt/v electrolyte concentration, the conductivity of NaOH solution is low comparatively to other concentrations which again lead to slower electrolysis and less amount of hydrogen gas bubbles generations and lower discharges leading to narrow slots

Table 5 Computed utility function values

Exp. no.	MRR (mg/min)	
	R_1	R_2
1	5.21054	5.92585
2	5.16232	5.73128
3	5.27371	6.33312
4	6.50063	6.95510
5	5.49204	6.01316
6	5.88537	7.01253
7	5.40034	5.89327
8	4.95325	5.52675
9	4.62914	5.10368
Mean	5.7223	

or smaller width of cut. As wire travel speed increases, it sweeps hydrogen gas bubbles with it making the gas film flimsy and reducing the volume of space in which discharge will occur in turn reducing the intensity of spark causing reduction of width of slots making narrow cuts. In this case, the minimum WOC is obtained at 140% wire speed.

At maximum interelectrode distance, the resistance in ionic movement increases resulting in low electrolysis generating less amount of hydrogen bubbles causing weak discharge resulting in narrow slots. Here, the minimum WOC is obtained at 13 cm interelectrode gap. Results for WOC were analyzed using ANOVA to find significant factors which are affecting response factor. ANOVA table for width of cut is given in Tables 4 and 5 shows the values of computed utility functions.

Using Eqs. 8 and 9, confidence interval of confirmation experiments (CICE) comes shown in Eq. 9:

$$\mathrm{CI_{CE}} = \pm 0.0656 \tag{9}$$

As the Mean of all WOC reading, $T_\rho = 0.2334$

Using Eq. 10, the value of estimated mean μ_ρ:

$$\mu_\rho = 0.1537 \text{ mm} \tag{10}$$

Using Eq. 11, the predicted optimal range for 95% confidence interval for the confirmation of experiments is given by:

$$0.0881 < \mu_\rho < 0.2193 \tag{10}$$

Two possibilities arise, one is the prescribed combination of parameters level is identical to one of those in the experimental design and second is the prescribed combination of parameters level is not included in the experiment. In the present

Table 6 Response for means for utility function

Level	Voltage (V)	Electrolyte conc (%wt/v)	Wire speed (%)	Interelectrode gap (cm)
1	5.606	5.981	5.752	5.396
2	6.310	5.480	5.680	5.848
3	5.251	5.706	5.734	5.924
Delta	1.059	0.501	0.072	0.528
Rank	1	3	4	2

investigation, second situation exists and optimization of results has been carried out through the prescribed combination. The experiment performed on these input factors gives a value of 0.197 mm which comes well in this range.

3.3 *Multiresponse Optimization*

The output responses from machining using WECDM have been unified using utility approach as explained in previous section. MRR and WOC are included to make the utility response function thus converting them to utility data as given in Table 6.

3.3.1 Preference Scales

(a) *Material removal rate (PMRR)*

X^* = Predicted output value of MRR = 0.2420 mg/min.
$X_i^{'}$ = Minimum admissible value of MRR = 0.0036 mg/min, using the above values and Eq. 5, the preference scale for 'MRR' has been found using Eq. 11;

$$P_{\text{MRR}} = 4.92 \times \log\left(\frac{X_{\text{MRR}}}{0.0036}\right) \tag{11}$$

(b) Width of Cut (PWOC)

X^* = Predicted output value of WOC = 0.1537 mm.
$X_i^{'}$ = Maximum admissible value of WOC = 0.354 mm. Using the above values and Eq. 5, the preference scale for 'WOC' has been found as shown in Eq. 12;

$$P_{\text{WOC}} = -24.84 \times \log\left(\frac{X_{\text{WOC}}}{0.354}\right) \tag{12}$$

Table 7 ANOVA for utility function

Source	DF	Seq SS	Contribution (%)	Adj SS	Adj MS	*F*-value	*P*-value
Voltage (V)	2	3.4842	46.54	3.4847	1.7421	6.96	0.015
Electrolyte conc (%wt/v)	2	0.7557	10.10	0.7557	0.3779	1.51	0.272
Wire speed (%)	2	0.0168	0.23	0.0168	0.0084	0.03	0.967
Interelectrode gap (cm)	2	0.9774	13.06	0.9774	0.4887	1.95	0.197
Error	9	2.2517	30.08	2.2517	0.2501		
Total	17	7.4860	100.00				

3.3.2 Utility Value Calculation

Though it is a characteristic of WECDM process to yield a low MRR, various techniques have been used in previous sections to obtain an optimized setting of input parameter to attain maximum MRR. On the other hand, slots of minimum possible WOC are desirable. The indirect effect of achieving a higher MRR is getting a higher WOC which compromises the situation. The aim of this paper is to achieve such a combination of input factors that the available MRR may be the highest possible along with minimum WOC without compromising each other characteristics. So, keeping in mind the value of both the output responses equal priority weightage has been assigned to both the response measures. The utility function dependent on the experiment runs was computed as shown in Eq. 13:

$$U(n, R) = \text{PMRR}(n, R) \times \text{WMRR} + \text{PWOC}(n, R) \times \text{WWOC} \qquad (13)$$

where WMRR = ½ and WWOC = ½ are the priority weights assigned to the output responses. *n*&*R* are the trial run and replication, respectively ($n = 1, 2, 3, \ldots.9$; $R = 1, 2$). The calculated utility values are given in Table 7.

Using Minitab software, graphs are obtained to show the effect of input factors on utility value. The ranking of the input factors can be observed, and it can be seen that voltage is again dominating to affect the result and it can be observed that to obtain optimum utility value, the optimized combination is A2, B1, C1, and D3, that is, 33 V DC supply, 10% wt/v concentration, 100% wire speed, and 13 cm interelectrode gap. Results for utility value were analyzed using ANOVA to find the significant factors. ANOVA table for utility value is given in Table 7. Table is generated for 95% confidence level. For 95% confidence interval, a factor is said to be significant if F-value is greater than table value and *P*-value is less than 0.05. Table *F*-value for numerator 2 and denominator 17 is 19.43. *F*-value shows that voltage is the most significant parameter in width of cut response whereas wire speed is the least.

4 Conclusions

Based on the study through DOE techniques, ANOVA, and optical microscopic images, following conclusions can be made:

1. The optimized combination to obtain maximum MRR is 40 V DC supply, 20% wt/v concentration, 120% wire speed, and 3 cm interelectrode gap.
2. To obtain minimum WOC, the optimized combination is 25 V DC supply, 10% wt/v concentration, 140% wire speed, and 13 cm interelectrode gap.
3. In case of multiresponse optimization, the optimal setting to obtain larger MRR and smaller WOC simultaneously is 33 V DC supply, 10% wt/v concentration, 100% wire speed, and 13 cm interelectrode gap.
4. From optical microscopic images, it is clear that the inner surface texture of the cut slots is smooth when NaOH electrolyte solution is used.

References

1. Kumar M, Vaishya RO, Suri NM (2020) Machinability study of zirconia material by micro-ECDM. In: Sharma VS, Dixit US, Sørby K, Arvind Bhardwaj RT (eds) Manufacturing engineering, pp 195–209. Springer, Singapore. https://doi.org/10.1007/978-981-15-4619-8_15
2. Chen X, Ji L, Bao Y, Jiang Y (2012) Improving cutting quality by analysis of microstructure characteristics and solidification behaviour of recast layer formation on laser cut ceramic. J Eur Ceram Soc 32:2203–2211. https://doi.org/10.1016/j.jeurceramsoc.2012.03.020
3. Singh J, Vaishya R, Kumar M (2019) Fabrication of micro features on quartz glass using developed WECDM setup. ARPN J Eng Appl Sci 14:725–731
4. Singh M, Singh S (2020) Sustainable electrochemical discharge machining process: Characterization of emission products and occupational risks to operator. Mach Sci Tech 1:19. https://doi.org/10.1080/10910344.2020.1752238
5. Arab J, Dixit P (2020) Influence of tool electrode feed rate in the electrochemical discharge drilling of a glass substrate. Mater Manuf Processes 00:1–12. https://doi.org/10.1080/10426914.2020.1784936
6. Arab J, Mishra DK, Kannojia HK, Adhale P, Dixit P (2019) Fabrication of multiple through-holes in non-conductive materials by electrochemical discharge machining for RF MEMS packaging. J Mater Process Technol 271:542–553. https://doi.org/10.1016/j.jmatprotec.2019.04.032
7. Singh T, Dvivedi A, Arya RK (2019) Fabrication of micro-slits using W-ECDM process with textured wire surface: An experimental investigation on Kerf overcut reduction and straightness improvement. Precis Eng 59:211–223. https://doi.org/10.1016/j.precisioneng.2019.05.008
8. Wang J, Fu C, Jia Z (2018) Research on oil film-assisted wire electrochemical discharge machining. Int J Adv Manuf Tech 96:2455–2461. https://doi.org/10.1007/s00170-018-1781-8
9. Yadav P, Yadava V, Narayan A (2018) Experimental investigation of Kerf characteristics through wire electrochemical spark cutting of alumina epoxy nanocomposite. J Mech Sci Technol 32:345–350. https://doi.org/10.1007/s12206-017-1234-6
10. Rattan N, Mulik RS (2017) Improvement in material removal rate (MRR) using magnetic field in TW-ECSM process. Mater Manuf Process 32:101–107. https://doi.org/10.1080/10426914.2016.1176197
11. Paul L, Hiremath SS (2016) Improvement in machining rate with mixed electrolyte in ECDM process. Proc Tech 25:1250–1256. https://doi.org/10.1016/j.protcy.2016.08.218

12. Gupta PK, Dvivedi A, Kumar P (2015) Developments on electrochemical discharge machining: a review of experimental investigations on tool electrode process parameters. Proc Inst Mech Eng Part B: J Eng Manuf 229:910–920. https://doi.org/10.1177/0954405414534834
13. Singh A, Jawalkar CS, Vaishya R, Sharma AK (2014) A study on wire breakage and parametric efficiency of the wire electro chemical discharge machining process. In: 5th International and 26th all India manufacturing technology, design and research conference AIMTDR 272(1)–272(6)
14. Kuo KY, Wu KL, Yang CK, Yan BH (2013) Wire electrochemical discharge machining (WECDM) of quartz glass with titrated electrolyte flow. Int J Mach Tools Manuf 72:50–57. https://doi.org/10.1016/j.ijmachtools.2013.06.003
15. Jain VK, Rao PS, Choudhary SK, Rajurkar KP (1991) Experimental investigations into traveling wire electrochemical spark machining (TW-ECSM) of composites. J Eng Ind 113:75–84. https://doi.org/10.1115/1.2899625
16. Oza AD, Kumar A, Badheka V (2020) Improving quartz micro-machining performance by magnetohydrodynamic and zinc-coated assisted traveling wire-electrochemical discharge machining process. Mater Today: Proc 28:970–976. https://doi.org/10.1016/j.matpr.2019.12.334
17. Kumar M, Vaishya RO, Oza AD, Suri NM (2020) Experimental investigation of wire-electrochemical discharge machining (WECDM) performance characteristics for quartz material. Silicon 12:2211–2220. https://doi.org/10.1007/s12633-019-00309-z
18. Oza AD, Kumar A, Badheka V, Arora A (2019) Traveling Wire electrochemical discharge machining (TW-ECDM) of quartz using zinc coated brass wire: investigations on material removal rate and Kerf width characteristics. Silicon 11:2873–2884. https://doi.org/10.1007/s12633-019-0070-y
19. Dhiman P, Vaishya R, Kumar M (2019) A review on machining by electrochemical discharge phenomena. Int J Tech Innov Mod Eng Sci (IJTIMES) 5:2017–2020
20. Rajput V, Goud M, Suri NM (2019) Study on effective process parameters: toward the better comprehension of ECDM process. Int J Mod Manuf Tech 11:105–118
21. Verma R, Kant S, Suri NM (2019) Process optimization of slurry spray technique through multi-attribute utility function. Arab J Sci Eng 44:919–934. https://doi.org/10.1007/s13369-018-3337-5
22. Verma R, Kant S, Suri NM (2014) Adhesion strength optimization of slurry sprayed mullite-based coating using Taguchi method. Proc Inst Mech Eng Part E: J Process Mech Eng 230:87–96. https://doi.org/10.1177/0954408915595948

Effect of Process Parameters on Etch Depth of Aluminium Material in Photochemical Machining

Abhay Utpat, N. D. Misal, B. P. Ronge, and B. A. Kamble

1 Introduction

The emerging non-traditional process PCM finds wide scope in miniaturization of dimensions in mechanical components. PCM is also known as etching, photochemical machining, wet etching, photoetching, photochemical milling, etc. The photoresist used PCM has two types given below.

- Positive photoresist
- Negative photoresist

The process is started with by creating required design called as phototool in AutoCAD software and printed it on transparent sheet of plastic for etching. The photoresist is applied on metal surfaces. The design is kept on photoresist and is exposed by using high-intensity ultraviolet light. The pattern is transferred on photoresist, and it will be visible after development. Davis et al. [1] etched varieties of precision parts and decorative components to develop microfluidic channels, etc. Allen et al. [2, 3] demonstrate the three-dimensional etching process and gave characterization of ferric chloride etchant. The application of PCM has widely been used in the manufacturing of thin, flat and complex metal parts (i.e. lead frames, colour

A. Utpat (✉) · N. D. Misal · B. P. Ronge · B. A. Kamble
SVERI's College of Engineering, Pandharpur 413304, India
e-mail: aautpat@coe.sveri.ac.in

N. D. Misal
e-mail: ndmisal@coe.sveri.ac.in

B. P. Ronge
e-mail: bpronge@coe.sveri.ac.in

B. A. Kamble
e-mail: banduakamble@coep.sveri.ac.in

H. K. Dave and D. Nedelcu (eds.), *Advances in Manufacturing Processes*, Lecture Notes in Mechanical Engineering, https://doi.org/10.1007/978-981-15-9117-4_7

TV masks, sensors, heat plates, printed circuit boards) in electronics, precision engineering and decorative industries in the past forty years. Cakir et al. [4] etched copper by cupric chloride and suggested suitable process for regenerated waste treatment. Jang et al. [5] developed a 3D system in package (SIP) process. It has been considered as a superb microelectronics packaging system. Barron et al. [6] studied the negative and positive type photoresist chemistry. In negative photoresist, the UV light incident part becomes hard and in positive type, the light incident part becomes soluble in developer solution. Bruzzone et al. [7] generated simulation model of 2D photochemical machining process. The simulation was carried out in six steps. An experimental analysis of the process parameters on microgeometry is reported. Saraf et al. [8] presented mathematical model of PCM and optimized conditions of PCM process. Zhang et al. [9] investigated surface texturing of carbon steel by PCM process. The study shows the effect of process parameters on surface texturing. Mudigond et al. [10] show the surface texturing of Monel 400 and studied effect of process parameters on response measure. Bhasme et al. [11] investigated response surface methodology on SS316L steel. The selection of etchant for machining of parts is depending on the surface topology and hardness of material. The surface finishing of etched part is varying with etchant to etchant.

The major steps involved on 3D PCM are given below

- Preparation of phototool
- Selection of base metal
- Preparation of workpiece
- Photoresist coating
- Developing
- Etching
- Stripping and inspection.

As stated from Gao et al. [12], PCM is best suitable process for manufacturing microchannels. Misal et al. [13, 14] show the surface topography of Inconel 718 and responses are measured at different conditions and effect of process parameters on response measure is checked for hard materials like Monel and Inconel 718. Wangikar et al. [15] optimized process parameters of brass and German silver and concluded with brass has a better etching performance. Agrawal et al. [16, 17] create microfluidic channels, microholes feature by dissolving away the metal parts in etching process and optimized process. There is no use of cutting tool for manufacturing of components, due to this the stresses and other defects normally arise from metal cutting which is absent in the final part of PCM. Kamble et al. [18] show variable etching process for different colours. In three-dimensional PCM variable, depth can be achieved by less and more exposing of samples. The energy content of light is varying with colour by colour, which results in less and more harden of photoresist material. The 3D etching process is mainly depending on a development of photoresist material. Jadhav et al. [19] have focused the etching depth variation of brass material for different operating conditions. The objective of this study is to achieve depth on two-dimensional surfaces. The effect of process parameters on a etching depth on a aluminium strip were studied. The objective of this study is to achieve depth on two-dimensional surfaces. The effect of process parameters on a etching depth on a aluminium strip was studied.

2 Material and Method

The aluminium is the material used for this study. Nowadays, this material is most used in industries. The size of specimen used is 30 mm × 30 mm × 0.5 mm ($L \times B \times T$). Photochemical machining process is carried out by using a phototool. The phototool is an AutoCAD drawing with detailed dimensions of the required shape which is then printed on a tress paper or transparent paper. The phototool used in PCM is made up by printing different colours on transparent sheet. The 2400 × 2400 dpi printer was used to print phototool as shown in Fig. 2a. The accuracy of PCM is mainly depending on the quality of printing images. The thickness of phototool is 80 μm, the less thickness means less power loss and vice versa.

After preparing the phototool, the aluminium material is cut into required size. The first task is to clean the workpiece. The cleaning of the surface is done firstly by polishing the surface with a sand paper and then cleaning it by thinner (trichloroethylene or acetone). This step is done to make the surface of workpiece free from foreign particles, debris, dust, grease, oil and other contaminants. Because of polishing and cleaning with thinner, it gets easier to have a good adhesion of photoresist with the metal surface. The improper cleaning leads to inaccurate etching and bad surface finish. After cleaning the workpiece, a thin layer of photoresist is applied on the metal surface. The negative-type photoresist was used for experimentation. The characteristic of negative photoresist is it gets hardened when UV light is incident on it. It depends on the energy content of light coming out from phototool. It is necessary to bake photoresist after coating for better adhesion of film onto the metal surface. The baking time is 2 min and photoresist thickness range is 5–8 μm. The dip coating method is applied to coat photoresist onto the base metal. The 1 kW power UV light bulb is used for exposing the sample. It radiates the intensity 1800 μW per centimetre square at 25 cm distance. In 2D PCM process, the samples are exposed in only one direction.

The generated phototool is put on the coated workpiece and exposed to UV source for 90 s. The 2D exposing unit is shown in given Fig. 2b. The developing solution is the mixture of sodium carbonate and water. After UV exposure, the specimen is held in the developer for 90 s, which results in the formation of impression on workpiece. After that specimen is washed with water. The final step of the machining is to put the developed metal workpiece into the etchant. The etchant is the solution of $FeCl_3$ and water.

In etching process, the phototool is black and white, in which only etching of black colour is happened and there is no effect on white colour. The etching process set-up is shown Fig. 2c. Figure 1 shows the PCM flow chart, which explains the PCM process and its different steps.

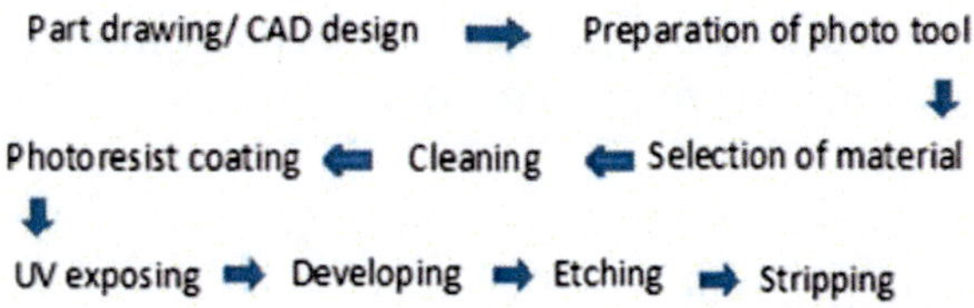

Fig. 1 PCM flow chart

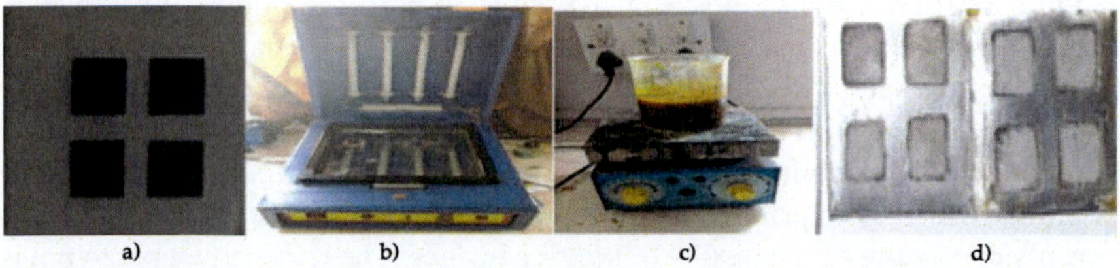

Fig. 2 **a** Phototool, **b** 3D exposure unit, **c** 3D etching unit, **d** finished component

The finished samples with different etching depths on 2D surface and 3D surface are shown in Fig. 2d.

3 Design of Experiments

The numbers of preliminary experiment were performed to analyse etching behaviour of aluminium material. The input parameters are etching time, temperature, etc., and response parameter is etching depth. The depth of etch was measured by using digital micrometre having least count 0.001 mm. The concentration of etchant is constant throughout the experiments which is 40 Baume scale measured by hydrometer. The level of experiments is selected by studying the etching behaviour of aluminium material at different temperature. The etching depth increases by increasing temperature and time.

The levels of experiments are shown in Table 1.

The coating thickness of photoresist is 4–5 μm, exposure time of UV machine is 85 s and developing time is 90 s. These parameters were taken constant throughout the experimentation.

The experimental values are shown in Table 2.

Table 1 Control parameters with their levels

Control parameters	Level 1	Level 2	Level 3
Temperature (°C)	43	46	49
Time (min)	2, 4,6	2, 4, 6	2, 4, 6

Table 2 Etching parameters and etching depth relation

Etching temperature (°C)	Time (min)	Etching depth (μm)
43	2	29
	4	44
	6	80
46	2	115
	4	139
	6	360
49	2	135
	4	365
	6	545

4 Results and Discussion

The performance of photochemical machining of aluminium has been evaluated by etching depth at different time and temperature. The etching depth of each specimen was recorded at three locations, and average value is taken for analysis. The etching depth of etching material increases by changing process parameters such as etching temperature and etching time. The depth of etching increases by increase in etching time and temperature. At the low temperature, the depth of etch is low due to, i.e., less molecules are available to etching reaction, i.e., less collision of molecules on the aluminium surface. As temperature increases the collisions in etchant molecules is increasing, it results in higher depth of etch.

The etching depth of etching material is changed by changing process parameters such as etching temperature and etching time. Figure 3 shows the relation between etching depth and process parameters. The depth of etching increases by increase in etching time and temperature.

The depth of etching at low temperature, i.e. for 43 °C is low at time 2 min and it increases with time its highest at time 6 min which is 80 μm. Similarly, for other temperatures, the depth of etching is highest at higher temperature and time.

4.1 Comparison Between Etching Temperatures and Etching Depth

The depth of etching is maximum at 49 °C, i.e. 545 μm due to more collision of etchant molecules on the surface of aluminium. The etchant species develop more reactive molecules at this temperature compared to other. Low temperature means less reactive species are collide on aluminium surface which results in less etching depth. The energy of reactive species increases as increase in temperature.

The etching time also has significant effect on depth of etch, at minimum time, the interaction of molecules on surface is less which results in minimum depth of etch.

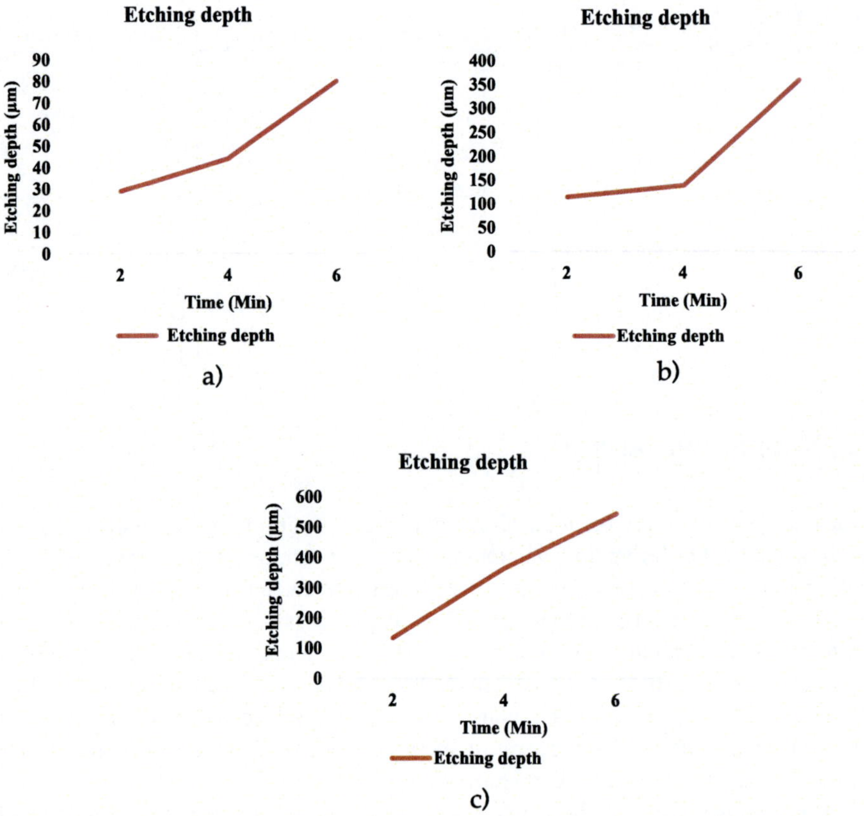

Fig. 3 **a** etching depth versus time at 43 °C, **b** etching depth versus time at 46 °C, **c** etching depth versus time at 49 °C

Similarly, at maximum time, the interaction of etchant species on surface is high which results in higher depth of etch. Figure 4 compared etching depth for different temperature and time. At highest temperature, the reaction between etchant species on surface happens rapidly compared to other temperature.

5 Conclusions

The present study investigates the depth of etching for different temperatures and time. As per this study, following conclusions were made.

- In this study, effect of process parameters like etching temperature and time on a etching depth is studied. The temperature has significant effect on a etching depth.
- The higher temperature results in more etching depth, due to which the rate collision of etchant species on the surface of aluminium is more at higher temperature.

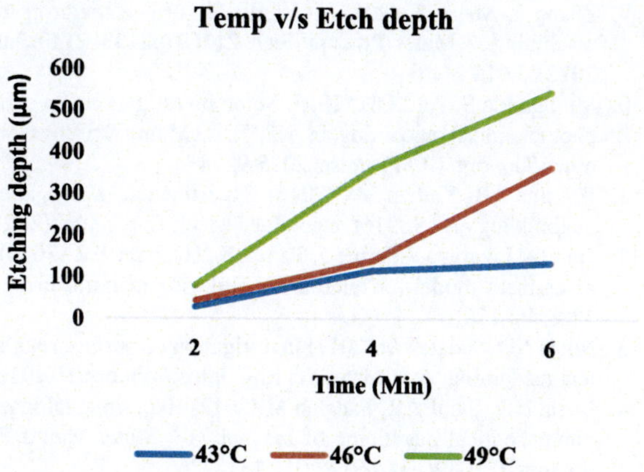

Fig. 4 Etching temperature versus etching depth

Similarly, at low temperature, the etchant molecules are less reactive and less in number which give less depth of etch.

- Depth of etching is goes on increasing by increase in temperature and etching time.
- The more etching depth obtained at 49 °C temperature, i.e. 545 µm in 6 min.
- The lowest depth obtained at 40 °C temperature, i.e. 29 µm in 2 min.

References

1. Davis PJ, Overturf GE (1986) Chemical machining as a precision material removal process. Precis Eng 8(2):67–71. https://doi.org/10.1016/0141-6359(86)90088-7
2. Allen DM, Brightman JRC, Gillbanks PJ, Smith PM, Crookall JR (1987) Three-dimensional photochemical machining. CIRP Ann 36(1):91–94. https://doi.org/10.1016/s0007-8506(07)62561-x
3. Allen DM, Almond HJ (2004) Characterisation of aqueous ferric chloride etchants used in industrial photochemical machining. J Mater Process Tech 149(1–3):238–245. https://doi.org/10.1016/j.jmatprotec.2004.02.044
4. Cakir O (2006) Copper etching with cupric chloride and regeneration of waste etchant. J Mater Process Tech 175(1–3):63–68. https://doi.org/10.1016/j.jmatprotec.2005.04.024
5. Jang DM, Ryu C, Lee KY, Cho BH, Kim J, Oh TS, Yu J (2007) Development and evaluation of 3-D SiP with vertically interconnected through silicon vias (TSV). In: 2007 proceedings 57th electronic components and technology conference, pp 847–852. IEEE. https://doi.org/10.1109/ectc.2007.373897
6. Barron AR (2009) Composition and photochemical mechanisms of photoresists. Creat Commons 1:1–7
7. Bruzzone AAG, Reverberi AP (2010) An experimental evaluation of an etching simulation model for photochemical machining. CIRP Ann 59(1):255–258. https://doi.org/10.1016/j.cirp.2010.03.070
8. Saraf AR, Misal ND, Sadaiah M (2012) Mathematical modelling and optimization of photochemical machining. Adv Mater Res 548:617–622. https://doi.org/10.4028/www.scientific.net/amr.548.617

9. Zhang J, Meng Y (2012) A study of surface texturing of carbon steel by photochemical machining. J Mater Process Tech 212(10):2133–2140. https://doi.org/10.1016/j.jmatprotec.2012.05.018
10. Mudigonda S, Patil DH (2015) Some investigations on surface texturing on Monel 400 using photochemical machining. In ASME. Int Manuf Sci Eng Conf Am Soc Mech Eng Digit Collect. https://doi.org/10.1115/msec2015-9294
11. Bhasme AB, Kadam MS (2016) Experimental investigation of PCM using response surface methodology on SS316L steel. Int J Mech Eng Tech 7(2):25–32
12. Gao Q, Lizarazo-Adarme J, Paul BK, Haapala KR (2016) An economic and environmental assessment model for microchannel device manufacturing: part 2–application. J Clean Prod 120:146–156
13. Misal ND, Sadaiah M (2017) Investigation on surface roughness of Inconel 718 in photochemical machining. Adv Mater Sci Eng. https://doi.org/10.1016/j.jclepro.2015.04.142
14. Misal ND, Saraf AR, Sadaiah M (2017) Experimental investigation of surface topography in photochemical machining of Inconel 718. Mater Manuf Process 32(15):1756–1763. https://doi.org/10.1080/10426914.2017.1317786
15. Wangikar SS, Patowari PK, Misra RD (2017) Effect of process parameters and optimization for photochemical machining of brass and German silver. Mater Manuf Process 32(15):1747–1755. https://doi.org/10.1080/10426914.2016.1244848
16. Agrawal D, Kamble D (2019) Effect and optimization of photochemical machining process parameters for manufacturing array of micro-hole. J Braz Soc Mech Sci Eng 41(4):178. https://doi.org/10.1007/s40430-019-1678-y
17. Agrawal D, Kamble D (2019) Optimization of photochemical machining process parameters for manufacturing microfluidic channel. Mater Manuf Process 34(1):1–7. https://doi.org/10.1080/10426914.2018.1512115
18. Kamble B, Utpat A, Misal ND, Ronge BP (2019) 3D photochemical machining of copper by using colored phototools. Int J New Tech Res (IJNTR) ISSN: 2454-4116 5(7): 28–32. https://doi.org/10.31871/ijntr.5.7.3
19. Jadhav Saurabh M, Karatkar Onkar V, Bangale Kamesh N, Choudhari Deepak B, Utpat AA, Kamble Banduraj K (2019) Etching depth variation of brass material for different operating conditions. Int J New Tech Res (IJNTR) ISSN: 2454-4116 5(4): 93–96 https://doi.org/10.31871/ijntr.5.4.42

Quartz Micro-machining Using Wire Electrochemical Spark Machining Process

Ankit D. Oza, Abhishek Kumar, Vishvesh Badheka, Aditya Nema, and Manoj Kumar

1 Introduction

In the past decades, various engineering insulating materials such as polymers, ceramics, composites, glasses, polydimethylsiloxane (PDMS), and Pyrex have incredible industrial applications in the defense, aviation, micro-electromechanical systems, optical, micro-opto-electromechanical, robotics, biomedical, electronics field, and space [1–9].

Nowadays, various micro-fluidic devices are widely taken for study at depth, and miniaturization of it will enable favorable applications for biological and chemical research. Mostly, micro-fluidic channels are made of glass and polymers materials. However, some major machining issues have been observed with the uses of a polymer material such as low solidity and permeability to moisture and meager chemical resistance. [8, 10–13]. Quartz is an untainted form of silica have over 99.9% SiO_2 contain and used widely in opto-electronics, computers, smartphones, micro-optics, etc. Further, it has high transmission, thermal chemical, and electrical resistivity, and due to these properties, it is hard to machine for micro-applications. However, few non-conventional micro-machining processes already used to machine quartz in micro-region. This process includes ultrasonic machining [14], Nd: Yag laser

A. D. Oza (✉) · A. Kumar
Department of Industrial Engineering, Pandit Deendayal Petroleum University, Gandhinagar 382007, Gujarat, India
e-mail: ankit.ophd15@sot.pdpu.ac.in

V. Badheka · A. Nema
Department of Mechanical Engineering, Pandit Deendayal Petroleum University, Gandhinagar 382007, Gujarat, India

M. Kumar
Department of Production and Industrial Engineering, Punjab Engineering College (Deemed to Be University), Chandigarh 160012, India

H. K. Dave and D. Nedelcu (eds.), *Advances in Manufacturing Processes*, Lecture Notes in Mechanical Engineering, https://doi.org/10.1007/978-981-15-9117-4_8

machining[15], abrasive slurry jet [16], rotary ultrasonic drilling [17], laser-assisted dry micro-grinding [18], ultrashort laser pulses [19], ultrashort laser machining [20], etc. However, these processes used highly sophisticated and costly instruments, and as a result machining and maintenance cost increases. Furthermore, higher HAZ, tool wear, and low material removal rate are key limitations [21–26]. Moreover, some well-known non-traditional micro-machining processes like W-EDM, ECM, and EDM cannot machine non-conductive materials like quartz used in this case. Utilizing the benefits of two or more non-conventional machining processes, we can develop a hybrid machining procedure which can assist in machining of electrically non-conductive materials. One such process is ECDM and the method of which this content would be is wire electrochemical spark machining (WECSM) process.

Varieties of high-strength, brittle materials, which are electrically non-conducting in nature, can be machine by wire electrochemical spark machining (WECSM) process [27–30]. WECSM process is a recently developed and hybrid non-conventional micro-machining process, which combines the principle of wire EDM and ECM process. WECSM process is also identified as traveling wire electro-chemical discharge machining process, wire spark-assisted chemical engraving, wire electro-erosion machining process, and wire electrochemical arc machining.

Figure 1 gives you an idea about the of WECSM process. In the WECSM process, mainly two electrodes are used: auxiliary (graphite) and active (wire). Both electrodes are dipped in the electrolyte (acidic or alkaline). To machine a workplace, the wire is used as an active electrode and it is considered as cathode and graphite electrode as the anode which is far from 25 to 80 mm from the quartz material (workpiece). In WECSM process, wire has been used to machine or slice the workpiece and therefore it continuously travels along the workpiece as shown in Fig. 1. Electrolysis process increases with an increase in applied voltage which is responsible to form a gas bubbles formation phenomena. These gas bubbles created a stable gas film

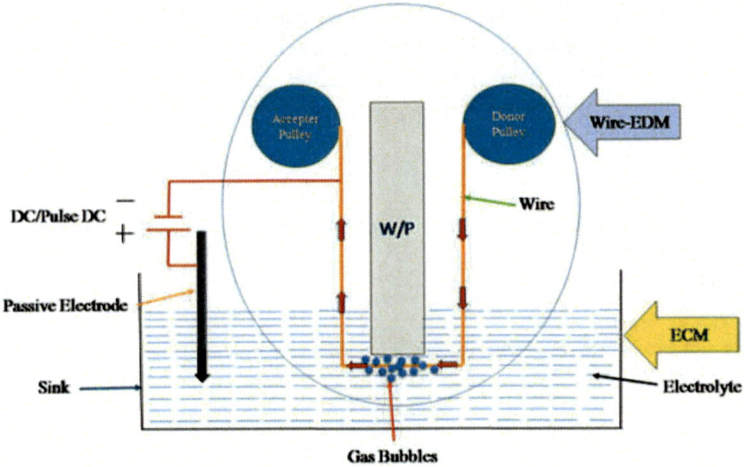

Fig. 1 Schematic diagram of WECSM process

around the wire surface which provides dielectric medium to produce spark. While the applied (input) voltage reached beyond the critical voltage, gas film breaks and sparking observed and due to this spark, high energy streamline of electrodes moves to the workpiece with tremendous speed and produced the high thrust shock wave toward the quartz (workpiece) as a result, workpiece melts and material is removed from the workpiece [31].

This process utilizes the energy of discharge to vaporize the workpiece thus causing micro-machining. WECSM has a wire as cathode and an auxiliary electrode extraordinarily larger in size than the wire as anode. Anode is dipped and cathode just touches the electrolyte placed in an electrolytic cell. The electrolyte is used as a working fluid here to complete the electrical circuit between them and also for other purposes like cooling. DC power is connected to the electrodes with supplied voltage varying with electrolyte concentration to be and the workpiece used. The work-piece used in this process is essentially non-conducting, brittle, and hard like glass, ceramics, etc., which are difficult to machine by other unconventional processes like electric discharge machining process (EDM), electrochemical machining process (ECM), and wire electrical discharge machining process (W-EDM) [32].

A DC voltage is applied between the electrodes which cause electrolysis and the formation of H_2 gas bubbles at cathode (wire) and O_2 gas bubbles at anode. When the number of H_2 gas bubbles formed at wire becomes substantial, these coalesce to form an insulating layer around the wire which causes huge resistance between wire electrode and electrolyte interface. This heightened resistance causes more heat, which vaporizes the electrolyte causing in turn an electrically non-conducting layer of H_2 and vapor mix layer around the wire. When the applied voltage overcomes this resistance, that is, when it crosses a critical level, discharge of electrons occurs in the form of spark at the wire electrolyte interface creating radiation powerful enough to vaporize the workpiece. Energy of that spark varies proportionally to the applied voltage between the two electrodes. In order to machine the workpiece, it has to be placed in a zone close to the discharge so that spark energy can be utilized to melt, vaporize, and erode it. The basic block diagram of WECSM setup is shown in Fig. 1.

The electrolyte chamber in the WECSM setup consists of electrolyte, anode, and cathode. The use of NaOH and KOH solutions must be free in water and act as electrolyte. Graphite rod acts as an inert anode, and zinc-coated brass wire acts as cathode. Graphite is completely dipped in electrolyte, whereas the wire just touches to electrolyte surface. When DC power supplied from a DC power source, it causes electrolysis of water breaking it down to its ionic form. The electricity completes the circuit with the cathode giving electrons creating an ionic movement in electrolyte and the anode receiving electrons. At cathode, that is, at wire, a reduction process always occurs in which a positive ion receives an electron to form a neutral molecule meaning positive ions will be attracted toward it to gain electron to get reduced. The ion that will get reduced at cathode depends on the energy authority which will be required to reduce it. Lower the energy required, more easily the ion will get reduced. In other words, only that ion will get reduced at cathode which has a lower position in electrochemical series. Here, in WECSM process, two ions are present near cathode ready to get reduced which is H^+ from water and Na^+ from NaOH electrolyte solution, H^+ being lower in electrochemical series gets reduced to form H_2 gas rather than Na^+. At anode, oxidation process occurs which attracts negative ions toward it to make the ion lose its electron to form a neutral molecule depending on its oxidation potential. Only OH^- available from both water and basic electrolyte is present near anode to get oxidized into water and O_2 gas [2, 33–35].

The graphite anode is an inert electrode which does not form any deposits and also does not interfere in reactions and only gives connection between the circuit and electrolyte. The ions involved in this cell are the dissociated ions available from water after giving electricity—H^+ and OH^- as given in Eq. 1 and the ions of the electrolyte that dissociate when dissolved in water. For example, if NaOH is taken as electrolyte, the available ions will be Na^+ and OH^- as shown in Eq. 1. The movement of these ions toward oppositely charged electrode helps in flow of current and completes the circuit.

If a basic electrolyte like NaOH is used, the hydrogen ions (H^+) will come from the water that is used to make the electrolytic solution and the hydroxide ions (OH^-) will come from both the basic electrolyte as well as water.

$$H_2O + 2e^- \rightarrow 2(OH)^- + H_2 \uparrow \tag{1}$$

The chemical reactions which occur with the supply of DC power evolve hydrogen gas around cathode which in turn helps in the discharge causing the machining which is given in Eq. 2 and the evolution of oxygen gas bubbles at anode is represented in the chemical reaction given in Eq. 3.

At cathode:

$$2H_2O \rightarrow O_2 \uparrow +4H^+ + 4e^- \tag{2}$$

At anode:

$$4(OH)^- \rightarrow 2H_2O + O_2 \uparrow +4e^- \tag{3}$$

Many researchers proposed few approaches to improve the WECSM process and its performance. In WECSM process, generally pulse (pulsating) DC and simple DC supply are used. Research shows that fine surface finish with minimum HAZ has been noticed with the use of pulse DC supply and therefore during the study, pulsating power supply has been used. Furthermore, the use of a reciprocating-type mechanism along with the addition of abrasive particles in an electrolyte provides a better surface finish and reduced the Kerf width [36]. Wang et al. [37, 38] made a vertically operating WECDM setup and machined 10 mm thick quartz material with aid of oil film insulation. They provided a round shape cut to quartz glass using vertical WECDM setup.

During the process, debris is stored in the machining zone, and therefore it reduced the quality of machined slots. To defeat these problems, magnetohydrodynamic convention approach has been used. This MHD approach helps out to flush the wreckage from the machining zone (distance between workpiece and wire). Furthermore, it improves the material removal rate and reduced the surface roughness and discharge current [39]. Another research study proposed titrated electrolytic flow to reduce the electrolyte waste and improve the machining performance. SiC abrasive particles were used in conjunction with titrated flow of electrolyte to machine a quartz material additionally, electrolyte consumption reduced, and therefore, it reduced the less machining cost and pollution [40, 41]. Mostly in WECSM process, NaOH and KOH are used as electrolyte. However, poor surface finish has been observed with NaOH. [23]. On the other hand, the use of KOH electrolyte reduced the slit expansion as compared to NaOH [36] and therefore in the present work, KOH electrolyte has been selected as an electrolyte. In the WECSM process, the wire breakage phenomenon has been observed and it reduced the machining efficiency and precision [2] and therefore it is required to minimize the wire breakage for stable machining operation. A recent study of W-EDM shows that the use of zinc-coated brass wire provides better cooling and surface finish and provides flushing as compared to plain wire. Moreover, it improves the conductivity, strength, and sparking ability and therefore zinc-coated brass wire selected as a wire material [42] and therefore during the study, 0.15 mm sized zinc-coated brass wire has been used to study the process performance.

From the above discussion, it can be concluded that WECSM is an advanced and cost-effective alternative to machine non-conducting materials irrespective of their physical properties. Different studies were proposed to improve the process. For better performance, effective material removal, wire breakage problem, effective electrolyte circulation to machining zone, and gas film formation are critical parameters.

2 Experimental Methodology

The WECSM setup in-house designed and fabricated to conduct the experiments is shown in Fig. 2. WECSM setup consists of various subcomponents like pulse DC power supply unit, electrode sample, machining chamber, machine structure, tool, workpiece holder, tool and auxiliary electrodes, and wire feeding mechanism. The ultra-high-molecular-weight (UHMW) polyethylene material has been used in the fabrication of the machining chamber that sustains the chemical erosion effect of electrolyte. To maintain the proper tension in the wire, Teflon materials pulleys were used. The DC motor was used to run the left-hand pulley to provide the wire feed in machining operations. The Zn-coated brass wire having a diameter of 150 μm, the auxiliary electrode of a graphite material, and NaOH electrolyte is used in the experimentations. The size of the auxiliary electrode is 80 mm in length and 20 mm in diameter. Pulsating power supply has been designed with the 0–200 V and 0–6 A range with 1 kHz. Table 1 shows the quartz (workpiece) properties.

The experiments were designed using Taguchi hybrid methodology along with an analysis of variance (ANOVA) using the MINITAB 17 package. The governing input

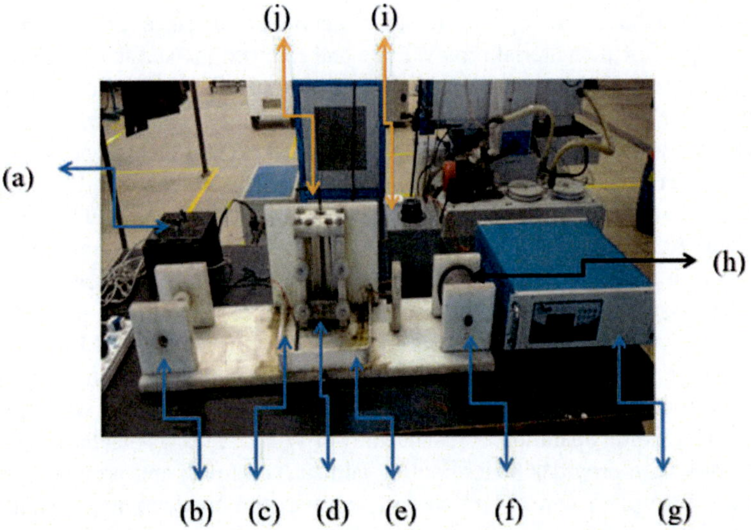

Fig. 2 WECSM setup [27], where, **a** wire feed controller, **b** accepter pulley, **c** auxiliary electrode, **d** workpiece, **e** electrolyte tank, **f** donor pulley, **g** power supply unit, **h** wire, **i** dimmer, and **j** workpiece feed controller

Table 1 Properties of quartz material

Parameters	Values
Size	75(L) × 25(W) × 2(T) mm
Young's modulus	7.2×10^{10} Pa
Density	2.2×10^3 kg/m^3
Tensile strength	50×10^7 Pa (N/m^2)
Thermal conductivity (at 20 °C)	1.4 W/m °C
Coefficient of thermal expansion (20–320 °C)	5.11×10^{-7} cm/cm °C
Hardness	5.5–6.5 Mohs' scale (N/mm^2)

parameters were selected using trial experiments after reviewing the literature for the process. Taguchi L_9 orthogonal has been used to evaluate the process performance. Input process parameters: applied voltage (V), %age of electrolyte concentration (g/L), and wire feed rate (m/min) and response parameters: average surface roughness (R_a) and material removal rate (MRR). The governing input parameters were selected using trial experiments after reviewing the literature for the process. For voltage 30–40 (V), electrolyte concentration—25–35% (g/L) and wire speed 0–15 (m/min) were selected for study as shown in Table 2. An analysis of variance (ANOVA) study was carried out to measure the percentage contribution and significance of each input

Table 2 Factors with levels

Levels	Factors		
	A	*B*	*C*
	Voltage (V)	Percentage concentration (g/L)	Wire speed (m/min)
1	30	25	3
2	35	30	8
3	40	35	13

parameter. *S*/*N* graphs representing the signal-to-noise ratio be plotted to determine and analyze the desirable signal value to undesirable noise.

The response parameter, i.e., MRR (mg/min) was evaluated with expression shown in Eq. 4, where M_b and M_a symbolize the mass of the quartz glass before and after the (mg) experimenting, respectively, and t is the total time (min) taken while machining [19, 21]. The machining operation was carried for a fixed interval of time, i.e., 10 min. The METTLER TOLEDO model weighing machine was used to measure the weight of the quartz glass sample (before and after experiments). A scanning electron microscope (SEM) has been used to investigate the profile of micro-slits. Average surface finish has been measured with surface tester (model no—ISR S400).

$$\text{MRR} = \left(\frac{M_b - M_a}{t}\right) \tag{4}$$

3 Results and Discussions

All the experiments were carried out with Taguchi orthogonal L_9 array as shown in Table no. 2 with all three input factors and each factor has three levels. The *S*/*N* ratio was calculated considering different desirability criteria for both response parameters. In *S*/*N* ration, higher the better for MRR and lower the better for R_a, selected [22]. The *S*/*N* ratio calculated using Eq. 5. The *S*/*N* ratio value after calculation for L_9 orthogonal array is shown in Table 3.

$$\frac{S}{N} = -10 \times \log(\text{mean square deviation}) \tag{5}$$

Table 3 *S/N* values for MRR and R_a

Experiment no.	MRR	R_a
1	−12.590	−17.527
2	−11.795	−17.889
3	−11.116	−18.121
4	−11.588	−18.842
5	−10.906	−19.042
6	−10.417	−19.528
7	−10.675	−19.859
8	−10.026	−20.456
9	−9.358	−21.055

3.1 Material Removal Rate (MRR)

The MRR plot of the *S/N* ratio is shown in Fig. 3 and it can be concluded that the MRR, *S/N* value increases with a raise in the applied voltage and electrolyte concentration. This might exist owing to the high sparking energy at higher-end levels, which helps in gas films, which aid in material removal. When the voltage increases, electrolysis reaction increases, and therefore more dense bubbles formed

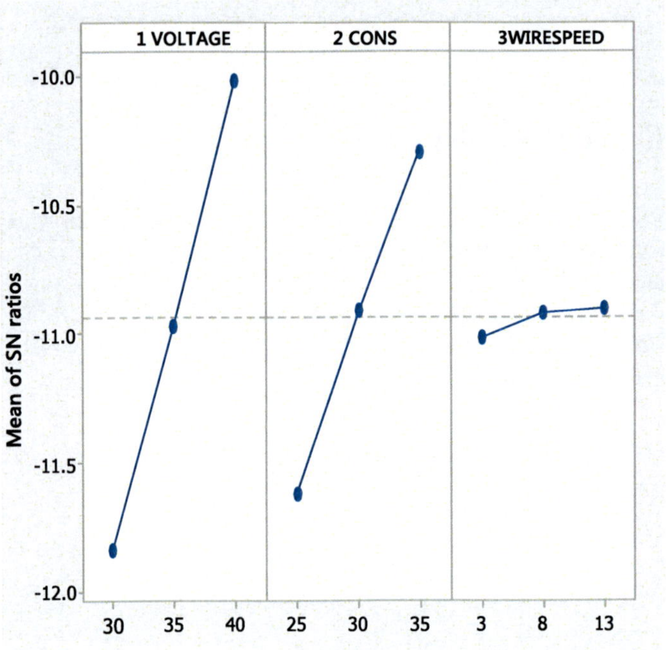

Fig. 3 *S/N* plot for MRR

Table 4 ANOVA for MRR and R_a

Source	DOF	Adj SS	Adj MS	*P*-value	Percentage age of contribution
MRR					
X	2	0.0053245	0.0026622	0.003	**65.402**
Y	2	0.0027873	0.0013937	0.005	**34.237**
Z	2	0.0000158	0.0000079	0.461	0.194
Error	2	0.0000135	0.0000067		0.165
Total	8	0.0081411			100
R_a					
X	2	11.3762	5.68809	0.006	**88.525**
Y	2	1.2232	0.61159	0.052	**9.5184**
Z	2	0.1849	0.09243	0.265	1.438
Error	2	0.0666	0.0333		0.518
Total	8	12.8508			100

resulted in high sparking observed at the wire electrode. Thus, the rate of material removal is increased. Dissolution rate has been enhancing with applied voltage. Furthermore, with increase in supply voltage, the temperature of the electrolytes also raises resulting in the mobility of ions that tremendously increases and this might be one reason for the improvement in material removal rate (MRR). Additionally, by increase in electrolyte concentration, the conductivity value of electrolyte also increases, resulting in a higher amount of ions formed during the process, and thus material removal rate has been increased. Wire speed plays a negligible role during machining operation for material removal rate, only 0.194% contributing. Table 4 represents the ANOVA table, applied voltage contributing more than 65%, followed by %age of electrolyte contributes 34.23% and wire speed affects less (0.194%). The best possible parametric set for MRR is A3, B3, C3.

Results for MRR and R_a were analyzed using ANOVA to find significant factors which are affecting response factor. ANOVA table for material removal is given in Table 4. Table is generated for 95% confidence level.

3.2 *Surface Roughness* (R_a)

Figure 4 represents the *S/N* ratio plot for R_a and from the plot it has been observed that surface roughness decreases by an increase in the input voltage and %age of the electrolyte concentration. As discussed in MRR section, when the voltage and % age of the electrolyte concentration increase, high dense and stable gas film generated and resulted in high intensity sparking phenomenon observed at the workpiece and therefore rough machining surface observed. Another reason, electrolytic nature, with increases in conductivity, more free ions with huge potential bombarded on the

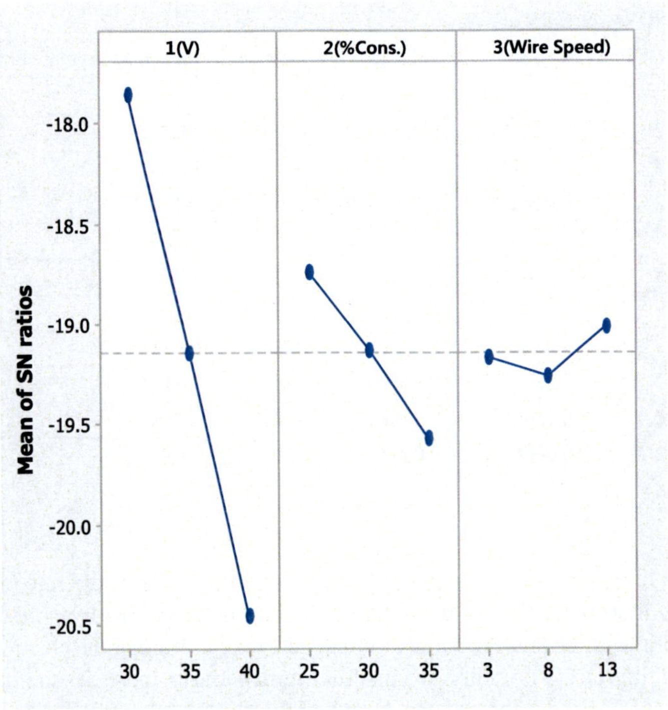

Fig. 4 *S/N* plot for R_a

workpiece surface, and therefore, the crater formed on the surface. Wire speed plays minor role (1.43%) on surface roughness.

Table 4 represents the ANOVA table, which shows the voltage (input) has the most significant parameter for surface roughness contributing more than 88%. Table 4 is generated for 95% confidence level. The most favorable level of input parameters is A1, B1, C3.

4 Analysis of Machined Quartz by Scanning Electron Microscope (SEM)

The quartz micro-machined profile is depicted in Figs. 5 and 6. Figure 5 illustrates the machined profile of quartz with the input parameters: voltage 30 V, electrode and 25% of concentration (250 g/L), wire speed 3 m/min and in Fig. 6 the input process parameters: voltage 40 V, secondary electrode and 35% of electrolyte concentration (350 g/L), wire speed 13 m/min. Supplementary electrode kept 30 mm from the wire electrode. Figure 5 shows the asymmetrical shape and HAZ at end side of the slit. It may be possible that higher wire speed and tension are accountable for uneven

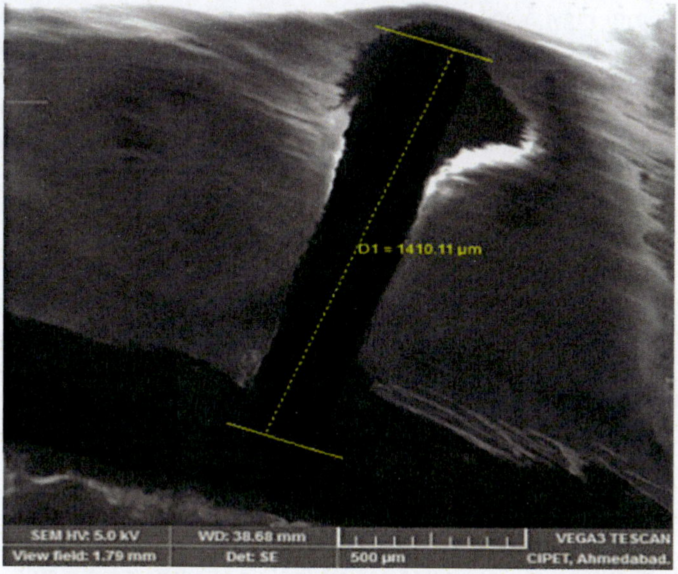

Fig. 5 Rough slit

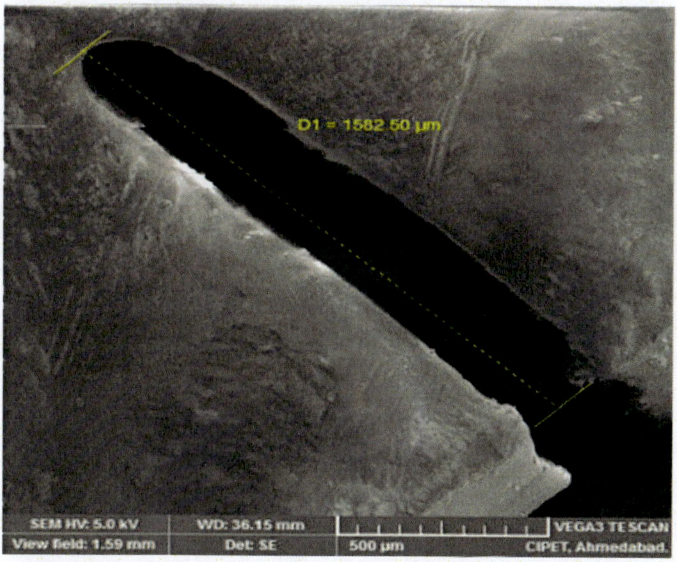

Fig. 6 Straight and smooth slit

micro-slit. From Figs. 5 and 6, it has been observed that some irregular machined profiles generated at lower input process parameters whereas at a higher voltage and concentration level (Fig. 6), the smooth surface profile is produced and the length of the machined cut is more in comparison with Fig. 5. Fine scattered machining surface has been observed because no flashing arrangement is provided during the operation. Furthermore, micro-burrs are detected along the slit, due to non-uniform gas film formation.

5 Conclusion

In the present work, it was demonstrated that the fabrication of micro-slits can be possible on quartz material with zinc-coated brass wire-assisted WECSM process. Based on experimental results in this study, we conclude the following.

(1) Smooth cutting machining operation has been observed with zinc-coated brass wire, reduced wire breakage problem, and improved material removal rate.
(2) Applied voltage and percentage of electrolyte concentration play a vital role during the machining operation and they govern the WECSM performance.
(3) Intended for higher material removal rates, DC supply voltage and % age of electrolyte concentration are the most major parameters which affect the process performance significantly. The best parametric set for material removal rate (MRR): A3, B3, C3 combination of the input parameter.
(4) For lower surface roughness response, the applied DC voltage is the most significant parameter having more than 88% contribution. The best parametric set for the surface finish: A1, B1, C3 combination of the input parameter.
(5) The SEM depicts the irregular surface finish at higher-end input parameters, due to the presence of crater wear. This might be because of poor flushing action across the machining zone.

Based on the experiments, results, analysis, and surface characterization, it can be concluded that the hard-to-machine quartz glass can be machined with the WECSM process at the micro-level by suitably selecting the combination of parameters. WECSM has tremendous potential to machine other non-conducting materials, for example, composites, glasses, FRP, diamond, fiberglass, foam, and ceramics materials. The process can be further modified to a hybrid process using ultrasonic vibrations, abrasive powders, and magnetic fields.

Acknowledgements The authors fully acknowledge the financial support of "Office of Research and Sponsored Program (ORSP)," Pandit Deendayal Petroleum University (P.D.P.U.), Gandhinagar—Gujarat, India. Project number: ORSP/R&D/SRP/2017/AOAK.

References

1. Arab J, Mishra DK, Kannojia HK, Adhale P, Dixit P (2019) Fabrication of multiple through-holes in non-conductive materials by electrochemical discharge machining for RF MEMS packaging. J Mater Process Technol 271:542–553. https://doi.org/10.1016/j.jmatprotec.2019.04.032
2. Rolf Wuthrich JDAZ (2015) Micromachining using electrochemical discharge phenomenon fundamentals and application of spark assisted chemical engraving. https://doi.org/10.1016/b978-0-323-24142-7.12001-3
3. Kumar M, Vaishya RO, Suri NM (2020) Machinability study of zirconia material by micro-ECDM. In: Sharma VS, Dixit US, Sørby K, Arvind Bhardwaj RT (eds) Manufacturing engineering, pp 195–209. Springer, Singapore. https://doi.org/10.1007/978-981-15-4619-8_15
4. Singh M, Singh S, Kumar S (2020) Experimental investigation for generation of micro-holes on silicon wafer using electrochemical discharge machining process. Silicon 12:1683–1689. https://doi.org/10.1007/s12633-019-00273-8
5. Gupta PK, Dvivedi A, Kumar P (2015) Developments on electrochemical discharge machining: a review of experimental investigations on tool electrode process parameters. Proc Inst Mech Eng Part B: J Eng Manuf 229:910–920. https://doi.org/10.1177/0954405414534834
6. Kumar N, Mandal N, Das AK (2020) Micro-machining through electrochemical discharge processes: a review. Mater Manuf Processes 00:1–42. https://doi.org/10.1080/10426914.2020.1711922
7. Jain VK, Chak SK (2000) Electrochemical spark trepanning of alumina and quartz. Mach Sci Tech 4:277–290. https://doi.org/10.1080/10940340008945710
8. Jain VK, Balasubramaniam R, Mote RG, Das M, Sharma A, Kumar A, Garg V, Kamaliya B (2020) Micromachining: an overview (Part I). J Micromanuf. 251659841989582..https://doi.org/10.1177/2516598419895828
9. Jain VK, Sidpara A, Ravisankar M, Das M (2016) Micromanufacturing: an introduction
10. Jahan MP, Perveen A, Rumsey AM (2019) A review on the conventional, non-conventional, and hybrid micromachining of glass. Mach Sci Tech 23:264–338. https://doi.org/10.1080/10910344.2019.1575411
11. Srivastava A, Yadav SKS (2020) Machining issues on electrochemical spark machining—a review. Mater Today: Procee. https://doi.org/10.1016/j.matpr.2020.02.593
12. Mishra DK, Arab J, Pawar K, Dixit P (2019) Fabrication of deep microfeatures in glass substrate using electrochemical discharge machining for biomedical and microfluidic applications. 2019 IEEE 21st Electron Packag Tech Conf EPTC 2019:263–266. https://doi.org/10.1109/EPTC47984.2019.9026714
13. Arab J, Dixit P (2020) Influence of tool electrode feed rate in the electrochemical discharge drilling of a glass substrate. Mater Manuf Process 00:1–12. https://doi.org/10.1080/10426914.2020.1784936
14. Sindhu D, Thakur L, Chandna P (2019) Parameter optimization of rotary ultrasonic machining on quartz glass using response surface methodology (RSM). Silicon. https://doi.org/10.1007/s12633-019-00160-2
15. Huang Z, Feng CL (2011) Study of machining microchannels on quartz by the laser-induced plasma. Adv Mater Res 154–155:1655–1658. https://doi.org/10.4028/www.scientific.net/AMR.154-155.1655
16. Qi H, Fan J, Wang J (2015) A study of the micro-machining process on quartz crystals using an abrasive slurry jet. Proc Inst Mech Eng Part B: J Eng Manuf 229:421–434. https://doi.org/10.1177/0954405414528167
17. Kumar V, Singh H (2019) Rotary ultrasonic drilling of silica glass BK-7: microstructural investigation and process optimization through TOPSIS. Silicon 11:471–485. https://doi.org/10.1007/s12633-018-9933-x
18. Hu MF, Xie J, Su HH, Liu JN (2018) Study on laser-assisted dry micro-ground surface of difficult-to-cut materials. Int J Adv Manuf Tech 94:2919–2928. https://doi.org/10.1007/s00170-017-1093-4

19. Varel H, Ashkenasi D, Rosenfeld A, Wähmer M, Campbell EEB (1997) Micromachining of quartz with ultrashort laser pulses. Appl Phys A Mater Sci Process 65:367–373. https://doi.org/10.1007/s003390050593
20. Lin Y-C, Lee C-C, Lin H-S, Hong Z-H, Hsu F-C, Hung T-P, Lyu Y-T (2017) Fabrication of microfluidic structures in quartz via micro machining technologies. Microsyst Tech 23:1661–1669. https://doi.org/10.1007/s00542-015-2717-y
21. Kumar M, Vaishya RO, Oza AD, Suri NM (2020) Experimental investigation of wire-electrochemical discharge machining (WECDM) performance characteristics for quartz material. Silicon 12:2211–2220. https://doi.org/10.1007/s12633-019-00309-z
22. Singh T, Dvivedi A, Arya RK (2019) Fabrication of micro-slits using W-ECDM process with textured wire surface: an experimental investigation on Kerf overcut reduction and straightness improvement. Precis Eng 59:211–223. https://doi.org/10.1016/j.precisioneng.2019.05.008
23. Oza AD, Kumar A, Badheka V, Arora A (2019) Traveling wire electrochemical discharge machining (TW-ECDM) of quartz using zinc coated brass wire: investigations on material removal rate and Kerf width characteristics. Silicon 11:2873–2884. https://doi.org/10.1007/s12633-019-0070-y
24. Kumar U, Singh M, Singh S (2020) Wire-electrochemical discharge machining of SiC reinforced Z-pinned polymer matrix composite using grey relational analysis. Silicon. https://doi.org/10.1007/s12633-020-00484-4
25. Singh M, Singh S (2020) Electrochemical discharge machining: fumes generations, properties and biological effects. Int J Adv Manuf Technol 106:357–370. https://doi.org/10.1007/s00170-019-04632-y
26. Oza AD, Kumar A, Badheka V (2015) Electrochemical discharge machining—a future prospect. In: International conference on precision, meso, micro and nano engineering (Copen-9) 10–12 Dec 2015
27. Oza AD, Kumar A, Badheka V (2020) Improving quartz micro-machining performance by magneto hydrodynamic and zinc-coated assisted traveling wire-electrochemical discharge machining process. Mater Today: Proc 28:970–976. https://doi.org/10.1016/j.matpr.2019.12.334
28. Singh J, Vaishya R, Kumar M (2019) Fabrication of micro features on quartz glass using developed WECDM setup. ARPN J Eng Appl Sci 14:725–731
29. Yadav P (2019) Experimental investigation for performance study of wire electrochemical spark cutting of silica epoxy nanocomposites. Silicon. https://doi.org/10.1007/s12633-019-00197-3
30. Oza AD, Kumar A, Badheka V (2017) Traveling wire electrochemical discharge machining: principle and possibilities. In: International conference on "advances in materials and processing: challenges and opportunities (AMPCO 2017) 30 Nov—2 Dec 2017
31. Oza AD, Kumar A, Badheka V (2019) Micro-machining characteristics of quartz using travelling wire-electrochemical discharge machining (TW-ECDM) process. In: International conference on precision, meso, micro and nano engineering (Copen-11) 12–14 Dec 2019
32. Kannojia HK, Arab J, Pegu BJ, Dixit P (2019) Fabrication and characterization of through-glass vias by the ECDM process. J Electrochem Soc 166:D531–D538. https://doi.org/10.1149/2.0141913jes
33. Singh M, Singh S (2019) Electrochemical discharge machining: a review on preceding and perspective research. Proc Inst Mech Eng Part B: J Eng Manuf 233:1425–1449. https://doi.org/10.1177/0954405418798865
34. Dhiman P, Vaishya R, Kumar M (2019) A review on machining by electrochemical discharge phenomena. Int J Tech Innov Modern Eng Sci (IJTIMES) Impact 5:2017–2020
35. Malik A, Manna A (2016) An experimental investigation on developed WECSM during micro slicing of e-glass fibre epoxy composite. Int J Adv Manuf Technol 85:2097–2106. https://doi.org/10.1007/s00170-016-8858-z
36. Yang CT, Song SL, Yan BH, Huang FY (2006) Improving machining performance of wire electrochemical discharge machining by adding SiC abrasive to electrolyte. Int J Mach Tools Manuf 46:2044–2050. https://doi.org/10.1016/j.ijmachtools.2006.01.006

37. Wang J, Fu C, Jia Z (2018) Research on oil film-assisted wire electrochemical discharge machining. Int J Adv Manuf Technol 96:2455–2461. https://doi.org/10.1007/s00170-018-1781-8
38. Wang J, Jia Z, Guo YB (2018) Shape-cutting of quartz glass by spark discharge-assisted diamond wire sawing. J Manuf Process 34:131–139. https://doi.org/10.1016/j.jmapro.2018.06.001
39. Rattan N, Mulik RS (2017) Improvement in material removal rate (MRR) using magnetic field in TW-ECSM process. Mater Manuf Process 32:101–107. https://doi.org/10.1080/10426914.2016.1176197
40. Kuo KY, Wu KL, Yang CK, Yan BH (2015) Effect of adding SiC powder on surface quality of quartz glass microslit machined by WECDM. Int J Adv Manuf Technol 78:73–83. https://doi.org/10.1007/s00170-014-6602-0
41. Kuo KY, Wu KL, Yang CK, Yan BH (2013) Wire electrochemical discharge machining (WECDM) of quartz glass with titrated electrolyte flow. Int J Mach Tools Manuf 72:50–57. https://doi.org/10.1016/j.ijmachtools.2013.06.003
42. Maher I, Sarhan AAD, Hamdi M (2014) Review of improvements in wire electrode properties for longer working time and utilization in wire EDM machining. Int J Adv Manuf Technol 76:329–351. https://doi.org/10.1007/s00170-014-6243-3

Stress Relaxation Study of Ultrafine-Grained AA 6061 Alloy Processed Through Combined Constrained Groove Pressing and Cold Rolling

K. Changela, K. Hariharan, and R. K. Digavalli

1 Introduction

The use of aluminum alloys as a lightweight structural material has become inevitable in the modern world due to their higher strength to weight ratio when compared to steel [1]. Among various heat-treatable aluminium alloys, AA 6061 alloys are extensively used in automobile, aerospace, and marine industries because of their properties such as high fatigue strength and high corrosion resistance [2, 3]. Despite the conventional strengthening methods, the tensile strength of AA 6061 alloy is usually lower when compared to steel. Further improvement in strength can be attained using various thermo-mechanical processes leading to diverse microstructural evolution and property modification. Grain refinement in metallic materials to a very fine size of the order of a few hundred nanometers increases strength and toughness simultaneously [4]. In recent years, severe plastic deformation (SPD) has been found as an effective top-down method for the production of bulk ultrafine-grained (UFG) metallic materials [5–7]. Constrained Groove Pressing (CGP) is one of the SPD technique to produce a UFG metallic sheets in which the sheet is deformed by repetitive shear deformation under fully constrained conditions. More details about this

K. Changela (✉)
Department of Mechanical Engineering, G H Raisoni College of Engineering and Management, Pune, Maharashtra 412207, India
e-mail: kandarp.changela06@gmail.com

K. Hariharan
Department of Mechanical Engineering, Indian Institute of Technology Madras, Chennai 600036, India
e-mail: hariharan@iitm.ac.in

R. K. Digavalli
Department of Mechanical Engineering, Indian Institute of Technology Delhi, New Delhi 110016, India
e-mail: dravi@mech.iitd.ac.in

H. K. Dave and D. Nedelcu (eds.), *Advances in Manufacturing Processes*, Lecture Notes in Mechanical Engineering, https://doi.org/10.1007/978-981-15-9117-4_9

process could be found in our earlier work [8, 9]. The unique advantage of CGP process is that the sample dimensions do not change during the deformation and the process is not influenced by the crystal structure of deforming materials. However, characterization of the CGP samples is quite difficult due to its irregular surface. It is known from previous work [8] that CGP process impart non-uniform deformation, and it is one of the serious issues for industrial application [10]. Cold rolling (CR) of the CGPed sheet is one of the possible solutions to overcome these difficulties.

The strength of UFG materials usually depends upon the extent of grain refinement. Due to the intense plastic strain during CGP and cold rolling, traditional optical microscopy is not suitable to quantify the extent of refinement in UFG materials. Advanced electron microscopy methods such as transmission electron microscope (TEM), and electron backscatter diffraction (EBSD) are suitable methods to evaluate UFG and nanostructured materials. These techniques require tedious sample preparation and the inspection is confined to a tiny region of the macroscopic sample. An overall estimate of the degree of grain refinement imparted in the sample is challenging. Alternately, indirect estimates using stress relaxation can be used to verify the grain refinement in UFG materials. Moreover, macroscopic transient tests also help to understand the deformation mechanism in coarse grain and ultrafine-grained materials. Using this method, it can be quickly estimated the average degree of grain refinement for a larger length scale. However, the present method does not focus on the local changes in the grain refinement and rather gives an average estimate, which is essential for practical applications.

1.1 Stress Relaxation and Related Phenomena

Stress relaxation is a reliable method to understand the plastic flow behavior of metals, particularly, their time-dependent deformation. In this test, the uniaxial tensile sample is deformed up to the predefined value of stress/strain and abruptly stopped without unloading. As shown in Fig. 1, the applied stress drops continuously with time. In repeated stress relaxation, multiple relaxation cycles are performed at the same value of stress (Fig. 1). This test can be used to understand the deformation mechanism and quantify the parameters like internal stress and activation volume [11]. The flow stress (σ) is composed of two components as the internal or a thermal stress component (σ_i) and the effective or thermal stress component (σ^*).

$$\sigma^* = \sigma - \sigma_i \tag{1}$$

Internal and effective stresses are related to long and short-range obstacles, respectively. The long-range obstacles correspond to the internal stress field created by other dislocations and incoherent precipitates acting on dislocation movement. The short-range obstacles are related to localized obstacles in the section of atomic distance such as solute atoms and forest dislocations. The Orowan equation ($\dot{\varepsilon}_p = \varphi\ \rho_m\ b$

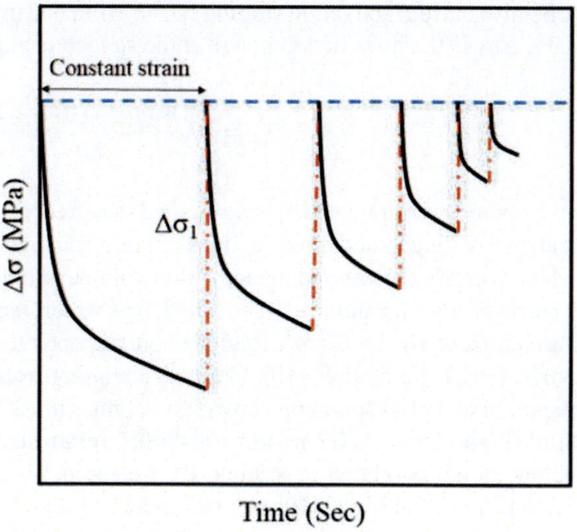

Fig. 1 Schematic representation of the repeated stress relaxation test [13]

v, where ϕ is a constant, ρ_m is the mobile dislocation density and v is the average dislocation velocity) can be used to model the stress relaxation behavior.

Assuming plastic deformation as a thermally activated process, the rate dependent velocity can be written as, $v = v_0 \exp(\frac{-\Delta G}{KT}))$, where ΔG is the Gibbs free energy, K is the Boltzmann constant, and T is the temperature. Substituting velocity equation in Orowan equation, and integrate the new equation led to the logarithmic model, as follows [12],

$$\Delta\sigma(\sigma_t - \sigma_0) = A\ln(1 + Bt) \tag{2}$$

where σ_t is the flow stress at time t and σ_0 is the stress when relaxation time (t) is zero, A and B are constants:

$A = \frac{KT}{V^*}$, $B = \frac{-E\varphi\rho_m b v_o V^*}{KT} \exp(-\frac{\Delta G - \sigma^* V^*}{KT})$

Equation (3) can be rewritten in the following form

$$\Delta\sigma = -\frac{KT}{V^*}\ln\left(1 + \frac{t}{C_r}\right) \tag{3}$$

where, V^* is the apparent activation volume and C_r is the time constant.

The apparent activation volume is estimated from a single relaxation using Eq. (3). The apparent activation volume is measured from the strain rate during relaxation and dependent on both mobile dislocation density and velocity of dislocation velocity. Whereas the actual activation volume characterizes the rate-dependent deformation mechanism measured using repeated relaxations. During repeated-stress relaxations, there is a sudden change in strain rate between first ($\dot{\varepsilon}_{(f,j)}$) and second ($\dot{\varepsilon}_{(\mathrm{i,j+1})}$) relaxation cycle under quasi-elastic condition leading to negligible change mobile dislocation density. The subscript 'j' refers to the stress relaxation number. The actual (effective or true) activation volume can be estimated using repeated-stress relaxation data. $\Delta\sigma_1$ is the flow stress change at the sudden change in strain rate.

$$V_e = kT\frac{\ln\left(\dot{\varepsilon}_{(i,j+1)}/\dot{\varepsilon}_{(f,j)}\right)}{\Delta\sigma_j} \tag{4}$$

Conrad and Narayan [14] have studied the effect of grain size on flow stress in Zn and they distinguished three-regimes of grain size (depicted in Figs. 1 and 2 of [14]) which identify corresponding activation volume and deformation behavior. Regime I corresponds to grain size 1 mm to 1 μm which has apparent activation volume varying from 10^2 to 10^3 b^3. Regime I corresponds to conventional deformation of polycrystalline materials [15]. Grain size ranging from 1 μm to 10 nm is related to regime II and corresponding activation volumes are $20b^3$ (grain size = 10–100 nm), $80b^3$ (grain size = 170 nm) and $100–200b^3$ (grain size = 5–50 μm). The grain size below 10 nm is related to Regime III. Lee et al. [16] found the activation volume of 10 b^3 for UFG Mn TRIP steel and indicated that the dislocation bow out was the possible deformation mechanism that was controlled by the grain boundary network. Mishra et al. [13] performed the single and multiple relaxations of AA 6061 alloy in different aging conditions and found that the deformation mechanism depends

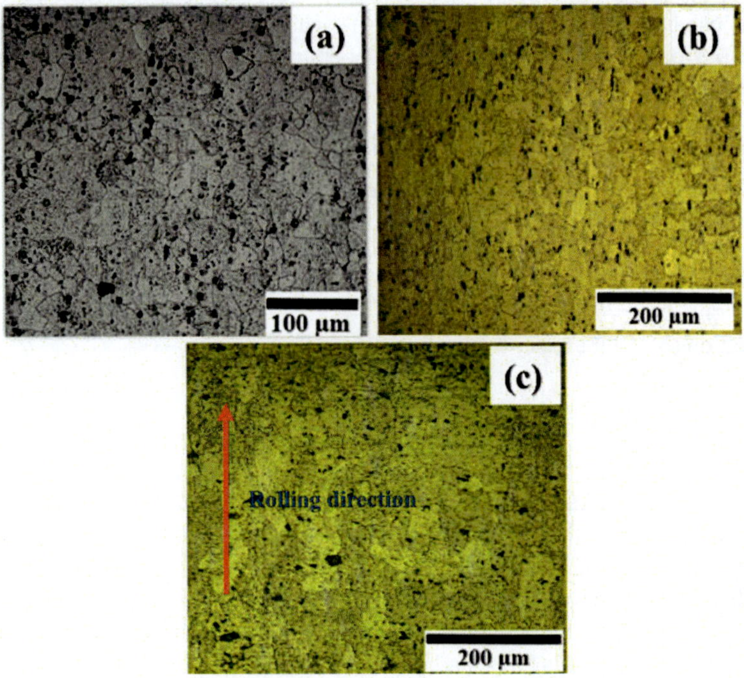

Fig. 2 Optical micrographs for AA6061 in different conditions, **a** solutionized, **b** 1 pass CGPed, **c** CGP + CR

on the type of solutes and precipitates developed during aging. Mohebbi et al. [17] obtained the apparent activation volume of UFG AA 1050 in the range of 50–70 b^3 suggesting grain boundary sliding contribution in relaxation.

The objective of the present study is to estimate the average grain refinement of UFG sheets using uniaxial tensile tests with single and repeated stress relaxations and understand the possible deformation mechanism based on the activation volume determined in deformed and undeformed conditions.

2 Experimental Procedure

A 3 mm thick AA6061 alloy sheets ('T6' temper) were purchased as the initial material condition in this work. Table 1 shows the chemical composition of AA 6061 alloy measured by standard spectroscopy method.

The length and width of the sheets were kept the same by 108 mm. The sheets were solutionized as per the ASTM B918M standard. The solutionizing can be done by slowly heating the sample at 530 °C in a muffle furnace, soaking for 2 h, and finally quenching in cold water. Solutionized (SL) sheets were severely deformed by one pass of constrained grove pressing (CGP). The CGP processed sheets were rolled at room temperature by keeping the grooves of the CGP sample parallel to rolling direction. During cold rolling, a 3 mm thickness of CGP sheets were reduced to 1 mm by applying true plastic strain of 1.09. The procedure of CGP and CGP + CR were discussed in our earlier work [8, 9].

The solutionized and all deformed samples were analyzed by optical microscopy. Samples were polished by standard metallographic procedures followed by electrochemically etching. The etching procedure was discussed in [9]. For monotonic and stress relaxation tests, the tensile specimens (prepared as per ASTM E8 standard) were extracted from SL, CGP, and CGP + CR samples. The tensile test samples with the gauge length of 25 mm were cut from the CGP sheet. The tensile tests were carried out in 100 kN capacity INSTRON 59OOR machine featuring video extensometer for strain measurement. Stress relaxation tests of SL and deformed samples were conducted at an initial strain rate of 10^{-3} s^{-1} with intermittent pausing of the uniaxial tensile test at predefined stress in the uniform elongation zone for a given interval of time. The constant relaxation time of 60 s was chosen for the present study. The predefined stress can be chosen by 97% of UTS to get more decrease in the stress value. To ensure the repeatability of results, three samples were tested for each processing condition. The apparent and actual activation volumes were calculated from Eqs. (3) to (4).

Table 1 Chemical compositions of AA 6061 alloy (wt%)

Mg	Si	Mn	Cr	Cu	Fe	Ti	Zn	Al
0.8	0.68	0.12	0.093	0.20	0.22	0.02	0.014	Balance

3 Results and Discussion

3.1 Microstructure

Figure 2 shows the optical micrographs of deformed and undeformed samples. Equiaxed grains with an approximate size of 88 μm was observed in the case of solutionized sample condition (Fig. 2a). As shown in Fig. 2b, the grain size reduced from 88 μm to approximately 22 μm after the severe deformation of sheet by one pass of CGP. Figure 2c shows the microstructure of thin sheets processed through CGP followed by cold rolling. It shows that grains are severely deformed and elongated in the rolling direction as shown in Fig. 2c. The optical microscope was not able to capture the severely fragmented grains due to diffused and ill-defined grain boundary structure in the material.

3.2 Monotonic Tensile Behavior and Mechanical Properties

Figure 3a shows the stress–strain response of the SL and deformed samples and the corresponding average mechanical properties are shown in Fig. 3b. The average hardness of the SL sample is 45 HV, which is 100% increased after applying the effective true strain (ε) of 1.16 by four stages of CGP process. The YS and UTS of the CGP samples are 185 MPa, and 202 MPa, respectively, which is higher than SL sample condition. The pure shear deformation under plane strain condition is responsible for significant improvement in the strength of the CGP sample. Work hardening and grain refinement are possible strengthening mechanisms during the CGP process.

Further improvement in strength and hardness is observed after the cold rolling of the CGPed sample. However, the total elongation sharply decreased from the

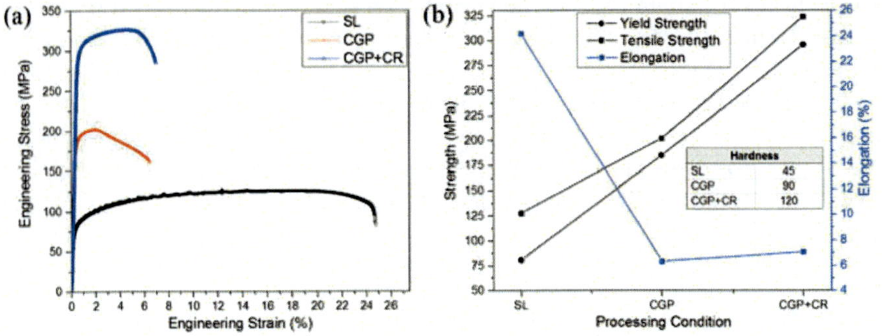

Fig. 3 **a** Engineering stress–strain curves and **b** mechanical properties of AA 6061 alloy samples in different conditions

SL condition. Important observation made from Fig. 3 that the CGP sample shows slightly lower elongation than the CGP + CR sample, and that is possibly due to remove the unevenness and achieve flat surface in CGP + CR sample. According to Kumar et al. [18], the geometric and strain in-homogeneities in the CGP samples are the responsible factors in lowering the total elongation.

3.3 Stress Relaxation

Figure 4a shows single relaxation behavior of AA 6061 alloy in SL, CGP, and CGP + CR conditions. Relaxation curves show excellent agreement of stress relaxation data between experimental and logarithmic model (Eq. 3). It is observed that relaxation is fast in the initial time of 10–12 s and then decreases at longer time in all the cases. Rate of relaxation is almost zero at longer time in SL and CGP samples while CGP + CR sample continues to relax. Amount of relaxation is significantly more in CGP, and CGP + CR samples compared to SL sample, which is concluded that the deformation in CGP and CGP + CR samples is more thermally activated, compared to SL sample. Figure 4b shows the comparison of apparent activation volume in SL, CGP and CGP + CR sample conditions which is measured from the single relaxation. The apparent activation volume in SL condition is 232 b^3 which is much larger than in CGP (133 b^3) and CGP + CR (57 b^3) samples. The smaller activation volume in CGP and CGP + CR samples is attributed to higher density of short-range obstacles compared to SL sample.

Figure 5a–c show stress-time cures of AA 6061 alloy for repeated relaxation. Repeated relaxation tests were performed at stress values of 97% of UTS, which are 124 MPa, 195 MPa, and 314 MPa for the SL, CGP, and CGP + CR samples, respectively. The stress drop decreases with relaxation number as distinctly shown for all sample conditions in Fig. 5a–c. It is due to reduction in mobile dislocation by annihilation and conversion to immobile (sessile) dislocation [19].

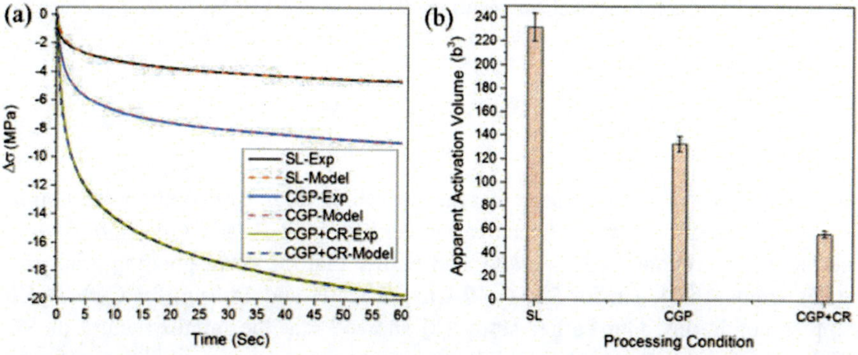

Fig. 4 **a** Stress relaxation behavior and **b** apparent activation volume of AA 6061 alloy in different conditions

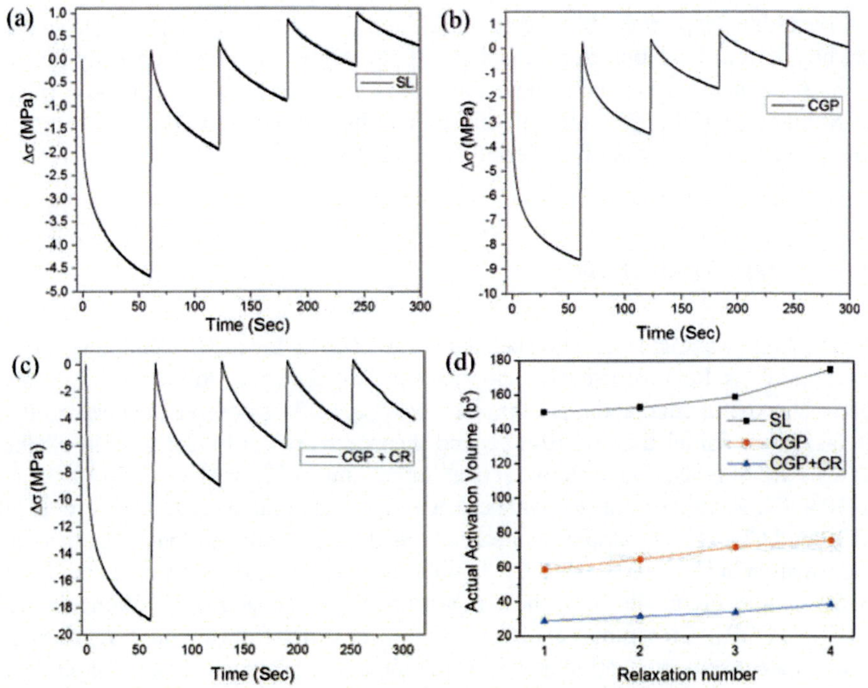

Fig. 5 **a**, **b**, **c** Repeated stress relaxation of AA 6061 alloy in SL, CGP, and CGP + CR conditions, respectively, and **d** actual activation volume with respect to stress relaxation number

The highest stress drop ($\Delta\sigma$) in each cycle is observed for CGP + CR sample compared to SL and CGP samples. The actual activation volume is determined from repeated relaxation. Figure 5d shows the actual activation volumes for all sample condition calculated using Eq. (4). It shows that the actual activation volume is lower in all the cases compared to apparent activation volume and that may be due to the effect of mobile dislocation density, which is additionally incorporated in the determination of actual activation volume.

3.4 Grain Refinement

From Figs. 4 and 5, it can be seen that the activation volume of solutionized sample lie within the regime I, while the activation volume in CGP and CGP + CR samples corresponds to regime II. It is mentioned earlier that the corresponding grain size ranges are 1 mm–1 μm for SL, 6–40 μm for CGP, and 180 nm for CGP + CR samples conditions. Our earlier study [8] showed that the microstructure of SL, CGP, and CGP + CR samples exhibits the grain size to be around 88 μm, 18 μm, and 277 nm, respectively. The activation volume decreases with increasing strain and decreasing grain size [16]. The low value of activation volume for UFG material is attributed to the fact that limited dislocations are involved in the thermal activation process.

3.5 Deformation Mechanism

It is believed that the rate-controlling mechanisms in coarse-grain structured materials are an intersection of forest dislocations and cross slip. However, it is not true for very fine or ultrafine-grained materials. According to Conrad and Narayan [14], dislocation glide and an intersection of dislocations is the suggested deformation mechanism corresponding to activation volume of 10^2–10^3 b^3 (Regime I). Activation volume of the present CGP + CR sample is far smaller than the SL material and also falls within the range of activation volume reported by Hayes et al. [20]. It is important to understand possible deformation mechanisms that occurs for very fine grain size with low activation volume. Due to large amount of dislocations are available at grain boundaries, diffusion occurs at very low strain rate during stress relaxation. Consequently, grain boundary sliding is a possible phenomenon that occurs in UFG materials. Kapoor et al. [21] also suggested the possibility of grain boundary deformation mechanism for UFG material at low activation energy and at low activation volume. The process involves the nucleation of dislocations at grain boundaries, glide through small grains, and absorbed at the opposite grain boundaries. Moreover, dynamic recovery also possibly occurs during stress relaxation at low strain rate in UFG materials. High stored energy available in very fine-grained materials act as driving force to promote the recovery in the dislocations during stress relaxation. It is also believed that dynamic recovery is more favorable than work hardening during relaxation due to continuously decreasing of plastic strain rate. Therefore, it is concluded that it is possible to estimate the grain refinement in the larger scale and understand the deformation mechanism by calculating the activation volume through simple stress relaxation experiments in uniaxial tension. It is suggested that the dislocation-dislocation interactions are the possible deformation mechanism in CGP and SL samples and grain boundary sliding in CGP + CR samples.

4 Conclusions

The effect of constrained groove pressing and cold rolling on single and repeated stress relaxation behavior of AA6061 alloy was investigated in the present work. The mechanical properties were also evaluated through the monatomic tensile test of SL, CGP, and CGP + CR samples. To understand the deformation mechanism of UFG materials, the apparent and actual activation volume was calculated from stress relaxation data.

1. The optical microstructures revealed severe grain refinement in CGP and CGP + CR samples due to intense plastic strain applied. However, the severely deformed structure with ill-defined grain boundaries makes the grain size determination difficult.
2. The YS of CGP and CGP + CR samples are 128 and 258% greater than solutionized sample. Increase in the strength of CGP sample is due to repetitive shear deformation during grooving and flattening. The ductility of CGP + CR sample is observed to be higher than CGP sample due to flat and smooth surface achieved in CGP + CR sample.
3. The activation volume of CGP + CR sample is quite smaller than that of SL sample. It suggests that small amount of dislocations are available in the grain interior and most of the dislocations are close to the grain boundaries. Smaller activation volume in CGP + CR sample corresponds to the regime II that shows the grain size is within the range of ultrafine-grain size.
4. The comparison of activation volumes indicated that the dislocation glide and interaction of dislocations are the possible deformation mechanisms for large grain size (solutionized) specimens, while grain boundary sliding could have occurred in UFG materials due to small grain size and large grain boundary network.

References

1. Lapovok R, Timokhina I, McKenzie PWJ, O'Donnell R (2008) Processing and properties of ultrafine-grain aluminium alloy 6111 sheet. J Mater Proccess Tech 200(1–3):441–450
2. Miller WS, Zhuang L, Bottema J, Wittebrood AJ, De Smet P (2000) Recent development in aluminium alloys for the automotive industry. Mater Sci Eng A 280:37–49
3. Starke EA, Staleyt JT (1996) Application of modern aluminium alloys to aircraft. Progr Aerosp Sci 32(2–3):131–172
4. Changela K, Naik HB, Desai KP, Raval HK (2020) Effecct of rolling temperatures on mechanical and facture behavior of AA 3003 alloy and pure Cu. SN Appl Sci 2:1109. https://doi.org/10.1007/s42452-020-2903-0
5. Changela K, Krishnaswamy H, Digavalli RK (2020) Mechanical behavior and deformation kinetics of aluminium alloys processed through cryorolling and subsequent annealing. Metallur Mater Trans A 51:648–666
6. Valiev RZ, Korznikov AV, Mulyukov RR (1993) Structure and properties of ultrafine-grained materials produced by severe plastic deformation. Mater Sci Eng A 168(2):141–148
7. Parmar V, Changela K, Srinivas B, Mani Kumar M, Mohanty S, Panigrahi SK, Hariharan K, Kalyanasundaram D (2019) Relationship between dislocation density and actibacterial activity of cryorolled and cold rolled copper. Materials. 12:200. https://doi.org/10.3390/ma12020200
8. Changela K, Krishnaswamy H, Digavalli RK (2019) Development of combined groove pressing and rolling to produce ultra-fine grained Al alloys and comparison with cryorolling. Mater Sci Eng 760:7–18
9. Changela K, Kumar S, Hariharan K, Ravi Kumar D (2019) Aging behavior of ultra-fine grained AA 6061 alloy subjected to constrained groove pressing followed by cold rolling. IOP Confe Ser Mater Sci Eng 651:012069. https://doi.org/10.1088/1757-899X/651/1/012069

10. Kumar S, Venkatachalam S, Hariharan K, Digavalli RK, Murthy HSN (2019) Influence of inhomogeneous deformation on tensile behavior of sheets processed through constrained groove pressing. J Eng Mater Tech 141(4):1–10
11. Varma A, Krishnaswamy H, Jain J, Lee MG, Barlat F (2019) Advanced constitutive model for repeated stress relaxation accounting for transient mobile dislocation density and internal stress. Mech Mater 133:138–153
12. Sargent GA (1965) Stress relaxation and thermal activation in niobium. Acta Meterialia. 13(6):663–671
13. Mishra S, Yadava M, Kulkarni KN, Gurao NP (2018) Stress relaxation behavior of an aluminium magnesium silicon alloy in different temper condition. Mech Mater 125:80–93
14. Conrad H, Narayan J (2002) Mechanisms for grain size hardening and softening in Zn. Acta Mater 50(20):5067–5078
15. Cheng S, Spencer JA, Milligan WW (2003) Strength and tension/compression asymmetry in nanostructured and ultrafine-grain metals. Acta Mater 51(15):4505–4518
16. Lee S, Lee SJ, De Cooman BC (2011) Work hardening behavior of ultrafine-grained Mn transformation-induced plasticity steel. Acta Mater 59(20):7546–7553
17. Mohebbi MS, Akbarzadeh A, Yoon YO, Kim SK (2015) Stress relaxation and flow behavior of ultrafine grained AA 1050. Mech Mater 89:23–34
18. Kumar S, Hariharan K, Ravi Kumar D, Paul S (2019) Accounting Bauschinger effect in the numerical simulation of constrained groove pressing process. J Manuf Process 38:49–62
19. Hariharan K, Dubey P, Jain J (2016) Time dependent ductility improvement of stainless steel SS 316 using stress relaxation. Mater Sci Eng A 673:250–256
20. Hayes RW, Witkin D, Zhou F, Lavernia EJ (2004) Deformation and activation volumes of cryomilled ultrafine-grained aluminum. Acta Mater 52(14):4259–4271
21. Kapoor R, Chakravartty JK (2007) Deformation behavior of an ultrafine-grained Al–Mg alloy produced by equal-channel angular pressing. Acta Mater 55(16):5408–5418

Effect of Friction Stir Welding Process Parameters on Tensile Strength and Forming Height of Tailor Welded Blanks

Sumit Patel, Shalin Marathe, Keyur Desai, and Harit Raval

1 Introduction

Tailor Welded Blanks (TWBs) are the best solution for the automobile and transportation industries as these sectors demands for lightweight products which ultimately will lead to fuel-efficient products. TWBs can lead to weight [1, 2] and cost reduction [3]. The major limitations of the TWBs are weld line shift [4, 5] and formability reduction [1, 6, 7] in comparison to the parent blanks involved in it. Because of the two mentioned limitations, TWBs are not used by many industries. There are many efforts have been made by researchers to study the limitation of weld line shift and formability reduction.

In order to decrease the weld line shift, deformation in strong and weak parent material should be the same. That can be done using properly design draw bead [8] or by controlling the Blank Holding Force (BHF) [9, 10]. A novel approach was used to decrease the weld line shift in which split punch technology [5] was used. Towards strong material side heating was provided which resulted in more deformation of strong material and ultimately the weld line shift was decreased. To improve the formability of the TWBs, die design [11] can be changed or lubrication can be a good solution [4]. To study the formability of the TWBs, Forming Limit Diagrams are required to be developed after experimentation. To achieve this objective, major and minor strains are required to be measured and for the grid marking method is required to be used [12]. There are many combinations of TWBs with different experimental conditions of different forming processes have been explored by researchers. Similar

S. Patel
Mechanical Engineering Department, RC Technical Institute, Ahmedabad, India

S. Marathe (✉) · K. Desai · H. Raval
Department of Mechanical Engineering, SV National Institute of Technology, Surat, Gujarat 395007, India
e-mail: shalin.marathe89@gmail.com

H. K. Dave and D. Nedelcu (eds.), *Advances in Manufacturing Processes*, Lecture Notes in Mechanical Engineering, https://doi.org/10.1007/978-981-15-9117-4_10

one combination has been tried which combines TWBs with Single Point Incremental Forming (SPIF) method [13–15]. In one such attempt, it was recommended that this combination can resolve many issues related to the TWBs. It was stated that by using appropriate position of forming tool, weld line shift can be controlled [16]. During SPIF of blanks, relationship of tool radius with the thickness of the blank plays an important role and this ratio should be nearly 2.2 for maximum formability [17].

Generally, TWBs are developed using steels and for that laser welding process is used [18–22]. But aluminum material offers more weight saving in comparison to the steel material and due to that use of aluminum material has increased in automobile industries.

The major problem associated with the aluminum material is regarding being joined using conventional welding processes [23]. FSW was specially developed for welding of aluminum and its alloys. It is a solid-state joining process in which welding takes place below the melting point and at plastic stage of parent material. The temperature during FSW welding is nearly 70–80% of the melting point of parent blanks. There are many process parameters involved in this process like tool feed, tool rotational speed, tool tilt angle, tool pin profile, type of tool, and tool indentation time. Out of these parameters tool rotational speed, tool transverse speed, and tool pin profile are major influencing factors to the weld quality of the FSW joints [24]. The prior art study also indicates that there has been very limited work on the orientation of the parent blanks.

Above reported efforts indicate that the formability of TWBs is of major concern. Formability of the TWBs has been investigated in this present investigation using Erichsen cup test. FSW has been adopted to fabricate the TWBs of dissimilar parent blanks.

2 Experimental Work

Experimental work for FSW process was done on a manual vertical milling machine (see Fig. 1). A special fixture was designed and developed to hold the plates being welded.

AA 6061 T6 (1.6 mm thick) and AA 3102 (2 mm thick) material were used as parent blanks in this study to develop the TWBs. AA 6061 T6 is a space-grade alloy and it is having applications in automobile sector as well. AA 3102 is having excellent strength and better corrosion resistance properties which is very important for automobile and shipbuilding industries. So, in order to develop TWBs, these two materials are selected having different thickness. H13 material was used to weld the faying surfaces [25]. During welding plunge depth of 0.4 mm was kept constant. Tool tilt angle was kept constant at 20 during FSW process [26]. Welding was followed by tensile testing and of TWBs. Tensile test specimen was prepared as per BM-557 M and during testing, the cross-head speed was kept as 1 mm/min. Table 1 represents the different levels of different parameters.

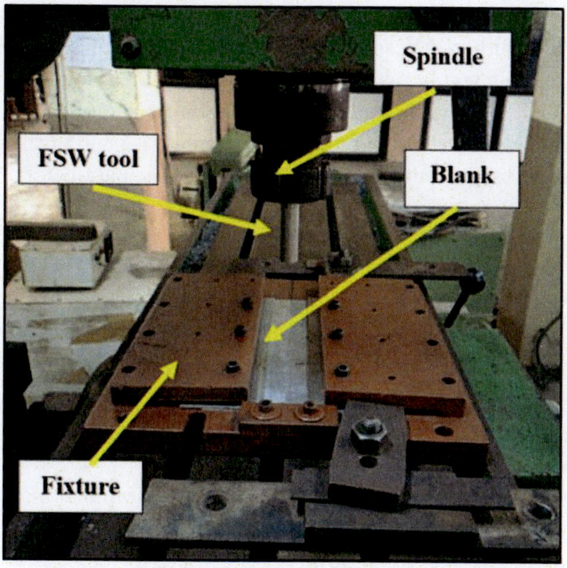

Fig. 1 Closeup view of spindle and Fixture on vertical milling machine

Table 1 Levels of different weld process parameters

Factor	Variable parameters	Level-1	Level-2	Level-3
A	Tool rotational speed (RPM)	1000	1400	2000
B	Feed (mm/min)	28	40	56
C	Tool (pin profile)	Taper	Threaded	–
D	Side (sheet orientation during welding)	Advancing side	Retreating side	–

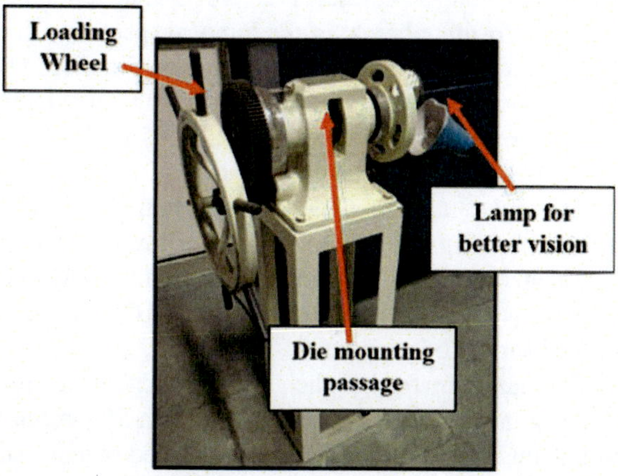

Fig. 2 Erichsen cup test setup (RET-20) model

Welded blanks were subjected to Erichsen cup test (see Fig. 2) and forming height was studied as response parameter. Standard model (RET-20) was used for Erichsen cupping test in which punch diameter is 20 mm.

Full factorial study has been adopted and effect of weld parameters and tooling on tensile strength and forming height has been investigated. ANOVA has also been done to find out the significance of all the parameters.

3 Result and Discussion

The result regarding tensile test and forming height before failure of welded blanks are represented in Table 2. The maximum welded strength of TWBs was found to be 142.4 MPa and maximum forming height was observed of 9.90 mm during Erichsen cup test.

The comparison of tensile strength of welded blanks with parent blanks is indicated in Fig. 3. The strength of the welded blanks is low in comparison to the AA 6061 T6 material while there is not much difference between parent blank and welded blanks regarding tensile strength. The resultant tensile strength of the welded blanks depends upon the tensile strength of the parent blanks. As AA 3102 is having strength of 135.3 MPa which is very low in comparison to the tensile strength of 301.1 MPa of AA 6061 T6. Due to this large variation of tensile strength of parent blanks, welded blanks are resulted in having low strength. In addition to that welding of thin sheet demands for more research because the welding of thick sheet (3–6 mm) is comparatively easy. Welding of this sheet also demanded for development considering the tool design. Improper tool design will lead to drilling action or inadequate heat generation. So, the low tensile strength of the welded blanks is result of combination of mentioned phenomenon.

The values of S/N ratios for different experimental condition are tabulated Table 3. Figures 4 and 5 represents the plot of interaction between different responses for UTS and forming height of the joints, respectively. It can be found that irrespective of types of tool, retreating side results in the more UTS of the joints. For the feed of 56 mm/min there is no significant change in UTS for two different tools. Similar observation can be made for 2000 RPM of welding tool. Maximum UTS is observed for 1400 RPM of tapered welding tool.

The tool transverse speed of 56 mm/min is not having much effect on UTS considering advancing or retreating side. Maximum UTS of joint is observed for retreating side at 40 mm/min of tool travel speed and 1400 RPM of tool rotational speed. Effect of feed also indicates that for maximum UTS of joints, 1000 RPM and 40 mm/min of feed should be adopted.

For maximum forming height, threaded tool, and retreating side were found better (see Fig. 5). Taper tool configuration with 56 mm/min of tool travel speed resulted in maximum forming height of the welded blanks. The interaction of tool and tool rotational speed revealed that 1400 RPM and threaded configuration of tool gives the maximum forming height. Maximum forming height resulted in interaction of

Table 2 Results of tensile test and formability test

Exp. run	Input parameters				Response characteristics	
	Tool rotational speed RPM	Feed mm/min	Tool pin profile	Side	UTS (MPa)	Formability height (mm)
1	1000	28	TH	A	111.2	8.60
2	2000	28	T	A	107.7	8.30
3	1400	56	TH	A	111.7	9.20
4	2000	40	T	R	123.0	8.50
5	1400	56	TH	R	126.2	9.85
6	2000	28	TH	R	122.2	8.60
7	1400	56	T	A	116.9	9.66
8	2000	28	T	R	131.0	8.30
9	2000	40	TH	A	110.0	9.50
10	1400	28	TH	R	109.3	9.90
11	1000	40	T	A	106.1	8.57
12	1400	40	T	R	126.8	8.80
13	1400	40	TH	R	124.6	9.50
14	1000	56	T	A	124.9	8.54
15	1000	28	T	R	117.6	8.94
16	1400	56	T	R	126.0	8.51
17	1000	56	TH	A	118.4	8.92
18	1400	40	TH	A	113.6	8.95
19	2000	56	TH	R	119.3	7.33
20	1000	56	T	R	109.5	9.65
21	1000	40	T	R	124.4	6.50
22	1000	40	TH	R	142.4	8.20
23	1400	28	T	A	120.6	6.43
24	1000	40	TH	A	118.2	8.53
25	1400	28	TH	A	110.5	9.10
26	1000	56	TH	R	111.2	8.25
27	1000	28	T	A	109.6	8.42
28	2000	56	T	A	104.0	9.35
29	2000	40	T	A	93.7	8.80
30	2000	56	TH	A	124.6	7.10
31	1000	28	TH	R	111.8	9.70
32	2000	28	TH	A	108.5	8.94
33	1400	28	T	R	135.4	8.31
34	1400	40	T	A	114.6	9.60
35	2000	40	TH	R	115.4	9.55

(continued)

Table 2 (continued)

Exp. run	Input parameters				Response characteristics	
	Tool rotational speed RPM	Feed mm/min	Tool pin profile	Side	UTS (MPa)	Formability height (mm)
36	2000	56	T	R	122.9	9.60

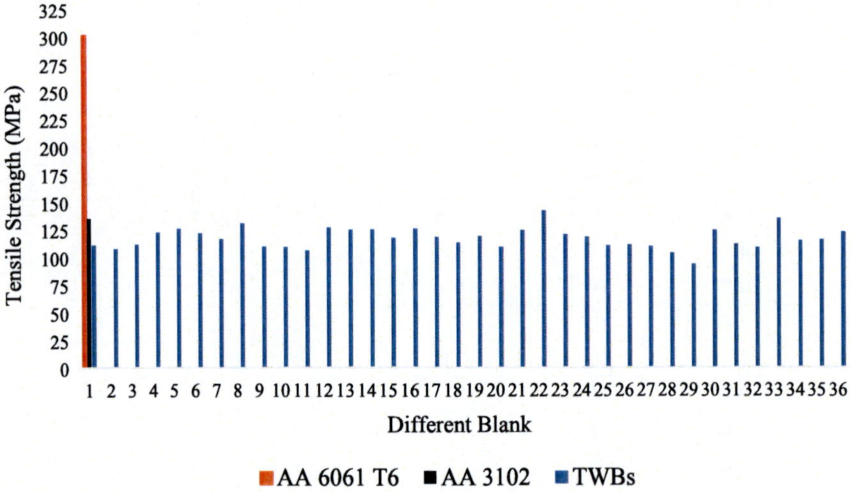

Fig. 3 Comparison of tensile strength of welded blanks with parent blanks

retreating side with 28 mm/min of tool travel speed and 1400 RPM of tool rotational speed. There is no significant change in forming height between 56 and 40 mm/min of tool travel speed at 1400 RPM of tool rotational speed.

3.1 Effect of Tool Rotational Speed

As the tool rotational speed increases beyond 1400 RPM, the meting of parent blank increases which reduces the coefficient of friction between tool and the parent material. This reduces to stirring effect and that ultimately reduces the UTS of joint.

The dynamic recrystallization is taken place due to viscoelastic deformation at high temperature, which reduces the flow stresses because of fine-strain free grains are nucleated [27]. This is not only reducing the dislocation hardening but also facilitates boundary rotation and sliding result in reduction of flow stress. Friction stir welding at higher rotational speed or lower welding speed results in an increase in both degree of deformation and peak temperature of thermal cycle. The increase in the intensity of deformation leads to reduction in the recrystallized grain size. On the other hand, the increase in peak temperature of thermal cycle results in a generation

Table 3 S/N ratio for different experimental run

Exp. run	Input parameters				S/N ratio of response parameters	
	Tool rotational speed	Feed	Tool	Side	UTS MPa	Formability height (mm)
1	1000	28	TH	A	40.922	18.690
2	2000	28	T	A	40.644	18.381
3	1400	56	TH	A	40.961	19.275
4	2000	40	T	R	41.798	18.588
5	1400	56	TH	R	42.021	19.868
6	2000	28	TH	R	41.741	18.690
7	1400	56	T	A	41.356	19.699
8	2000	28	T	R	42.345	18.381
9	2000	40	TH	A	40.827	19.554
10	1400	28	TH	R	40.772	19.912
11	1000	40	T	A	40.514	18.659
12	1400	40	T	R	42.062	18.889
13	1400	40	TH	R	41.910	19.554
14	1000	56	T	A	41.931	18.629
15	1000	28	T	R	41.408	19.026
16	1400	56	T	R	42.007	18.598
17	1000	56	TH	A	41.467	19.007
18	1400	40	TH	A	41.107	19.036
19	2000	56	TH	R	41.532	17.302
20	1000	56	T	R	40.788	19.690
21	1000	40	T	R	41.896	16.258
22	1000	40	TH	R	43.070	18.276
23	1400	28	T	A	41.626	16.164
24	1000	40	TH	A	41.452	18.619
25	1400	28	TH	A	40.867	19.180
26	1000	56	TH	R	40.922	18.329
27	1000	28	T	A	40.796	18.506
28	2000	56	T	A	40.340	19.416
29	2000	40	T	A	39.434	18.889
30	2000	56	TH	A	41.910	17.025
31	1000	28	TH	R	40.968	19.735
32	2000	28	TH	A	40.708	19.026
33	1400	28	T	R	42.632	18.392
34	1400	40	T	A	41.183	19.645
35	2000	40	TH	R	41.244	19.600
36	2000	56	T	R	41.791	19.645

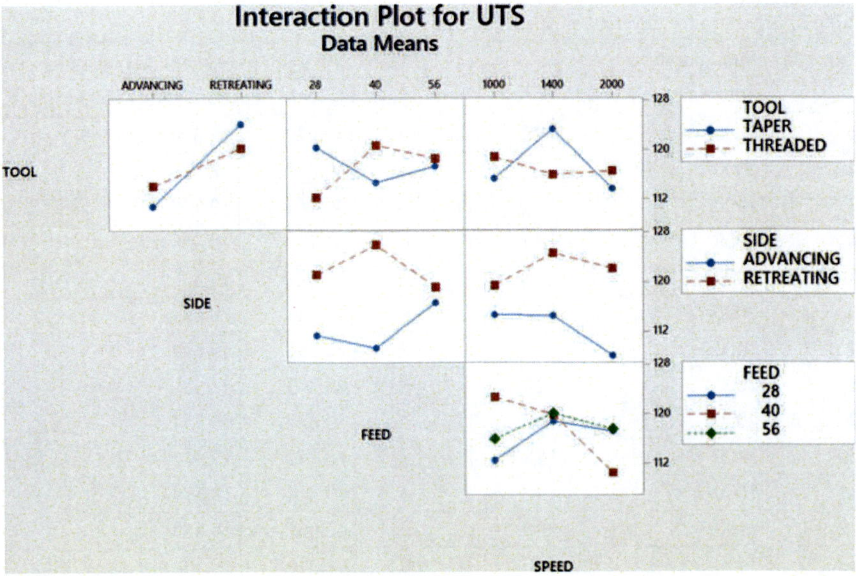

Fig. 4 Interaction plot for UTS (MPa) of the joints

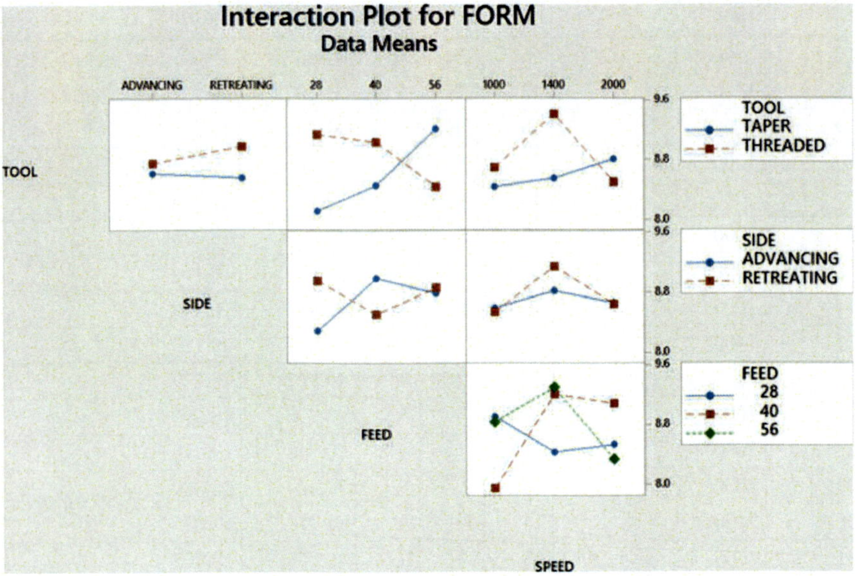

Fig. 5 Interaction plot for forming height (mm) of the joints

of coarse recrystallized grains and also leads to notable grain growth. Therefore, the variation of recrystallized grain size with welding speed in FSW depends on which factor is dominant, i.e., deformation or peak temperature. This trend indicates that the deformation is dominating up to 1400 rpm and then the peak temperature due the thermal cycle has dominant effect over deformation above 1400 rpm, which is considered for the reduction in formability of the weld. Also, the defect associated high rotation speed weaken the joint and result in to lower the formability of the joint.

3.2 *Effect of Tool Feed [24]*

It is well-known fact that the degree of material deformation and heat input are both decreased with increasing welding speed in the FSW when the tool rotation speed is constant. The decrease in the degree of material deformation during FSW accounts for coarser re-crystallized grains according to the general principles for re-crystallization. On the other hand, less heat input leads to finer re-crystallized grains, which gives rise to UTS during increase in rate of feed. Recrystallization temperature decrease with the dislocation density and annealing time, so with the increase in feed rate favor dynamic recrystallization due to sufficient time and temperature is available for the occurence of it, so the increase in the formability observed with increase in feed rate.

3.3 *Effect of Tool Pin Profile*

The tool geometry has a crucial role in flow of plasticized material and localized heating. During the initial stage of tool plunge, the heating results mainly due to the friction between pin and workpiece. When the tool shoulder touches the workpiece, the friction between the shoulder and workpiece produces the major component of the heat. From the heat generation aspect, the relative size of pin and shoulder is important, but stirring and movement of material are affected chiefly by the design features of tool pin profile.

Due to the friction of the shoulder, the metal underneath it first moves towards the tool pin and then moves downwards along the pin surface. After the metal arrives at the pin tip, it is forced outwards and moves back following a helical rotational path [28]. Thus, horizontal movement and vertical movement of the material orient the material in different direction which reduces the anisotropic behavior by mixing of the material [29].

In case of taper pin tool, the metal around pin undergoes simple extrusion, and there seems to be no downward movement along the pin surface, thus this type of tool pin yields less stirring action, which induces insufficient deformation of material and production of void in stir zone. Formability is the bidirectional property or depends upon anisotropy of the material, therefore, the threaded pin gives better formability than taper pin. In case of Taper pin tool, the metal around pin undergoes simple extrusion, and there seems to be no downward movement along the pin surface. Thus, this type of pin tool yields less stirring action, which induces insufficient deformation of material and production of void in stir zone.

3.4 Effect of Sheet Orientation

Stronger mechanical action pushes the soft material for a longer distance and an increase in the area involved by large deformation is found when the 6061 is in the AS (harder material) and 3102 material is on the retreating side. This phenomenon promotes solid diffusion process. The solid diffusion process promotes continuity in the material and reduces occurrence of the porosity kind of defect. So, both UTS and formability observed high when 3102 is considered on retreating side.

4 Conclusions

In the present article, experimental investigation on the TWBs developed using FSW process has been done. Erichsen cup test has been used to investigate the forming behavior of the TWBs. Effect of process parameters has also been investigated in detail in the present investigation. Following concluding remarks can be drawn from the experimental study.

- UTS increases as the Tool rotation speed increases up to 1400 rpm and then it decreases due to the dominance of thermal cycle over deformation cycle.
- The effect of static to dynamic volume ratio has negligible effect on the UTS considering both tool profile.
- Formability of the material depends upon the pin profile and threaded pin profile resulted in more formability than taper pin profile.
- The thick (2 mm) and low strength material (AA 3102) at retreating sides gives higher UTS and formability of the joint.

References

1. Parente M, Safdarian R, Santos AD, Loureiro A, Vilaca P, Jorge RN (2016) A study on the formability of aluminum tailor welded blanks produced by friction stir welding. Int J Adv Manuf Tech 83(9–12):2129–2141
2. Ebrahimzadeh P, Baseri H, Mirnia MJ (2018) Formability of aluminum 5083 friction stir welded blank in two-point incremental forming process. Proc Inst Mech Eng Part E: J Process Mech Eng 232(3):267–280

3. Riahi M, Amini A (2013) Effect of different combinations of tailor-welded blank coupled with change in weld location on mechanical properties by laser welding. Int J Adv Manuf Tech 67(5–8):1937–1945
4. Panda SK, Kumar DR (2008) Improvement in formability of tailor welded blanks by application of counter pressure in biaxial stretch forming. J Mater Process Tech 204(1–3):70–79
5. Suresh VS, Regalla SP, Gupta AK (2017) Combined effect of thickness ratio and selective heating on weld line movement in stamped tailor-welded blanks. Mater Manuf Process 32(12):1363–1367
6. Panda SK, Kumar DR (2008) Improvement in formability of tailor welded blanks by application of counter pressure in biaxial stretch forming. J Mate Process Technol 204(1–3):70–79
7. Rodrigues DM, Loureiro A, Leitao C, Leal RM, Chaparro BM, Vilaça P (2009) Influence of friction stir welding parameters on the microstructural and mechanical properties of AA 6016-T4 thin welds. Mater Des 30(6):1913–1921
8. Heo Y, Choi Y, Kim HY, Seo D (2001) Characteristics of weld line movements for the deep drawing with drawbeads of tailor-welded blanks. J Mater Process Tech 111(1–3):164–169
9. Zadpoor AA, Sinke J, Benedictus R (2007) Mechanics of tailor welded blanks: an overview. InKey Eng Mater 344:373–382
10. Kinsey B, Liu Z, Cao J (2000) A novel forming technology for tailor-welded blanks. J Mater Process Tech 99(1–3):145–153
11. Kesharwani RK, Panda SK, Pal SK (2015) Experimental investigations on formability of aluminum tailor friction stir welded blanks in deep drawing process. J Mater Eng Perf 24(2):1038–1049
12. Marathe S, Raval H (2019) Investigation on prediction of forming behaviour of blanks through laser and ultra violet (UV) grid marking methods during single point incremental forming (SPIF) process and Erichsen cup method. Int J Mod Manuf Tech 11(2):37–43
13. Silva MB, Skjødt M, Vilaça P, Bay N, Martins PA (2009) Single point incremental forming of tailored blanks produced by friction stir welding. J Mater Process Tech 209(2):811–820
14. Rattanachan K, Sirivedin K, Chungchoo C (2014) Formability of tailored welded blanks in single point incremental forming process. Adv Mater Res 979:339–342
15. Alinaghian I, Ranjbar H, Beheshtizad MA (2017) Forming limit investigation of AA6061 friction stir welded blank in a single point incremental forming process: RSM approach. Trans Ind Inst Metals 70(9):2303–2318
16. Marathe SP, Raval HK (2019) Numerical investigation on forming behavior of friction stir tailor welded blanks (FSTWBs) during single-point incremental forming (SPIF) process. J Braz Soc Mech Sci Eng 41(10):424
17. Al-Ghamdi KA, Hussain G (2015) Threshold tool-radius condition maximizing the formability in SPIF considering a variety of materials: experimental and FE investigations. Int J Mach Tools Manuf 88:82–94
18. Reis A, Teixeira P, Duarte JF, Santos A, Da Rocha AB, Fernandes AA (2004) Tailored welded blanks—an experimental and numerical study in sheet metal forming on the effect of welding. Comput Struct 82(17–19):1435–1442
19. Rojek J, Hyrcza-Michalska M, Bokota A, Piekarska W (2012) Determination of mechanical properties of the weld zone in tailor-welded blanks. Arch Civil Mech Eng 12:156–162
20. Cheng CH, Chan LC, Tang CY, Chow CL (2005) Determination of true stress-strain curve for the weldment of aluminum laser-welded blanks. J Laser Appl 17(3):159–170
21. Song Y, Hua L, Chu D, Lan J (2012) Characterization of the inhomogeneous constitutive properties of laser welding beams by the micro-Vickers hardness test and the rule of mixture. Mater Des 37:19–27
22. Spöttl M, Mohrbacher H (2014) Laser-based manufacturing concepts for efficient production of tailor welded sheet metals. Adv Manuf 2(3):193–202
23. Kulkarni N, Mishra RS, Yuan W (2015) Friction stir welding of dissimilar alloys and materials. Butterworth-Heinemann
24. Elangovan K, Balasubramanian V (2007) Influences of pin profile and rotational speed of the tool on the formation of friction stir processing zone in AA2219 aluminium alloy. Mater Sci Eng, A 459(1–2):7–18

25. Mishra RS, Ma ZY (2005) Friction stir welding and processing. Mater Sci Eng R: Rep 50(1–2):1–78
26. Elyasi M, Aghajani Derazkola H, Hosseinzadeh M (2016) Investigations of tool tilt angle on properties friction stir welding of A441 AISI to AA1100 aluminium. Proc Inst Mech Eng Part B: J Eng Manuf 230(7):1234–1241
27. Saeid T, Abdollah-Zadeh A, Assadi H, Ghaini FM (2008) Effect of friction stir welding speed on the microstructure and mechanical properties of a duplex stainless steel. Mater Sci Eng A 496(1–2):262–268
28. Li Y, Qin F, Liu C, Wu Z (2017) A review: effect of friction stir welding on microstructure and mechanical properties of magnesium alloys. Metals 7(12):524
29. Elangovan K, Balasubramanian V, Valliappan M (2008) Effect of tool pin profile and tool rotational speed on mechanical properties of friction stir welded AA6061 aluminium alloy. Mater Manuf Process 23(3):251–260

A Brief Review on Formability, Wall Thickness Distribution and Surface Roughness of Formed Part in Incremental Sheet Forming

Kiran R. More, Vikas Sisodia, and Shailendra Kumar

1 Introduction

Incremental sheet forming (ISF) process is one of the advanced sheet metal forming process which does not require dedicated punch and die for forming operation. A universal fixture for clamping the sheet metal blank, and forming tool is the only requirement. The foundation of the ISF process was laid down by Mason in the year 1978 [1]. In ISF process, blank is held between the blank holder and the backing plate, and a tool (or punch) is allowed to move with a small incremental step in negative z-axis on the blank, as shown in Fig. 1. ISF process is a flexible and innovative forming process in which sheet metal can be shaped into customized geometries without the need for a dedicated punch and die system [2].

ISF process can be performed using different types of forming tools, as shown in Fig. 2. Generally, two types of forming tools were used by researchers—(i) fluid tool [4], and (ii) rigid tool. The ISF process is classified according to the number of stages, forming method, and forming tool. Also, depending on the forming method, it is basically classified into three types such as single point incremental forming (SPIF) process, two-point incremental forming (TPIF) process, and hybrid incremental sheet forming (HISF) process.

The present paper describes a literature review on the influence of process parameters on response characteristics like formability, wall thickness distribution, and surface roughness in ISF process. Finally, the conclusion and recommendation for future research are drawn.

K. R. More (✉) · V. Sisodia · S. Kumar
Mechanical Engineering Department, Sardar Vallabhbhai National Institute of Technology, Surat, Gujarat, India
e-mail: kmore199@gmail.com

H. K. Dave and D. Nedelcu (eds.), *Advances in Manufacturing Processes*, Lecture Notes in Mechanical Engineering, https://doi.org/10.1007/978-981-15-9117-4_11

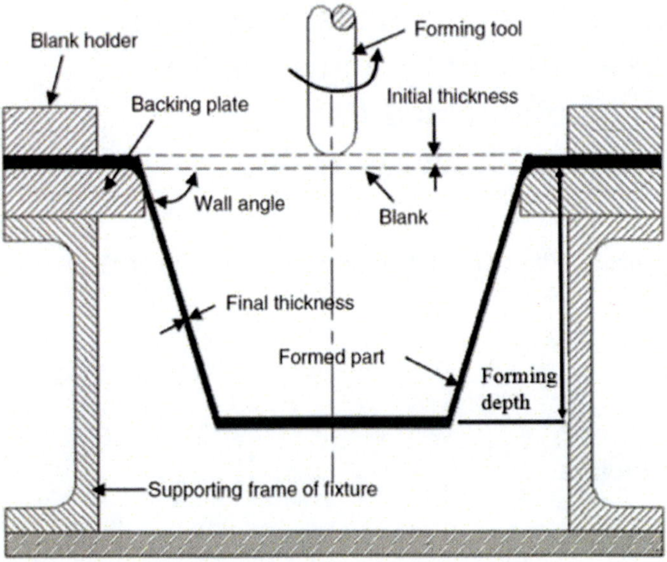

Fig. 1 Cross-sectional view SPIF process [3]

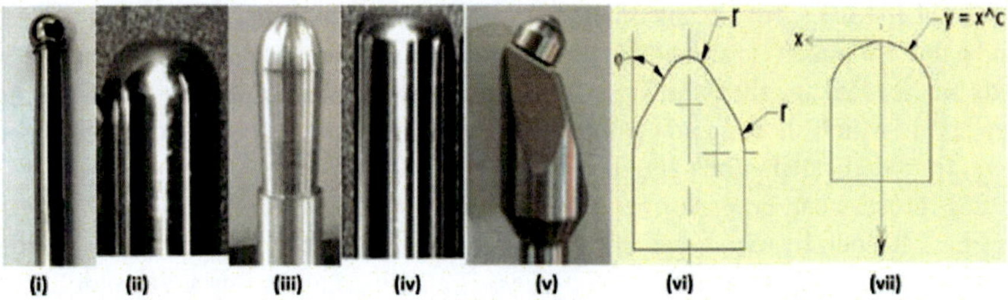

Fig. 2 Various tool shapes used in the ISF process [5–7]

2 Influence of Process Parameters on Formability and Wall Thickness Distribution in ISF

Due to some prevailing limitations of the ISF process like uneven wall thickness distribution, high forming time, and surface roughness of the formed part, it is not popular in sheet metal industries. Worldwide researchers have made efforts to improve the above characteristics of the ISF process to make it acceptable in sheet metal industries. Process parameters influence these responses. A brief literature review focused on the influence of process parameters on formability, wall thickness distribution and surface roughness of formed parts in the ISF process is discussed in subsequent sections.

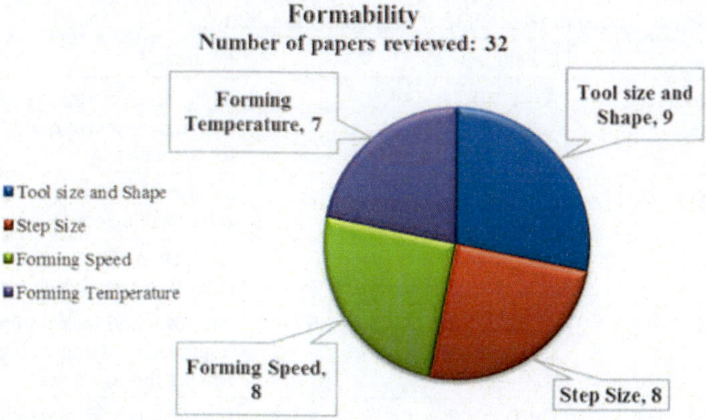

Fig. 3 Categorization of research papers reviewed on the influence of process parameters on formability

2.1 *Formability*

Formability in ISF process is more as compared to the conventional sheet metal forming because of the incremental nature of deformation. In ISF process, formability is generally measured in terms of maximum wall angle (also called maximum forming angle) attainable by the sheet metal [7, 8]. Forming depth is also one of the parameters used by researchers to investigate the formability [9]. Formability is influenced by various process parameters like tool size, step size, feed, etc. Total 32 research papers related to influence of parameters on formability have been reviewed. Categorization of these papers on the basis of process parameters is depicted in Fig. 3.

The summary of major research work related to the influence of process parameters on formability in ISF process is given in Table 1.

2.2 *Wall Thickness Distribution*

Wall thickness distribution in formed parts is influenced by process parameters. In this domain total 30 research papers have been reviewed. Categorization of these papers is shown in Fig. 4.

A summary of major research work related to the influence of process parameters on wall thickness distribution is given in Table 2.

Table 1 Summary of major research work related to formability

Author	Sheet material	Findings
Kim and Park [10]	AA 1050	High values of step size and tool size impose an adverse effect on formability
Ambrogio et al. [11]	AA 1050-O	Smaller step size as a measure of safer forming condition
Ham and Jeswiet [12]	AA3003-O	Interaction effect of tool size and sheet thickness is important
Duflou et al. [13]	Al 5152, AISI 5155, TiAl6V4 sheets	Accuracy and formability are improved by dynamic local heating through laser
Hussain et al. [14]	CP Ti	Lubricating with a paste of MoS_2 and grease improves formability and surface quality of the formed part
Ji and Park [15]	Magnesium AZ31	Formability of hard to form material increases at elevated temperature
Fan et al. [16]	AZ31 Magnesium	Hot incremental forming used hard to form material and they observed it is feasible and easy to control. Also, seen after hot forming asymmetric part shows more distortion than symmetric part
Ambrogio et al. [17]	Magnesium alloy AZ31	Warm forming condition is used for hard to form materials
Hussain et al. [18]	Titanium	Interaction between tool size and step size was found more significant than their individual effects
Decultot et al. [19]	ENAW - 5086-H111	High strain and spatial resolution of the 3D digital image correlation used for the development of strain field to study the formability of mentioned material
Hussain et al. [20]	AA 2024 preaged and annealed	Interaction of step size and tool size was found to be significant for formability
Nguyen et al. [21]	Cold-rolled Steel	Incremental forming of complex shape (human face). Used Taguchi's orthogonal array for FEM simulations

(continued)

Table 1 (continued)

Author	Sheet material	Findings
Bhattacharya et al. [22]	AL 5052	Tool diameter and sheet thickness interaction have a significant effect on material formability
Gottmann et al. [23]	Ti Grade 2 and TiAl6V4	They highlighted two main process limits, i.e., geometric accuracy and low formability at room temperature of hard to form material
Hussain et al. [24]	AA - 2024 - O	Interaction of sheet thickness and tool radius is found significant
Buffa et al. [25]	AA 1050-O, AA 1050-H24, AA 6082-T6	Improvement of formability through localized sheet heating
Fritzen et al. [26]	Brass (70-30)	With spiral (or helical) tool path improved formability is observed
Lu et al. [27]	AA1100, AA2024, AA5052 and AA6111	Developed oblique rolling ball tool which reduces friction and improves surface finish and formability
Adams and Jeswiet [28]	Al 6061-T6	Resistive heating (electrically assisted forming) which increases the sheet temperature is beneficial for hard to form materials
Liu et al. [29]	Ti6Al4V sheets	A new tooling concept that is electricity assisted incremental forming was introduced to provide localized heating to improve formability of hard to form material
Al-Ghamdi and Hussain [30]	Tri-Layered Cu-Steel-Cu Composite sheet metal	High spindle speed and feed rate has good influence on formability, irrespective of tool size and step size when the composite sheet is annealed at low temperature
Satish et al. [31]	Al-6061	They observed that overall formability and post-forming strength depends on the combined effect punch speed and temperature in the warm working temperature range
Baharudin et al. [32]	Al 6061-T6	Frictional stir-assisted SPIF was used for deformation
Barnwal et al. [33]	AA-6061 aluminum alloy	Plastic anisotropy has a strong effect on microstructure and texture development

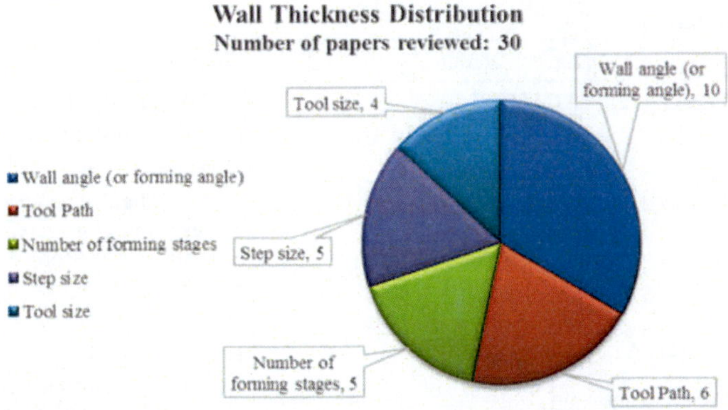

Fig. 4 Categorization of research papers reviewed on the influence of process parameters on wall thickness distribution

Table 2 Summary of major research work related to wall thickness distribution

Author	Sheet material	Findings
Young and Jeswiet [34]	Aluminum 3003-O	Double pass SPIF has effective technique to avoid excessive thinning in precise area
Hussain and Gao [35]	Aluminum alloy in annealed form	Proposed a method to test thinning limits
Ambrogio et al. [36]	Al 1050-o	Proposed new tool path strategy is known as a decremental slope to improve thickness distribution
Hamilton and Jeswiet [37]	Al 3003-H14	Feed is an insignificant parameter for wall thickness distribution
Manco et al. [38]	AA 1050-O	Improvement in minimum thickness was observed by decremental slope tool path strategy
Li et al. [39]	DC 56 steel	An equation to estimate the number of forming stages required for forming was proposed
Li et al. [40]	DC-04 steel	More number of forming stages leads to less thinning and more homogenous wall thickness distribution
Doss et al. [41]	Al 1050	Applied FEM to predict wall thickness distribution

(continued)

Table 2 (continued)

Author	Sheet material	Findings
Malwad and Nandedkar [42]	AA 8011	As wall angle increases stretching intensifies which results in non-uniform thickness distribution
Oleksik [43]	Dc-04	Thickness reduction is highly dependent on wall angle
Sisodia and Kumar [44]	Al-1050	Variation in wall thickness is small, and overall wall thickness distribution along the depth of formed parts is almost uniform in SPIF process with dummy sheet

2.3 *Surface Roughness*

Surface roughness is one of the important response parameters in the ISF process. The main reason behind high surface roughness is due to waviness in the sheet caused by the tool, also known as scallop height. Categorization of 31 research papers reviewed on the influence of process parameters on surface roughness depicted in Fig. 5.

A summary of major research work related to the influence of process parameters on surface roughness is given in Table 3.

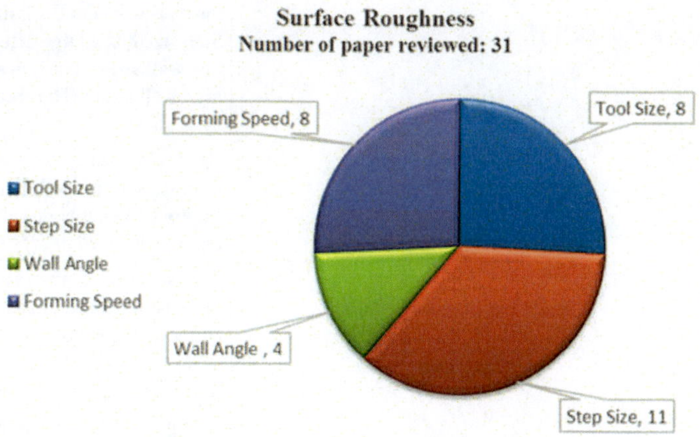

Fig. 5 Categorization of research papers reviewed on the influence of process parameters on surface roughness

Table 3 Summary of major research work related to surface roughness

Author/Year	Sheet material	Remarks
Hagan and Jeswiet [45]	Annealed Al 3003	Spindle speed is less significant parameter for surface roughness
Durante et al. [46]	AA 7075 T0	Interaction between step size and tool size was found. Low values of pitch and high value of tool size (in specified and acceptable range) results in a good surface finish
Hamilton and Jeswiet [37]	Al 3003-H14	Developed a predictive model for external non-contact surface roughness (orange peel effect) using measured surface roughness and using forming parameters
Bhattacharya et al. [22]	AL 5052	Surface roughness decreases with an increase in tool size at all step sizes. For a certain wall angle, the roughness increases with an increase in step size and then decreases
Ambrogio et al. [47]	AA-1050-O, AA 5754, AA 6082-T	Feed rate has less influence on surface roughness, so high feed can be given to minimize the forming time
Jason [48]	AA 5052	Interaction between step size and the wall was found significant. Step size has a large effect at a low wall angle, but less effect at a high wall angle. The feed rate has little effect
Radu et al. [49]	AA 1050	Surface roughness and microstructural modification has improved when high-value parameters used except step size
Echrif and Hrairi [50]	AA 1050-O	Tool size and step are most influencing parameters for surface roughness
Jagtap et al. [51]	Al 1050	The type of tool path also affects the surface quality. Profile tool path gives higher surface roughness as compared to helical tool path

(continued)

Table 3 (continued)

Author/Year	Sheet material	Remarks
Bastos et al. [52]	AA1050-H11 and dual-phase steel (DP600, DP780, and DP1000)	Increasing feed-rate deteriorates surface finish for dual-phase steels. Aluminum1050-H111 appears to be insensitive to feed rate variations. So, the material type may be a parameter which can influence the surface roughness
Najafabady and Ghaei [53]	Ti–6Al–4V titanium alloy	They worked on electric hot incremental forming of Ti–6Al–4V titanium alloy and observed that lubrication plays a vital role in this particular process
Radhika et al. [54]	AA5052	The minimal surface roughness can be obtained with minimal processing time when using oil as the lubricant in the multipoint incremental forming process
Slota et al. [55]	DC04 steel	The interaction between the work piece and tool-tip shows small linear grooves on the inner surface and the source of the premature crack is outer surface. Residual stresses are also measured by X-ray diffraction method
Sisodia and Kumar [56]	AA 1050	Vital role of dummy sheet in improvement of surface finish of target sheet

3 Applications of the Incremental Forming Process

Various researchers have made efforts to use ISF process to form a variety of industry-specific free-form shapes like vehicle headlight reflector, exhaust manifold, etc. A total of 18 research papers are reviewed related to applications of the ISF process. Various areas of application include automotive and aerospace, biomedical, architectural, households and kitchenware, and other industrial sheet metal parts. This is depicted in Fig. 6.

A summary of the literature review related to applications of the ISF process is given in Table 4.

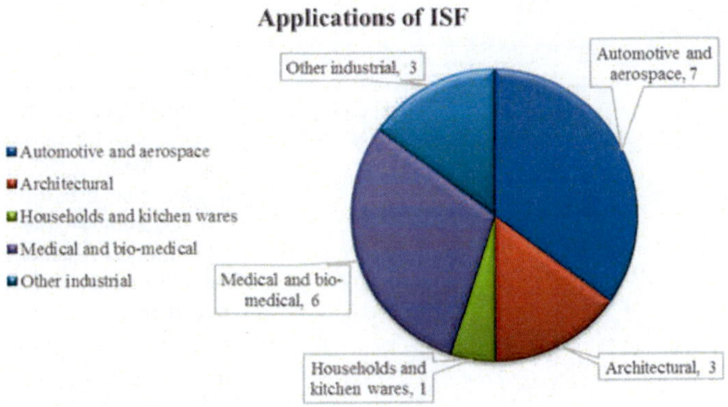

Fig. 6 Various applications of ISF process

Table 4 Application of ISF process

Broad application area	Particulars	Reference
Automotive and aerospace	Automobile headlight reflector, exhaust manifolds, motorbike seat and gas tank	Jeswiet and Hagan [57]
	Bus sheet metal body parts	Liu [58]
	Housings and fairings of aerospace and ship structures, Powder bed fusion (PBF) of metal components	Jeswiet et al. [6], Diaz [59]
	1/8 scale model of Shinkansen (Bullet Train)	Lora et al. [60]
	Automotive service parts like hood and side panels of the car	Amino et al. [61]
	Car tail light bracket	Li et al. [40]
Architectural	Free-form architectural envelopes	Castaneda et al. [62]
	Self-supporting architectural Structures	Bailly et al. [63]
	bespoke formwork, decorative panels	Jeswiet et al. [6]
Households and kitchen wares	Solar cooker/oven	Jeswiet et al. [64]
Medical and biomedical	Cranial	Duflou et al. [65], Lu et al. [66]
	Knee	Eksteen and Merwe [67]
	ankle	Ambrogio et al. [68]
	Denture base	Milutinovic et al. [69], Sbayti et al. [70]

(continued)

Table 4 (continued)

Broad application area	Particulars	Reference
Other industrial components	Hole flange	Silva et al. [71], Bambach et al. [72]
	Dies and molds	Allwood et al. [73]

4 Conclusion

Worldwide researchers have made research efforts in the domain of the ISF process. In the present study, more than 70 research papers focused on formability, wall thickness distribution, surface roughness, and applications of ISF process have been reviewed. The following are the conclusions drawn from the literature review:

1. Very few efforts are made in studying the effect of forming speed, forming temperature, and material anisotropy on formability. There is a requirement of a strategy that can clearly define and relate the influencing parameters relative to material type with formability. So research efforts are required to develop a strategy that can predict and define the formability limit of specific material and capable of recommending the optimum set of forming variables.
2. Also, very fewer research efforts have been reported on studying the influence of tool path, the number of forming stages, and step size on wall thickness distribution of formed part. Although very comprehensive literature is available related to the ISF process, still there is the absence of strategy which can predict and define the thinning limit of sheet. Also, no such effective strategy has been proposed to achieve a uniform thickness distribution or to identify or predict the location of minimum thickness. Thus research efforts are required to quantify the wall thickness distribution in the domain to the ISF process.
3. Very less research efforts have been applied for the influence of wall angle and forming speed on surface roughness. Therefore efforts are required to study the influence of said process parameters on surface roughness. Also, no such effective strategy has been proposed to achieve minimum surface roughness. Thus, research efforts are required to quantify the surface roughness in the domain to the ISF process.

Although very comprehensive literature is available related to ISF process but still there is deficiency of work regarding computational tools and technique which can properly predict, generate and define the various quality characteristics of formed part. Also, no such effective strategy has been proposed till date to achieve minimum surface roughness and uniform wall thickness distribution. A dedicated theoretical model is required to establish a better understanding about the process behavior. Also, in order to improve the performance of the process a dedicated computational tool needs to be developed. Such tools should be able to predict and simulate the process. For example, development of such algorithm is required which can predict behaviour of formed part characteristics taking the materials' properties into consideration.

The ISF process has great potential to meet the continuously changing demands of the customers as far as the small batch of parts is concerned. Various researchers have successfully applied the ISF process in various field of engineering. Any sort of late customization is possible in the formed part due to process flexibility. The process is also found effective in deforming the tailor welded blanks. Due to layer by layer deformation characteristics, forming forces required to deform the sheet metal is less as compared to conventional sheet metal forming processes. Also, this allows difficult for form materials like titanium, magnesium, etc. can be deformed relatively with less lead and setup time.

References

1. Emmens WC, Sebastiani G, van den Boogaard AH (2010) The technology of incremental sheet forming—a brief review of the history. J Mater Process Technol 210(8):981–997
2. Kurra S, Swetha N, Vinodh Reddy C, Regalla SP (2018) Experimental and finite element studies of single stage incremental forming process: effect of process parameters on maximum wall angle and thickness distribution. Adv Mater Process Technol 4(2):322–334
3. Kurra S, Rahman NH, Regalla SP, Gupta AK (2015) Modeling and optimization of surface roughness in single point incremental forming process. J Mater Res Technol 4(3):304–313
4. Jurisevic B, Kuzman K, Junkar M (2006) Water jetting technology: an alternative in incremental sheet metal forming. Int J Adv Manuf Technol 31(1–2):18–23
5. Lingam R, Bansal A, Reddy NV (2016) Analytical prediction of formed geometry in multi-stage single point incremental forming. IntJ Mater Form 9(3):395–404
6. Jeswiet J, Micari F, Hirt G, Bramley A, Duflou J, Allwood J (2005) Asymmetric single point incremental forming of sheet metal. CIRP Ann 54(2):88–114
7. Jeswiet J, Young D (2005) Forming limit diagrams for single-point incremental forming of aluminium sheet. Proc Inst Mech Eng Part B J Eng Manuf 219(4):359–364
8. Hussain G, Gao L, Hayat N, Qijian L (2007) The effect of variation in the curvature of part on the formability in incremental forming: an experimental investigation. Int J Mach Tools Manuf 47(14):2177–2181
9. Attanasio A, Ceretti E, Fiorentino A, Mazzoni L, Giardini C (2009) Experimental test to study feasibility and formability in incremental forming process. Key Eng Mat 410:411
10. Kim YH, Park JJ. (2002) Effect of process parameters on formability in incremental forming of sheet metal. J Mater Process Technol 130:42–46
11. Ambrogio G, Filice L, Fratini L, Micari F (2004) Process mechanics analysis in single point incremental forming. AIP Conf Proc 712(1):922–927
12. Ham M, Jeswiet J (2006) Single point incremental forming and the forming criteria for AA3003. CIRP Annals Manuf Technol 55(1):241–244
13. Duflou JR, Callebaut B, Verbert J, De Baerdemaeker H (2007) Laser assisted incremental forming: formability and accuracy improvement. CIRP Ann Manuf Technol 56(1):273–276
14. Hussain G, Gao L, Zhang ZY (2008) Formability evaluation of a pure titanium sheet in the cold incremental forming process. Int J Adv Manuf Technol 37(9–10):920–926
15. Ji YH, Park JJ (2008) Formability of magnesiumAZ31 sheet in the incremental forming at warm temperature. J Mater Process Technol 201(1–3):354–358
16. Fan G, Gao L, Hussain G, Wu Z (2008) Electric hot incremental forming: a novel technique. Int J Mach Tools Manuf 48(15):1688–1692
17. Ambrogio G, Filice L, Manco GL (2008) Warm incremental forming of magnesium alloy AZ31. CIRP Ann Manuf Technol 57(1):257–260

18. Hussain G, Gao L, Hayat N (2009) Empirical modelling of the influence of operating parameters on the spifability of a titanium sheet using response surface methodology. Proc Inst Mech Eng Part B J Eng Manuf 223(1):073–081
19. Decultot N, Robert L, Velay V, Bernhart G (2010) Single point incremental sheet forming investigated by in-process 3D digital image correlation. EPJ Web Conf 6:11001
20. Hussain G, Gao L, Hayat N, Dar NU (2010) The formability of annealed and pre-aged AA-2024 sheets in single-point incremental forming. Int J Adv Manuf Technol 46(5–8):543–549
21. Nguyen DT, Park JG, Lee HJ, Kim YS (2010) Finite element method study of incremental sheet forming for complex shape and its improvement. Proc Inst Mech Eng Part B J Eng Manuf 224:913–924
22. Bhattacharya A, Maneesh K, Reddy NV, Cao J (2011) Formability and surface finish studies in single point incremental forming. J Manuf Sci Eng 133(6):061020
23. Gottmann A, Diettrich J, Bergweiler G, Bambach M, Hirt G, Loosen P, Poprawe R (2011) Laser-assisted asymmetric incremental sheet forming of titanium sheet metal parts. Prod Eng 5(3):263–271
24. Hussain G, Khan HR, Gao L, Hayat N (2013) Guidelines for tool-size selection for single-point incremental forming of an aerospace alloy. Mater Manuf Process 28(3):324–329
25. Buffa G, Campanella D, Fratini L (2013) On the improvement of material formability in SPIF operation through tool stirring action. Int J Adv Manuf Technol 66(9–12):1343–1351
26. Fritzen D, Daleffe A, Castelan J, Schaeffer L (2013) Brass 70/30 and incremental sheet forming process. Key Eng Mater 554–557:1419–1431
27. Lu B, Fang Y, Xu D, Chen J, Ou H, Moser N, Cao J (2014) Mechanism investigation of friction-related effects in single point incremental forming using a developed oblique roller-ball tool. Int J Mach Tools Manuf 85:14–29
28. Adams D, Jeswiet J (2014) Single-point incremental forming of 6061–T6 using electrically assisted forming methods. Proc Inst Mech Eng Part B J Eng Manuf 228(7):757–764
29. Liu R, Lu B, Xu D, Chen J, Chen F, Ou H, Long H (2016) Development of novel tools for electricity-assisted incremental sheet forming of titanium alloy. Int J Adv Manuf Technol 85(5–8):1137–1144
30. Al-Ghamdi KA, Hussain G (2016) Parameter-formability relationship in ISF of tri-layered Cu-Steel-Cu composite sheet metal: response surface and microscopic analyses. Int J Precision Eng Manuf 17(12):1633–1642
31. Raja Satish D, Ravi Kumar D (2019) Formability of AA6061 alloy sheets in warm forming temperature range. Proc Inst Mech Eng Part L J Mater Des Appl 233(3):413–425
32. Baharudin BTHT, Azpen QM, Sulaima S, Mustapha F (2017) Experimental investigation of forming forces in frictional stir incremental forming of aluminum alloy AA6061-T6. Metals 7(11):484
33. Barnwal VK, Chakrabarty S, Tewari A, Narasimhan K, Mishra SK (2018) Forming behavior and microstructural evolution during single point incremental forming process of AA-6061 aluminum alloy sheet. Int J Adv Manuf Technol 95(1–4):921–935
34. Young D, Jeswiet J (2004) Wall thickness variations in single-point incremental forming. J Eng Manuf 218:1453–1459
35. Hussain G, Gao L (2007) A novel method to test the thinning limits of sheet metals in negative incremental forming. Int J Mach Tools Manuf 47(3–4):419–435
36. Ambrogio G, Filice L, Gaudioso M, Manco GL (2010) Optimized tool-path design to reduce thinning in ISF process. Int J Mater Form 3(1):959–962
37. Hamilton K, Jeswiet J (2010) Single point incremental forming at high feed rates and rotational speeds: surface and structural consequences. CIRP Annals Manuf Technol 59(1):311–314
38. Manco L, Filice L, Ambrogio G (2011) Analysis of the thickness distribution varying tool trajectory in single-point incremental forming. Proc Inst Mech Eng Part B J Eng Manuf 225(3):348–356
39. Li J, Hu J, Pan J, Geng P (2012) Thickness distribution and design of a multi-stage process for sheet metal incremental forming. Int J Adv Manuf Technol 62(9–12):981–988

40. Li J, Geng P, Shen J (2013) Numerical simulation and experimental investigation of multistage incremental sheet forming. Int J Adv Manuf Technol 68(9–12):2637–2644
41. Doss QM, Abaas TF, Bedan AS (2013) An investigation study of thinning distribution in single point incremental forming using FEM analysis. Al-Khwarizmi Eng J 9(3):1–14
42. Malwad DS, Nandedkar VM (2014) Deformation mechanism analysis of single point incremental sheet metal forming. Proc Mater Sci 6(ICMPC):1505–1510
43. Oleksik V (2014) Influence of geometrical parameters, wall angle and part shape on thickness reduction of single point incremental forming. Proc Eng 81:2280–2285
44. Sisodia V, Kumar S (2019) Experimental study of single point incremental forming with dummy sheet. Int J Mater Eng Innov 10(1):60–82
45. Hagan E, Jeswiet J (2004) Analysis of surface roughness for parts formed by computer numerical controlled incremental forming. Proc Inst Mech Eng Part B J Eng Manuf 218(10):1307–1312
46. Durante M, Formisano A, Langella A (2010) Comparison between analytical and experimental roughness values of components created by incremental forming. J Mater Process Technol 210(14):1934–1941
47. Ambrogio G, Filice L, Gagliardi F (2012) Improving industrial suitability of incremental sheet forming process. Int J Adv Manuf Technol 58(9–12):941–947
48. Lasunon OU (2013) Surface roughness in incremental sheet metal forming of AA5052. Adv Mater Res 753:203–206
49. Radu C, Herghelegiu E, Cristea ION, Schnakovszky C (2013) Analysis of the surface quality of parts processed by single point incremental forming. J Eng Stud Res 19(3):76–79
50. Echrif SBM, Hrairi M (2014) Significant parameters for the surface roughness in incremental forming process. Mater Manuf Process 29(6):697–703
51. Jagtap R, Kashid S, Kumar S, Hussein HM (2015) An experimental study on the influence of tool path, tool diameter and pitch in single point incremental forming (SPIF). Adv Mater Process Technol 1(3–4):465–473
52. Bastos RN, de Sousa RJ, Ferreira JA (2016) Enhancing time efficiency on single point incremental forming processes. IntJ Mater Form 9(5):653–662
53. Najafabady SA, Ghaei A (2016) An experimental study on dimensional accuracy, surface quality, and hardness of Ti-6Al-4 V titanium alloy sheet in hot incremental forming. Int J Adv Manuf Technol 87(9–12):3579–3588
54. Radhika T, Venkatesan R, Sathiyanarayanan C, Kumar MP (2019) Investigations and optimization of wall angle and surface roughness in incremental forming of AA5052 aluminum alloy. Carib J Sci 53:707–720
55. Slota J, Krasowski B, Kubit A, Trzepiecinski T, Bochnowski W, Dudek K, Neslušan M (2020) Residual stresses and surface roughness analysis of truncated cones of steel sheet made by single point incremental forming. Metals 10(2):237
56. Sisodia V, Kumar S (2020) Study of AA-1050 sheet metal parts processed by single point incremental forming with dummy sheet. Int J Mater Eng Innov 11(2):105–126
57. Jeswiet J, Hagan E (2001) Rapid prototyping of a headlight with sheet metal. Canadian Institute of Mining, Metallurgy and Petroleum (Canada), pp 109–114
58. Liu G (2011) The application of single point incremental forming technology of sheet metal in bus prototyping. Adv Mater Res 338:46–55
59. Diaz A (2019) Surface texture characterization and optimization of metal additive manufacturing-produced components for aerospace applications. In: Additive manufacturing for the aerospace industry, pp 341–374
60. Lora FA, Boff U, Yurgel CC, Folle LF, Schaeffer L (2013) Validation of the computer simulation process applied to the incremental forming process for the evaluation of strain paths. Key Eng Mater 554:2453–2461
61. Amino M, Mizoguchi M, Terauchi Y, Maki T (2014) Current status of "Dieless" Amino's incremental forming. Proc Eng 81:54–62
62. Castaneda E, Lauret B, Lirola JM, Ovando G (2015) Free-form architectural envelopes: digital processes opportunities of industrial production at a reasonable price. J Facade Des Eng 3(1):1–13

63. Bailly D, Bambach M, Hirt G, Forming M, Aachen R, Pofahl T, and Della G (2015) Investigation on the producibility of freeform facade elements made of sheet metal as self-supporting structures by means of incremental sheet forming. In: The METEC and 2nd European steel technology and application days
64. Jeswiet J, Duflou JR, Szekeres A, Lefebvre P (2005) Custom manufacture of a solar cooker—a case study. Adv Mater Res 6:487–492
65. Duflou JR, Lauwers B, Verbert J, Gelaude F, Tunckol Y (2005) Medical application of single point incremental forming: cranial plate manufacturing. In: Virtual modelling and rapid manufacturing: advanced research in virtual and rapid prototyping proc. 2nd international conference on advanced research in virtual and rapid prototyping. Leiria, Portugal, p 161
66. Lu B, Xu DK, Liu RZ, Ou HA, Long H, Chen J (2015) Cranial reconstruction using double side incremental forming. Key Eng Mater 639:535–542
67. Eksteen PD, Van der Merwe, AF (2012) Incremental sheet forming (ISF) in the manufacturing of titanium-based plate implants in the bio-medical sector. In: CIE42 proceedings, pp 15–18
68. Ambrogio G, Denapoli L, Filice L, Gagliardi F, Muzzupappa M (2005) Application of Incremental forming process for high customised medical product manufacturing. J Mater Process Technol 162:156–162
69. Milutinovic M, Lendel R, Potran M, Vilotic D, Skakun P, Plancak M (2014) Application of single point incremental forming for manufacturing of denture base. J Technol Plast 39(2):15–24
70. Sbayti M, Bahloul R, BelHadjSalah H, Zemzemi F (2018) Optimization techniques applied to single point incremental forming process for biomedical application. Int J Adv Manuf Technol 95(5–8):1789–1804
71. Silva MB, Martinho TM, Martins PAF (2013) Incremental forming of hole flanges in polymer sheets. Mater Manuf Process 28(3):330–335
72. Bambach M, Voswinckel H, Hirt G (2014) A new process design for performing hole-flanging operations by incremental sheet forming. Proc Eng 81:2305–2310
73. Allwood JM, Bramley AN, Ridgman TW, Mileham AR (2006) A novel method for the rapid production of inexpensive dies and moulds with surfaces made by incremental sheet forming. Proc Inst Mech Eng Part B J Eng Manuf 220(2):323–327

Parametric Study of Non-axisymmetric Stretch Flanging Process on AA-6061-T6 Sheet

Sai Dhruthi Killi, Keyur Desai, and Harit Raval

1 Introduction

Simply to define, flanging is a contour bending operation that is performed on a sheet which requires compound curvatures. It is encountered in many stretching processes say for, e.g. cups, panels and boxes and also in many deep drawing processes. Flanges are especially used for enhancing aesthetic appeal, for hidden joints and for rigidity strengthening of edges of sheet parts such as complex panels and automobile front fendors, etc., Of all the three types of flanges available (say shrink, stretch and straight), stretch flanging falls under that class where flange curvature appears to be concave (when seen from front side of the flanged surface). Hence, in this case, the metal in the flange is subjected to tension. In the stretch flanging, the sheet metal is forced locally near curved edge of die (around it) to form flanges of given angle and length. Edge crack phenomenon which encounters at die-profile radii along width of the sheet is the most dominant failure in NASF process.

Various researchers have focussed on different geometrical parameters concerning to stretch flanging operations. Wang et al. [1] carried out flanging of V-shaped metal blank wherein they found the effects of ratio of initial flange length and flange radius, flange angle on maximum free edge strain. Xu Feng et al. [2] have studied the formability of stretched curved flanging where the influence of punch profile radius and length to radius ratio on circumferential strain is studied. Li et al. [3] also focussed their study on stretch flanging of V-shaped sheet metal by enlightening the effect of flange length and flange angle on circumferential strain. Yogesh Dewang et al. [4] is one of those who has focussed his research on "NASF" process and his parameters of study are punch-die clearance, initial flange length, punch-profile radius, die-profile

S. D. Killi (✉) · K. Desai · H. Raval
Department of Mechanical Engineering, Sardar Vallabhbhai National Institute of Technology, Surat, Gujarat 395007, India
e-mail: killi.dhruthi@gmail.com

H. K. Dave and D. Nedelcu (eds.), *Advances in Manufacturing Processes*, Lecture Notes in Mechanical Engineering, https://doi.org/10.1007/978-981-15-9117-4_12

radius, Blank holding force and coefficient of friction. There exist many methods for carrying out this analytical study of which 'Finite element method, is proved out to be one of the finest methods. In this method, the component on which a particular analysis has to be done is alienated into number of finite elements (or number of nodes) and then the effect of particular load (force) on each and every element is carried out making it a highly versatile process. Earlier many formulas have been formulated so as to carry out this analytical study, but later on with globalisation new computer softwares have been invented for this FE analysis. Many authors have utilised various softwares for carrying out FE simulations. Sartkulvanich et al. [5] carried out the FE analysis to analyse the quality of blanked edge formed after stretch flanging on AHSS sheet by making use of a DEFORM-2D software. Butcher et al. [6] developed a 3D symmetrical finite element model to analyse the stretch flange forming process of AA 5182 sheet using LS-DYNA code. Bahloul [7] focussed on optimisation of the bending process of High Strength Low Alloy sheet metal by making use of ABAQUS-3D. Of all these, the more recent studies have used ABAQUS-2D and ABAQUS-3D. ABAQUS/CAE also provides a user-friendly and interactive environment to define motion of different components as well as procedure for geometric modelling and problem definition. Hence the present study is aimed to carryout FE simulation of NASF process on AA-6061-T6 metal blank and find out the effects of two geometrical parameters say clearance and punch profile radius on maximum circumferential strain, by using ABAQUS/CAE 6.13.

2 Finite Element Modelling

The different parts included in this NASF process are punch, die, blank holder and blank. The die is modelled in semi-circular shape or else it turns out to be hole flanging process. Instead of modelling a blank holder, the sheet is partitioned into 2 parts along periphery of die and BHF is distributed uniformly on portion of sheet that rests on die while free portion goes under punch during operation. The punch and die are modelled as rigid entities and hence their meshing is done by free meshing technique. The blank is modelled as deformable entity, considering it as solid element (C3D8R element type) for carrying out circumferential strain analysis. The global mesh size of sheet is taken as 1 mm after carrying out mesh convergence study.

Figure 1 shows FE model of sheet wherein Fig. 2 shows assembly of parts.

The material which is assigned to the blank is AA-6061-T6. The strength and damage components of this material have been taken from Lesuer et al. [8] in which Johnson–cook (JC) material model has been used and the values of its constants are as shown in Table 2.

The mechanical properties of AA-6061-T6 taken from manufacturing processes for engineering materials 5th ed. Kalpakjian et al. [9] are shown in the Table 3.

The interactions between different surfaces are shown in Table 4. Since punch and die are considered as discrete rigid entities, they are considered as master surfaces wherein sheet being modelled as deformable entity is considered as slave surface.

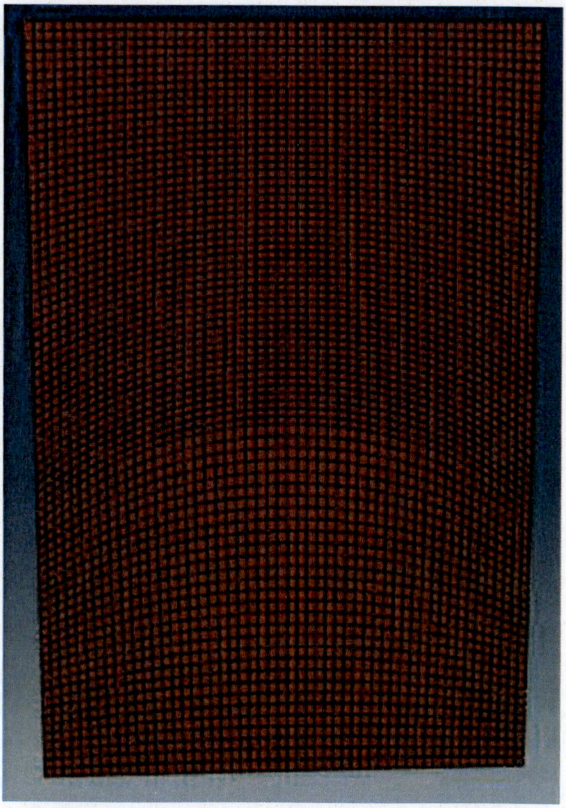

Fig. 1 FE model of blank

The boundary conditions employed in this analysis are as shown in Table 5.

As the mass scaling range can vary from 10 to 10^7, for the present work, a mass scaling of 100 has been employed.

3 Result and Discussion

The circumferential strain plots obtained for different values of clearances are shown in Fig. 3.

Hence, from the simulations done it has been observed that as clearance increases the maximum value of circumferential strain decreases and the values of maximum values of circumferential strain for different values of clearances are shown in the Table 6.

The circumferential strain goes on increasing from edge of the sheet to mid-section of sheet along die-profile radius and if the clearance is kept less, then an edge crack begins from a point at sheet edge and propagates towards centre of sheet. But the edge crack phenomenon does not encounter because of only circumferential strain

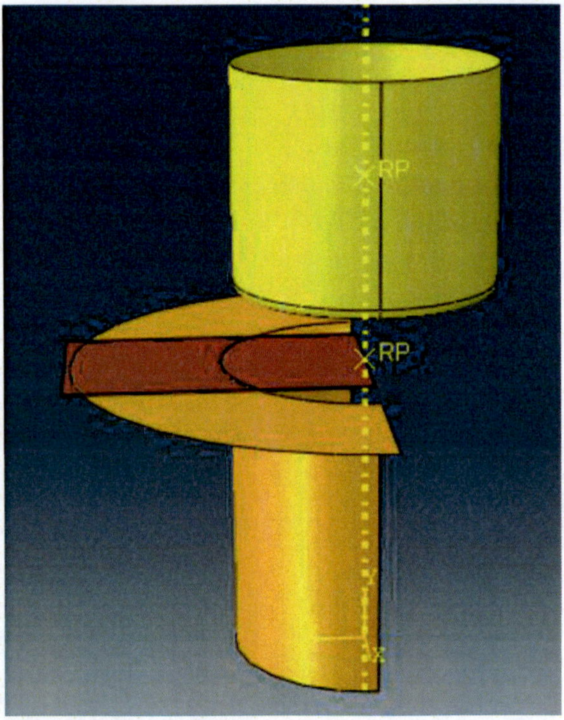

Fig. 2 Assembly of part models on **ABACUS**

Table 1 Dimensions of the designed parts

S. No.	Part	Dimension
1	Die	(a) Radius of die = 35.5 mm (b) Height of die = 80 mm
2	Punch	(a) Radius of punch = 35 mm (b) Height of punch = 60 mm
3	Sheet blank	(a) Length = 80 mm (b) Thickness = 0.5 mm (c) Width = 60 mm

Table 2 The values of constants of JC model for AA-6061 material

AA-6061-T6	
A = 324	B = 113.8
C = 0.002	M = 1.34
n = 0.42	$D_1 = -0.77$
$D_2 = 1.45$	$D_3 = -0.47$
$D_4 = 0$	$D_5 = 1.6$

Table 3 The mechanical properties of AA-6061 material

Yield strength (σ_y) in MPa	270
Ultimate tensile strength (σ_t) in MPa	339
Young's modulus (E) in MPa	68,900
Density (ρ) in kg/mm^3	2.7×10^{-6}
Poisson's ratio (μ)	0.33

Table 4 Master-slave assignments

Interaction no	Interaction 1	Interaction 2
Master surface	Die	Punch
Slave surface	Bottom surface of blank	Top surface of blank

Table 5 The Boundary conditions employed

S. No.	Part on which condition is assigned	Boundary condition
1	Die	Encaster
2	Punch	$U_2 \neq 0$
3	Punch	$V_2 = -100$ mm/s

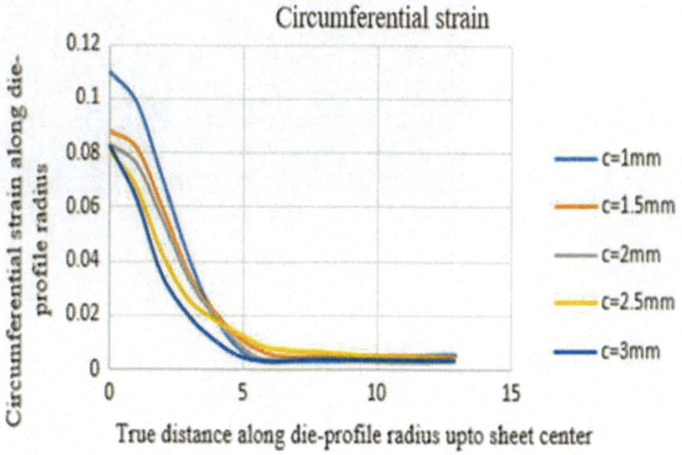

Fig. 3 Circumferential strain plots for different values of clearances

Table 6 Maximum circumferential strain observed for different values of clearances

Clearance (mm)	Maximum value of C.S
1	0.110105
1.5	0.087392
2	0.082725
2.5	0.080170
3	0.082477

and it is found to occur due to the combined effect of both maximum circumferential and radial strains Yogesh Dewang et al. [4]. The image obtained in the simulation is as shown in Fig. 4.

Similar studies have been carried out for different values of punch-profile radius and the plots obtained are shown in Fig. 5.

Fig. 4 Image showing propagation of crack for clearance of 0.6 mm

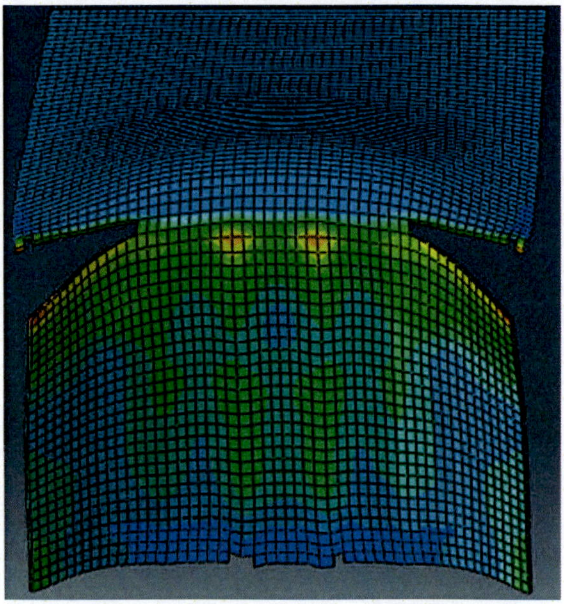

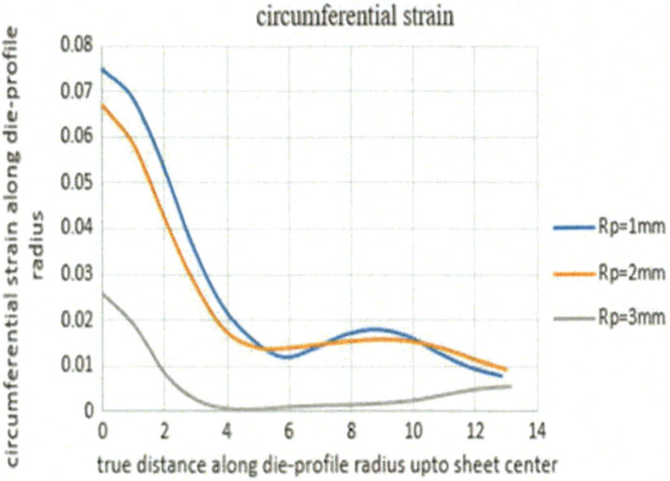

Fig. 5 Circumferential strain plots for different values of punch profile radius

Table 7 Maximum circumferential strain observed for different values of punch profile radii

Punch-profile radius (mm)	Maximum value of Circumferential strain
1	0.07465
2	0.06683
3	0.02560

Hence it has been observed that as punch profile radius increases the maximum value of circumferential strain decreases and the Table 7 shows the maximum values of circumferential strain for given punch-profile radii.

4 Conclusion

The ABAQUS studies carried out on non-axisymmetric stretch flanging of AA-6061-T6 metal blank can be summarised as:-

1. As the clearance has increased from 1 to 3 mm the maximum value of Circumferential strain has reduced by around 25.09%. Smaller clearances result in larger shearing at the extreme nodes thereby result in tearing of sheet.
2. Similarly, as punch-profile radius has increased from 1 to 3 mm the maximum value of circumferential strain has reduced by around 65.7% because sharp corners result in cutting operation rather than forming operation.
3. For lower values of clearances, edge crack phenomenon encounters because of which a crack begins at edge of the metal blank and propagates towards the centre of sheet.

References

1. Wang NM, Johnson LK, Tang SC (1984) Stretch flanging of "V"-shaped sheet metal blanks. J Appl Metalwork 3(3):281–291
2. Feng X, Zhongqin L, Shuhui L, Weili X (2004) Study on the influences of geometrical parameters on the formability of stretch curved flanging by numerical simulation. J Mater Process Technol 145(1):93–98
3. Li D, Luo Y, Peng Y, Hu P (2007) The numerical and analytical study on stretch flanging of V-shaped sheet metal. J Mater Process Technol 189(1):262–267
4. Dewang Y (2015) Analysis of non-axisymmetric stretch flanging process to investigate deformation behaviour of AA 5052. PhD thesis
5. Sartkulvanich P, Kroenauer B, Golle R, Konieczny A, Altan T (2010) Finite element analysis of the effect of blanked edge quality upon stretch flanging of AHSS. CIRP Ann Manuf Technol 59(1):279–282
6. Butcher C, Chen Z, Worswick M (2006) A lower bound damage-based finite element simulation of stretch flange forming of Al–Mg alloys. Int J Fract 142(3–4):289–298

7. Bahloul R, Dal Santo P, Potiron A (2008) Optimisation of the bending process of high strength low alloy sheet metal: numerical and experimental approach. IntJ Mater Form 1(1):113–116
8. Lesuer DR, Kay GJ, LeBlanc MM (1999) Modelling large-strain, high-rate deformation in metals. In: Paper presented at third biennial tri-laboratory engineering conference modeling and simulation, Pleasanton, CA, November 3–5, 1999
9. Kalpkjian S, Schmid S (2009) Manufacturing processes for engineering materials, 5th edn

Experimental Investigation and Effects of Process Parameters on Forming Time and Forming Accuracy in Incremental Sheet Forming

Bhavin Desai, Shalin Marathe, Keyur Desai, and Harit Raval

1 Introduction

Working principle of incremental sheet forming process (ISFP) is similar to the layered manufacturing principle of rapid prototyping. Complex 3D sheet metal parts are formed layered by layered on CNC machines without using dedicated punch and dies. Due to less lead time compared to conventional forming processes, ISFP provides economical solution for manufacturing of sheet metal components. Qualities of formed parts are also at par with components formed with conventional punch and dies. ISFP setup consists of a simple support or a fixture, upper clamping plate and sometimes backup plate as shown in Fig. 1. In ISFP, tool and workpiece are in direct contact, and tool moves along the predetermined path along the workpiece material. Hence, it takes higher forming time for unit production compare to conventional forming processes [1]. A rigid forming tool with flat, hemispherical or spherical end is generally used to perform incremental forming of sheet metals. A hemispherical end forming tool gives better results in term of forming accuracy [2, 3]. During forming, sheet deformation is localized and confined the close vicinity of the contact area. This may result in higher formability [4].

In the present scenario, ISFP can be divided mainly into three categories: single-point forming (SPF), two-point forming (TPF) and hybrid incremental sheet forming [5, 6]. Ambrogio et al. [7] carried out study of profile error by measuring it at the bottom of the formed part at two different points. Few researchers carried out parametric investigation to improve geometric accuracy [8, 9]. It was reported that better geometric accuracy may achieve with smaller step depth. However, their work is limited to pyramidal geometry. A comprehensive knowledge of effects of various incremental sheet metal forming parameters on forming efficiency and quality will

B. Desai (✉) · S. Marathe · K. Desai · H. Raval
Department of Mechanical Engineering, S V National Institute of Technology, Surat 395007, India
e-mail: desai628@gmail.com

H. K. Dave and D. Nedelcu (eds.), *Advances in Manufacturing Processes*, Lecture Notes in Mechanical Engineering, https://doi.org/10.1007/978-981-15-9117-4_13

Fig. 1 Work holding device used during experimentation

allow end users to select the appropriate forming method. Hence, in the present study experimental investigation and effects of various process parameters viz. incremental depth step (Tz), wall angle (Wa), forming speed (feed) (Ff), forming height (Th) and tool diameter (Td) on the forming time and forming accuracy are investigated.

2 Methodology

2.1 Experimentation

Experimenters have used several design of experiments (DOE) strategies [10]. It is also essential to determine simultaneously the individual and interactive effects of factors that could affect the process performance [11]. Conventional experimental design methods are too complex and difficult to use when number of input parameters is more. It requires large number of experimental trial run. In comparison with other design of experiment methods, Taguchi Method (TM) provides a significant reduction in the number of experimental runs [10]. Hence, in the present investigation, experimentations are planed with Taguchi's L16 orthogonal array. During experimentation, two replication per combination; by randomizing the experimental run; have been performed by keeping in mind an economic consideration without affecting the sensitivity of the experiment. Selected input process parameters and their levels for experimentation are presented in Table 1. Forming time and profile

Table 1 Forming parameters and their levels for experimentation

Factor	Parameters			Level 1	Level 2	Level 3	Level 4
A	Incremental depth step		(mm)	0.3	0.45	0.60	0.75
B	Wall angle		(deg)	40	50	60	70
C	Feed		(mm/min)	50	125	200	275
D	Target height		(mm)	12	15	–	–
E	Tool diameter		(mm)	8	10	–	–

accuracy are selected as response characteristics. Interaction graphs are plotted in order to study effect of process parameters on selected response characteristics.

During experimentations, frustum of cone geometry with 44 mm top (major) diameter is formed from 19-gauge AA1200-H14 on three-axis CNC milling machine; refer Fig. 2. AA1200-H14 is a lightweight aluminium alloy and normally used for industrial applications like interior panelling for trucks, cabins, etc. It may be clad with alloy 7072 for better protection against stress corrosion cracking. Design geometries are formed using contour tool path strategy [12] with hemispherical end tool; refer Fig. 3. Forming tool is manufactured from SS304 material in single piece to ensure the rigidity.

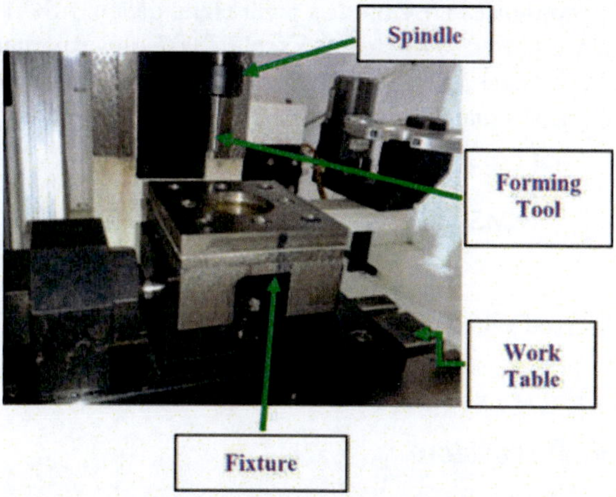

Fig. 2 CNC machine and work holding device utilized for ISFP

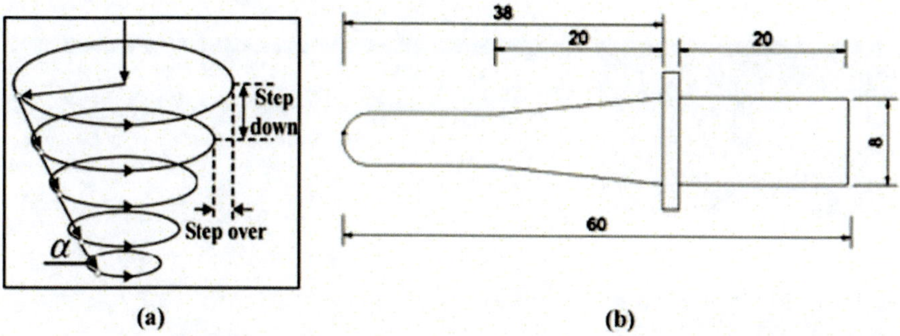

Fig. 3 **a** Contour tool path, **b** Hemispherical end forming tool

2.2 *Response Measurement*

In the present investigation, forming time and profile accuracy are investigated as response characteristics with aimed to increase the industrial applicability of ISFP and to quantify the quality of formed parts. In ISFP, the time required to travel the distance by forming tool to generate a tool path is considered as forming time and directly noted from CNC controller.

Profile accuracy is considered as the deviation in the shape of the formed geometry from the deigned geometry. A needle-type scanner PIX 30 provided with Roland Active PICZA Sensor technology is utilized to verify the profile accuracy. The formed components are tested at 106 equidistant points as shown in Fig. 4. After scanning the components, coordinates are stored as point cloud data in ASCII format. These scanned data are transferred to Microsoft Excel® 2007, and experimental profile is compared with the designed profile. The deviation between these two profiles are expressed as root mean square error (RMSE) and calculated with Eq. 1.

$$\mathrm{RMSE} = \sqrt{\frac{\sum_{i=1}^{n} (X_{\mathrm{obs},i} - X_{\mathrm{model},i})^2}{n}} \tag{1}$$

where X_{obs} is observed values, and X_{model} is modelled values at time/place *i*.

3 Result and Discussion

In the present work, forming time and forming accuracy are considered as the response characteristics. Moreover, profile error considered as forming accuracy.

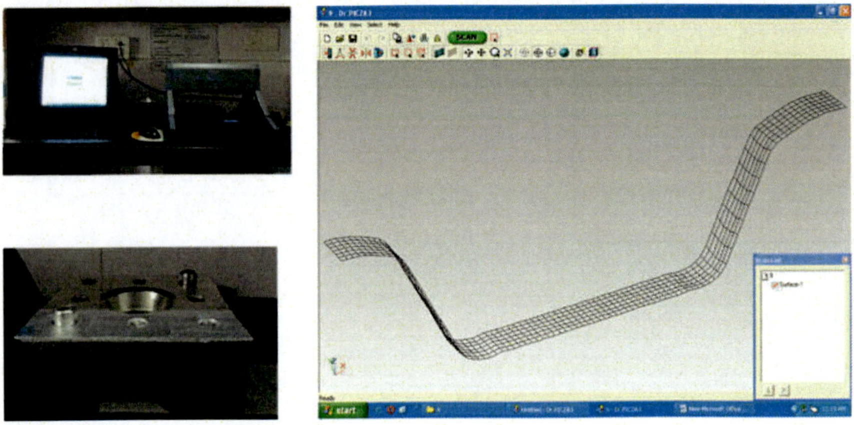

Fig. 4 Scanning of component with PICZA

Table 2 Observations and S/N ratios for forming time and profile accuracy

Exp. No.	Forming time (min)			Profile error (RMSQ—mm)		
	Trial 1	Trial 2	S/N ratio	Trial 1	Trial 2	S/N ratio
1	66.41	66.42	−36.445	0.748	0.745	2.539
2	31.20	31.24	−29.889	1.031	1.032	−0.269
3	25.41	25.41	−28.100	0.342	0.344	9.294
4	20.22	20.23	−26.118	0.647	0.645	3.795
5	19.44	19.43	−25.772	1.477	1.481	−3.399
6	60.18	60.17	−35.588	0.484	0.481	6.330
7	11.01	11.02	−20.810	0.626	0.625	4.075
8	16.07	16.06	−24.118	0.509	0.508	5.874
9	8.57	8.57	−18.660	0.529	0.528	5.539
10	7.38	7.38	−17.361	1.176	1.178	−1.416
11	51.39	51.39	−34.218	0.637	0.639	3.904
12	23.03	23.04	−27.248	0.702	0.705	3.055
13	5.49	5.47	−14.776	1.486	1.484	−3.435
14	9.31	9.30	−19.374	1.140	1.142	−1.146
15	14.36	14.35	−23.140	0.411	0.413	7.702
16	38.32	38.33	−31.670	0.539	0.535	5.400

So, minimum error between designed and formed component is always desirable. Hence, to analyse the results, S/N ratios are calculated with small-the-better quality characteristics [13] using Eq. (2). Calculated S/N ratios are reported in Table 2.

$$\eta_i = -10\log_{10}\left(\frac{1}{n}\sum_{j=1}^{n} y_{ij}^2\right) \tag{2}$$

ANOVA is performed to determine which factor significantly affects forming time and profile error in ISFP. It is known that irrespective of the type of quality characteristics, the maximum value of S/N ratio should always be preferred [13]. The mean values of S/N ratio of forming time and profile error for all parameters at all selected levels are calculated from Table 2 and are reported in Table 3 and Table 4, respectively.

3.1 Analysis of Results for Forming Time

Using the results reported in Table 3, a main effect plot for forming time is drawn as shown in Fig. 5. The optimal performance for the forming time can be obtained at incremental depth step in negative Z direction 0.75 (level 4), wall angle 40-degree

Table 3 Response table for mean of S/N ratio for forming time

Symbol	Parameters	Mean of S/N ratio				Max–Min	Rank
		Level 1	Level 2	Level 3	Level 4		
A	Incremental depth step	−34.48	−26.51	−22.56	*−19.77*	14.71	1
B	Wall angle	*−23.91*	−25.55	−26.57	−27.29	3.38	3
C	Feed	−30.14	−26.58	−24.37	*−22.24*	7.90	2
D	Forming height	*−25.27*	−26.40	–	–	1.13	4
E	Tool diameter	−25.86	*−25.80*	–	–	0.06	5

Table 4 Response table for mean of S/N ratio for profile error

Symbol	Parameters	Mean of S/N ratio				Max–Min	Rank
		Level 1	Level 2	Level 3	Level 4		
A	Incremental depth step	*3.840*	3.220	2.771	2.131	1.709	4
B	Wall angle	0.311	0.875	*6.244*	4.531	5.933	1
C	Feed	4.543	1.772	*4.890*	0.755	4.135	2
D	Forming height	*3.681*	2.300	–	–	1.380	5
E	Tool diameter	1.825	*4.156*	–	–	2.331	3

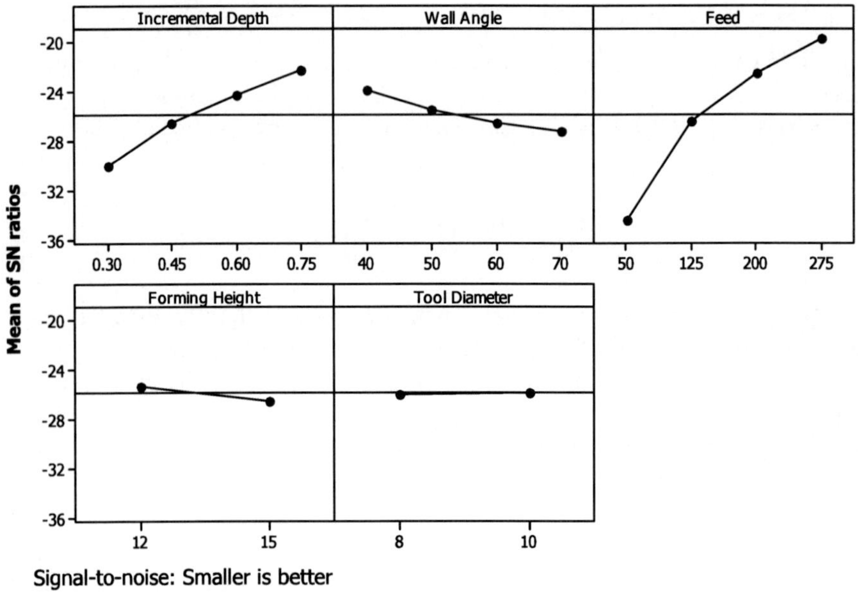

Fig. 5 Main effect plot for forming time

Table 5 ANOVA for forming time

Symbol	Factors	DOF	Sum of squares	Variance	F	P	% contribution
A	Incremental depth step	3	490.583	163.528	4217.28	0.000	74.54
B	Wall angle	3	25.727	8.576	221.16	0.000	3.91
C	Feed	3	136.544	45.515	1173.79	0.000	20.75
D	Forming height	1	5.143	5.143	132.64	0.000	0.78
E	Tool diameter	1	0.016	0.016	0.40	0.561	0.00
Error		4	0.155	0.039	–	–	–
Total		15	658.167	–	–	–	–

(level 1), feed 275 mm/min (level 4), forming (target) height of 12 mm (level 1) and tool diameter of 10 mm (Level 2). From ANOVA for forming time, it can be noted that p-values for factors, i.e. incremental depth step in negative Z direction, wall angle, feed and forming (target) height are less than 0.05 (almost zero); refer Table 5. The percentage contributions of the parameters are also presented in the same table. From the results, it could be concluded that incremental depth step and feed are the most significant parameters followed by wall angle and forming height to achieve minimum forming time for forming of the component with ISFP. This result is in agreement with the results reported in the literature [14].

3.2 *Analysis of Results for Profile Error*

The mean values of S/N ratios of profile error for each selected parameter are given in Table 4. Based on the best S/N ratio combinations, the main effect plot is plotted as shown in Fig. 6. The analysis of S/N ratio for profile error reveals that the optimal result can be obtained at lower incremental depth of 0.3 mm (level 1), wall angle 600 (level 3), feed 200 mm/min (level 3), forming height 12 mm (level 1) and tool diameter of 10 mm (level 2). Further, to find out the parameters that significantly affect the response characteristic, ANOVA is performed. The ANOVA result for S/N ratios of profile error is given in Table 6.

The comparisons among designed and formed profiles are also carried out as per the procedure explained in Sect. 2 (b) for all the formed components. It can be seen that geometrical error in formed parts are very less except at the top and bottom area; refer Figs. 7 and 8. Geometry specific backing plate can minimize the sheet bending in top area, i.e. near the initial tool contact location [15]. The observed pillow effect at the bottom; resulted due to spring back; can be eliminated by using larger tool radius and extended tool path at the bottom of component. These results are in good agreement with the work reported in the literature [8]. It was reported that the

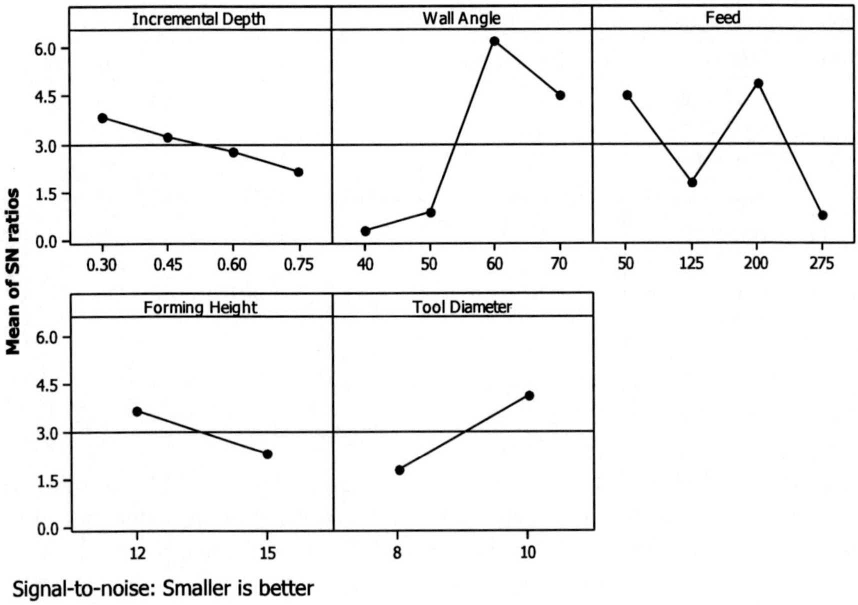

Fig. 6 Main effect plot for profile error

Table 6 ANOVA for profile error

Symbol	Factors	DOF	Sum of squares	Variance	F	P	% contribution
A	Incremental depth step	3	6.25	2.08	0.20	0.892	2.77
B	Wall angle	3	98.45	32.82	3.14	0.149	43.59
C	Feed	3	50.01	16.67	1.60	0.323	22.14
D	Forming height	1	7.63	7.63	0.73	0.441	3.38
E	Tool diameter	1	21.74	21.74	2.08	0.223	9.63
Error		4	41.76	10.44	–	–	–
Total		15	225.83	–	–	–	–

pillowing of sheet material in a part's bottom generally increases as the tool radius decreases.

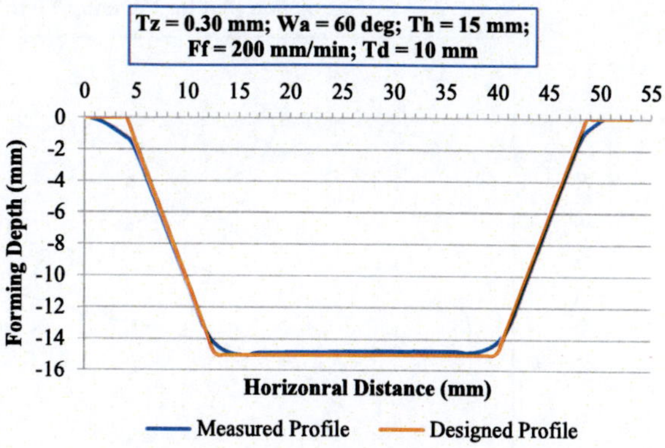

Fig. 7 Comparison of deigned and formed profile

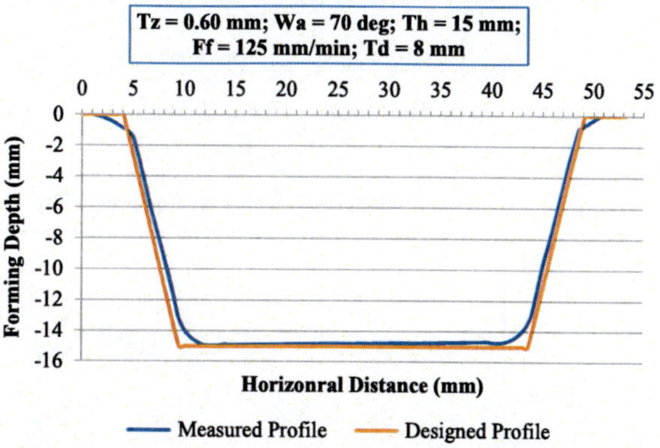

Fig. 8 Comparison of deigned and formed profile

3.3 *Interaction Effect Study*

In order to determine whether two process parameters are interacting or not, one can use a simple but powerful graphical tool called interaction graphs [16]. The interaction graph between incremental depth step and wall angle shows that the effect of wall angle on forming time at different level of incremental depth step is not the same. This implies that there is an interaction between these two process parameters. The forming time (in minute) is minimum when the wall angle is at low level and incremental depth step at high level (refer Fig. 9).

Similarly, the interaction graph between feed and incremental depth step is plotted as shown in Fig. 10. The graph shows that the effect of feed at different levels of incremental depth step is almost same. Moreover, the lines are almost parallel, which indicates that there is little interaction between these two factors. The interaction plot suggests that the forming time (in minute) is minimum when the feed is at high level and incremental depth step is also at high level.

Antagonistic interaction may present between forming height and incremental depth step. Because, the lines on the interaction plot cross each other (Fig. 11). In

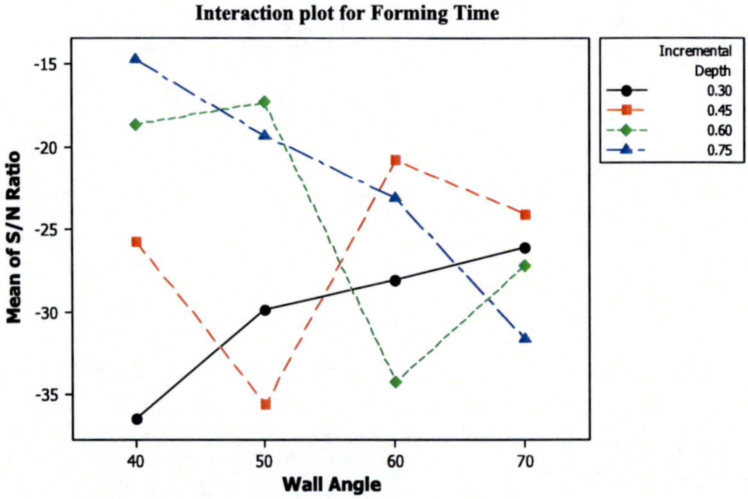

Fig. 9 Interaction plot between incremental depth step and wall angle

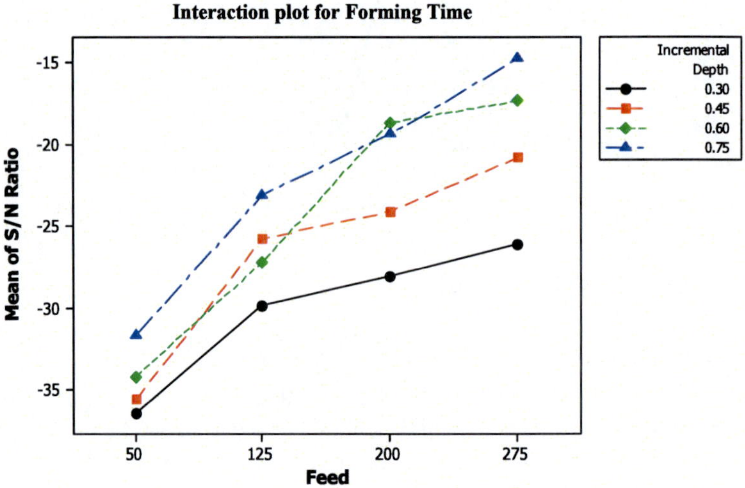

Fig. 10 Interaction plot between incremental depth step and feed

this case, it can be seen that forming height is less sensitive to variation in mean of S/N ratio for forming time at high level of incremental depth step.

Figure 12 clearly indicates the existence of interaction between the factors feed and wall angle as lines are not parallel in interaction plot. The interaction plot for feed and wall angle shows that forming time is minimum when the level of forming speed (feed) is kept at higher level and wall angle at lower level. Similarly, to interaction plot forming height and incremental depth step (Fig. 9), antagonistic interaction

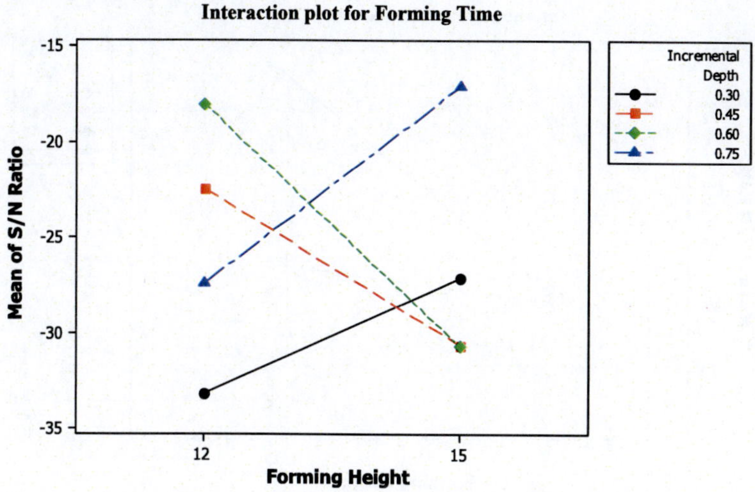

Fig. 11 Interaction plot between incremental depth step and forming height

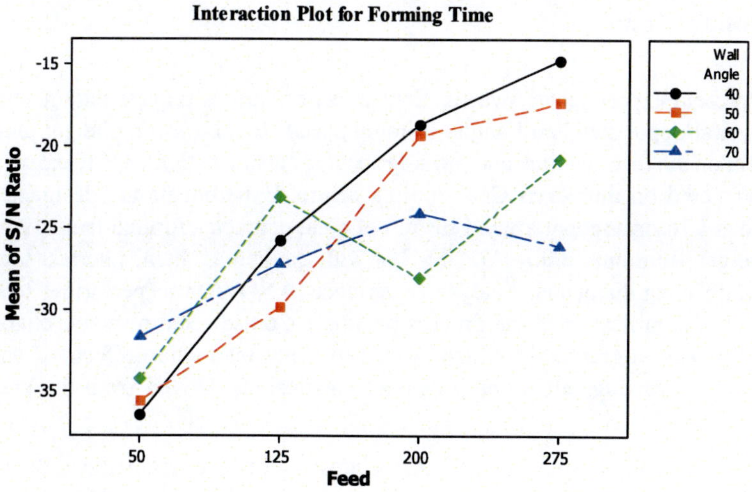

Fig. 12 Interaction plot between wall angle and feed

may present between the forming height and wall angle. Because, the lines on the interaction plot cross each other (Fig. 12). In this case, the change in mean of S/N ratio for forming height at higher level (represented by 15 mm) is noticeably high compared to lower level. In other words, forming height is less sensitive to variation in mean of S/N ratio for forming time at lower level of wall angle (Fig. 13).

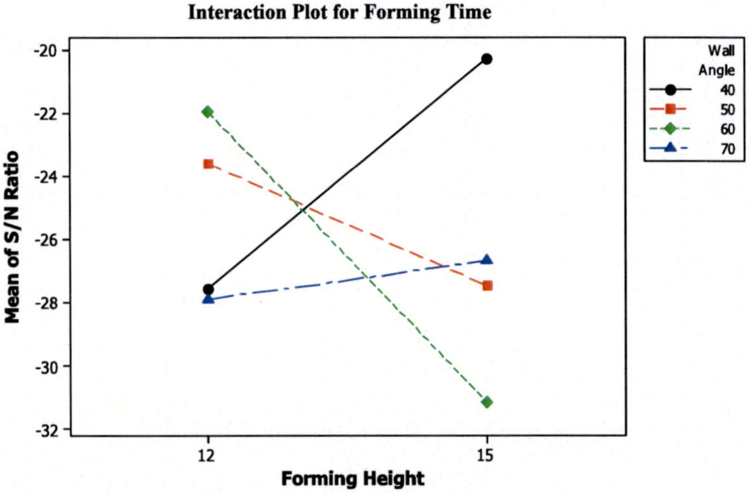

Fig. 13 Interaction plot between wall angle and forming height

4 Conclusions

In the present experimental investigations effect of various process parameters viz. incremental depth step, wall angle, forming speed (feed), forming height and tool diameter on the forming time and forming accuracy are investigated. All experiments are conducted on three-axis CNC milling centre. Experiments are designed with Taguchi's L16 orthogonal array. Frustum of cone geometry formed from 19-gauge lightweight aluminium alloy Al1200-H14 with the help of hemispherical forming tools of different diameters. Analysis of variance (ANOVA) was performed to know the level of significance of the process parameters on selected response characteristics. Interaction graphs were also plotted to study interaction effects of various processes. Followings are some important conclusions derived from the presents study.

1. Incremental depth step and feed are the most significant parameters followed by wall angle and forming height to achieve minimum forming time for forming of the component with ISFP. This result is in agreement with the results reported in the literature [14].
2. High forming accuracy may achieve at lower incremental depth of 0.3 mm (level 1), wall angle 60° (level 3), feed 200 mm/min (level 3), forming height 12 mm (level 1) and tool diameter of 10 mm (level 2).
3. To increase profile accuracy of the parts formed by ISFP, geometry specific backing plate could be used to minimise the sheet bending in top area, i.e. near the initial tool contact location.

4. The observed pillow effect at the bottom; resulted due to spring back; can be eliminated by using larger tool radius and extended tool path at the bottom of component. This result is in good agreement with the work reported in the literature [8].
5. Antagonistic interaction may observe from interactions plots between various process parameters for forming time. It can be seen that the forming time is minimum when the wall angle and forming height are at low level and feed and incremental depth step at high level.

References

1. Jeswiet J, Micari F, Hirt G, Bramley A, Duflou J, Allwood J (2005) Asymmetric single point incremental forming of sheet metal. CIRP Annals Manuf Technol 54(2):88–114
2. Hussain G (2014) Experimental investigations on the role of tool size in causing and controlling defects in single point incremental forming process. Proc Inst Mech Eng Part B J Eng Manuf 228:266–277
3. Li Y, Lu H, Daniel WJT, Meehan PA (2015) Investigation and optimization of deformation energy and geometric accuracy in the incremental sheet forming process using response surface methodology. Int J Adv Manuf Technol 79:2041–2055
4. Silva MB, Skjoedt M, Atkins AG, Bay N, Martins PAF (2008) Single-point incremental forming and formability–failure diagrams. J Strain Anal Eng Des 43(1):15–35
5. Allwood JM, Braun D, Music O (2010) The effect of partially cut-out blanks on geometric accuracy in incremental sheet forming. J Mater Process Technol 1, 210(11):1501–1510
6. Araghi BT, Manco GL, Bambach M, Hirt G (2009) Investigation into a new hybrid forming process: incremental sheet forming combined with stretch forming. CIRP Annals 1, 58(1):225–228
7. Ambrogio G, Cozza V, Filice L, Micari F (2007) An analytical model for improving precision in single point incremental forming. J Mater Process Technol 1, 191(1–3):92–95
8. Hussain G, Lin G, Hayat N (2011) Improving profile accuracy in SPIF process through statistical optimization of forming parameters. J Mech Sci Technol 1, 25(1):177–182
9. Lu H, Kearney M, Li Y, Liu S, Daniel WJ, Meehan PA (2016) Model predictive control of incremental sheet forming for geometric accuracy improvement. Int J Adv Manuf Technol 1, 82(9–12):1781–1794
10. Dave HK, Desai KP, Raval HK (2013) A Taguchi approach-based study on effect of process parameters in electro discharge machining using orbital tool movement. Int J Mach Mach Mater 1, 13(1):52–66
11. Myers RH, Montgomery DC, Anderson-Cook CM (2016) Response surface methodology: process and product optimization using designed experiments. Wiley, pp 378–380
12. Desai BV, Desai KP, Raval HK (2014) Die-less rapid prototyping process: parametric investigations. Proc Mater Sci 1(6):666–673
13. Phadke MS (1989) Quality engineering using robust design. Prentice-Hall, Englewood Cliffs, NJ, pp 13–40
14. Sarraji W, Hussain J, Ren WX (2012) Experimental investigations on forming time in negative incremental sheet metal forming process. Mater Manuf Process 1, 27(5):499–506
15. Desai BV, Desai KP, Raval HK (2017) The performance of tool shape on efficiency and quality of forming in incremental sheet-forming process. Int J Rapid Manuf 6(4):215–234
16. Antony J (2014) Design of experiments for engineers and scientists. Elsevier, p 22

Effect of Operating Parameters on Forces During Backward Flow Forming Process for AA6061

Jaydeep B. Bhatt, Ravi J. Bhatt, Harit K. Raval, and Keyur P. Desai

1 Introduction

The flow forming process has incremental nature. It incorporates point deformation at contact zone unlike bulk deformation in other forming processes that subsequently reduces force and power requirement. The production of near net shape part is of high demand in today's competitive industrial environment. In industries like aerospace and defense, parts require very high-dimensional tolerance and surface finish. To produce parts with low cost and with less material usage, chipless process like flow forming provide tremendous advantages. Especially, in aerospace industries parts that are needed generally not only requires high accuracy but parts have also very complex shape which also incorporates change in thickness within single part. To cost effectively produce this type of parts with increased mechanical properties, flow forming process is suitable. In this process, preform is compressed between mandrel and single or multiple rollers. During forming, material flow occurs in two directions axial and circumferential. Flow forming process can be performed by two strategies forward and backward. In backward flow forming strategy, the preform is held on head stock, and it is supported against fixture, the roller flows material in opposite direction of its travel, and advances toward fixed end. The preform is pressed against the mandrel and head stock. Backward strategy has advantage that roller does not have to travel over entire length of the work piece. In Fig. 1, diagram of backward flow forming is shown with the forces encountered during the process. In any forming

J. B. Bhatt · H. K. Raval · K. P. Desai
Mechanical Engineering Department, S V National Institute of Technology (SVNIT), Surat 395007, Gujarat, India
e-mail: bhattjaydeep25@gmail.com

R. J. Bhatt (✉)
Mechanical Engineering Department, C G Patel Institute of Technology, Uka Tarsadia University, Maliba Campus, Bardoli, Surat, Gujarat, India
e-mail: ravi.bhat1412@gmail.com

H. K. Dave and D. Nedelcu (eds.), *Advances in Manufacturing Processes*, Lecture Notes in Mechanical Engineering, https://doi.org/10.1007/978-981-15-9117-4_14

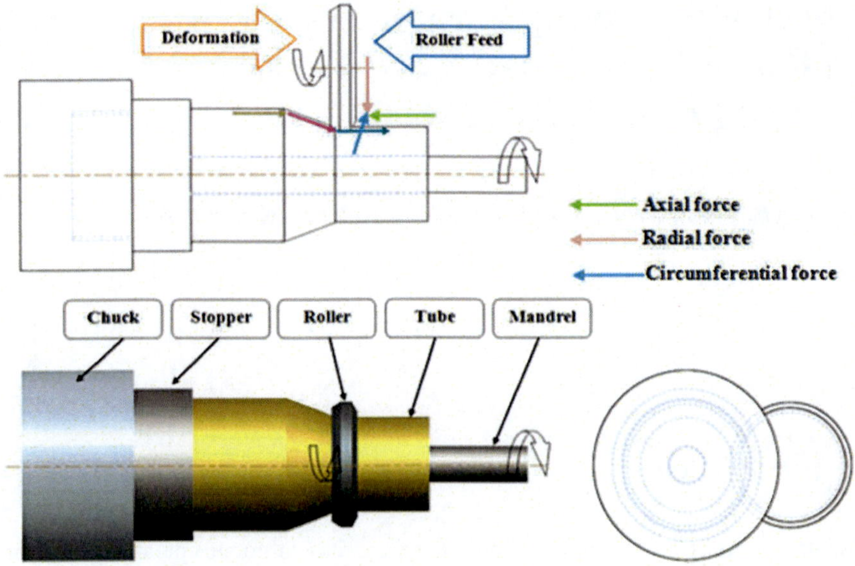

Fig. 1 Process schematic of backward flow forming

process, study of force required is of primary concern. By knowing force required in forming process, design of set up can be done, and appropriate energy requirement can be calculated.

Since last decade, majority of the researchers had studied power, axial force, and stress generation for flow forming process considering experiments and finite element analysis. Parsa et al. [1] investigated feed rate and attack angle on change of stress and strain distribution and also studied circumferential to axial length ratio change by finite element analysis. It was shown that circumferential to axial length ratio increases with feed rate and attack angle. Experiments and explicit ABAQUS simulation were used to study redundant strain in the process by Mohebbi et al. [2] for AA6061. It was found that high shear strains were occurring at both cross section and longitudinal directions of the formed work piece. Kim et al. [3] had put efforts to minimize axial force and to find optimal angle of attack by upper bound method. Simulation by elastic plastic model was also done by taking various thickness reduction and attack angle values. It was illustrated that angle with smallest value of axial force also resulted in the smallest magnitude of material build up and spring back. Xiao et al. [4] had studied metal flow model for cylindrical part by counter roller spinning process, and mathematical model for equivalent strain formula was proposed. Bhatt and Raval [5] investigated effect of operating parameters like feed, depth, speed, and friction coefficient on axial, circumferential, and radial force for AA6063 material by using finite element analysis. Axial feed and friction component have highest effect on the axial and circumferential forces. Bhatt and Raval [6] investigated effect of friction factor, strength coefficient K, strain hardening exponent n, material density, and temperature on force required using finite element analysis.

It was found that axial force is highest than radial and at last circumferential force. However, online force measurement is difficult task in commercial machines because of complex and bulky designs of machines. Hence, in the present research, in situ investigations have been conducted on forces for AA6061 material. Design of experiment (DOE)-based analysis technique was adopted after full factorial experimental study. It was observed that axial force is having major influence on the process compared to radial and circumferential components of the forces.

2 Materials and Methods

The design of experiment (DOE) is an effective tool to carry out experiments so that data can be analyzed, and important conclusion can be drawn out of it [7]. In this study, full factorial design was considered for experiments. Three parameters were selected for analysis, i.e., speed (rpm), % reduction, and feed (mm/rev). Each input parameters with its selected levels are shown in Table 1. Each parameter was taken at three levels. Full factorial method was used so required number of experiments were 27. The detailed experimental plan and observations of forces are shown in Table 2. Experimentations were performed by using single roller on engine lathe machine HMT-QETCOS. The levels of the parameters were selected based on the literature review [8–14] and the capacity of the available machine tool. Material selected for the study was AA6061. AA 6061 material has excellent corrosion resistance, moderate strength, high-plane strain fracture toughness, and it is easily available. Material was cut from extruded rod of AA6061, and then, de-burring operation was performed in order to remove burrs. After obtaining initial desired length, preforms were full annealed to temperature of 420 °C for one hour. Material had total tube length of 100 mm, forming length 50 mm (to avoid the collision of roller with chuck), outer diameter 30 mm, and inner diameter 20 mm. Major alloying elements of AA6061 material are Al–1Mg–0.6Si–0.25Cu–0.2Cr.

Forces were measured with the use of lathe tool dynamometer. Shank of dynamometer was clamped on carriage, and flow forming process was carried out. Roller was attached with dynamometer by taper roller bearing and tool holder. Dynamometer has accuracy of measuring 1 kg force, and it has scanning speed of 0.2 s. Figure 2 shows the experimental setup which includes roller used in this study and force dynamometer. The roller was made up of die steel D3 material with the roller geometry as roller attack angle 30°, nose radius 4 mm with roller outer

Table 1 Selected input parameters with their levels

Parameters	Levels		
	Level 1	Level 2	Level 3
A: Speed (rpm)	150	250	420
B: % reduction	20	35	50
C: Feed (mm/rev)	0.07	0.1	0.13

Table 2 Detailed experimental plan and observation table for 27 experiments

Exp. No.	Speed (rpm)	Reduction (%)	Feed (mm/rev)	Circumferential force (N)	Axial force (N)	Radial force (N)	Resultant force (N)
1	420	20	0.13	87.128	−915.238	−61.962	921.46
2	150	20	0.1	12.839	−671.238	−60.790	674.11
3	250	20	0.13	19.211	−896.741	−72.724	899.89
4	420	50	0.07	27.411	−1272.47	−63.051	1274.33
5	250	35	0.07	31.007	−1435.200	−66.621	1437.08
6	150	50	0.07	30.326	−1478.63	−82.039	1481.22
7	150	50	0.1	34.850	−1439.82	−93.699	1443.29
8	250	20	0.07	22.560	−1125.55	−84.894	1128.97
9	150	20	0.07	15.817	−871.744	−72.429	874.89
10	420	35	0.07	70.578	−815.936	−46.264	820.29
11	420	50	0.13	24.869	−1382.26	−94.058	1385.68
12	250	35	0.1	29.991	−1371.95	−110.27	1376.71
13	250	50	0.07	29.890	−1397.87	−114.68	1402.88
14	420	20	0.1	61.817	−731.717	−63.300	737.05
15	250	50	0.13	30.945	−1431.85	−87.056	1434.83
16	420	20	0.07	19.051	−958.615	−80.897	962.21
17	250	50	0.1	38.373	−1669.18	−109.71	1673.23
18	150	35	0.07	28.839	−1400.07	−84.414	1402.91
19	150	50	0.13	46.851	−1788.3	−93.276	1791.35
20	150	35	0.1	26.705	−1306.11	−83.144	1309.03
21	420	35	0.1	38.185	−1825.9	−103.70	1829.24
22	150	20	0.13	9.810	−883.488	−82.963	887.43
23	420	50	0.1	13.353	−1247.73	−87.369	1250.86
24	250	20	0.1	47.408	−579.994	−50.734	584.14
25	250	35	0.13	17.200	−1287.68	−94.708	1291.27
26	420	35	0.13	19.450	−1368	−102.73	1371.99
27	150	35	0.13	18.941	−1507.99	−116.66	1512.61

diameter of 64 mm and width of 32.12 mm. Mandrel was used for providing support at the internal diameter of the tube. Mandrel was manufactured with EN8 (AISI 1040) material. This lathe tool dynamometer was coupled with computer screen, and data acquisition was done with software of lathe tool dynamometer. Forces encountered in flow forming process had three mutually perpendicular components, i.e., axial, circumferential, and radial force.

In the present study, three forces were considered orthogonal, and resultant force of three can be found as per Eq. (1).

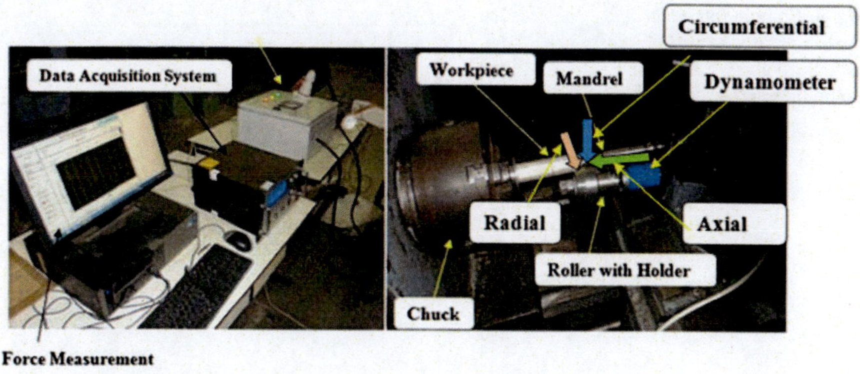

Fig. 2 Experimental setup with online data logging system

$$\text{Resultantforce} = \sqrt{F_x^2 + F_y^2 + F_Z^2} \tag{1}$$

where F_x = raidal force, F_y = circumferential force, and F_z = axial force. Observations of forces are shown in Table 2.

It can be observed from Table 2 that the axial force is having highest magnitude. The radial force values are intermediate, and circumferential force values are least among three. Also, Table 2 shows the highest values of the each individual force observed during experimentations. Further, the axial and radial forces are compressive in nature due to the axial feed and depth applied to the roller. The circumferential force is due to the instantaneous contact between roller and workpiece.

3 Result and Discussions

In flow forming process, nature of forces are shown in Fig. 3 for experiment no. 6 (5) reduction). It was observed from the graph of forces versus time that axial force was highest and had compressive nature. All three forces were first increases as process advances, and after some time, forces had stable value. Axial force had direction of axis of work piece. Axial force was highest due to main force that compresses the material ahead of roller and increases the length of the work piece. Material ahead of the roller tries to compress the roller and tries to displace it away in opposite direction of lathe feed so this force had compressive nature. Circumferential force had tensile nature, and circumferential force was due to instantaneous contact between work piece and roller. At any instance, roller was only driven due to friction between roller and work piece that will create force in tangential direction. Radial force had direction perpendicular to the axis of roller, and it had nature to separate the roller and work piece (nature to decrease depth of forming). Radial force first increases in tensile direction, and then, it stabilized in compressive direction after some time.

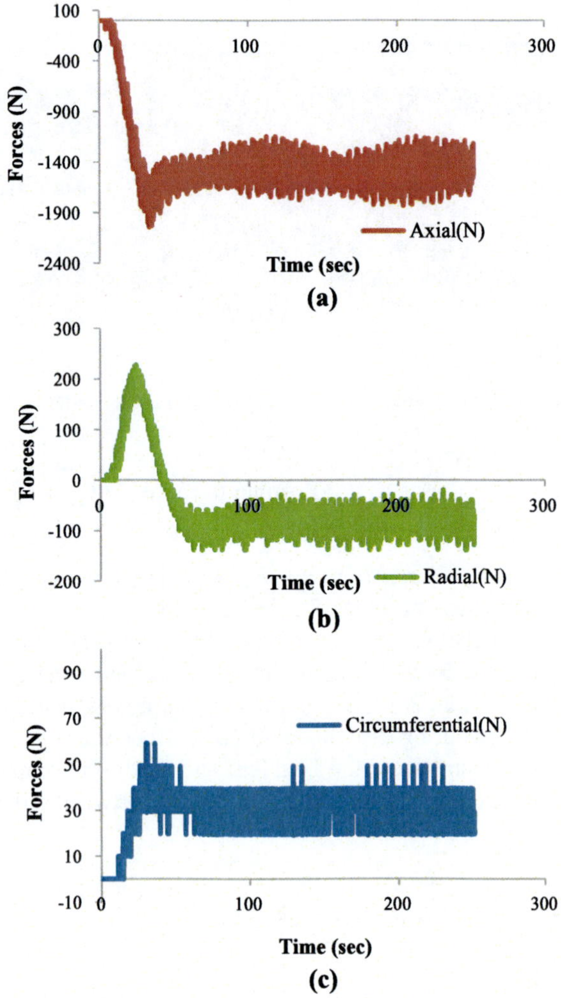

Fig. 3 Sample forces of experiment no. 6 (50% reduction) **a** axial, **b** radial, **c** circumferential

Bell-shaped curve was observed in the radial force direction, and the height of the bell shape curve changes with % reduction. In Fig. 3, sample nature of three forces had been shown for experiment number 6. It was observed that bell shape of the radial force had higher height for higher value of reduction. Direction change of radial force was attributed as elasticity of roller assembly in radial direction. At the start of process, roller deflects due to less rigidity of assembly of dynamometer and roller. Due to less rigidity of the assembly, at the start of forming process, roller first tries to climb on work piece material, and then, its position was stabilized after some time. This phenomenon of climbing over work piece caused the nature of radial force to change from tensile to compressive. This change in direction of radial force also causes the initial formed tube to have some radius of curvature. The formed samples

(a) Formed part of exp. no 3 (b) Formed part of exp. no 5 (c) Formed part of exp. no 6

Fig. 4 Sample photograph of flow formed part for initial radius of curvature

are shown in Fig. 4. It can be clearly seen that as % reduction increases initial radius of the part increases.

In the present study, it was observed that axial force is highest followed by radial and circumferential force. Similar observation reported by Wang and Long [15] and [16]. In the present study, single roller without separate external rotary actuation behaves like shear forming. In which, the shear forces are more dominating than compressive forces due to no external rotary motion of the roller as reported by Arai [17] and Hayama [18]. That could be the reason for the higher axial force (in deformation direction). If the roller has the external rotary motion (using hydraulic motors), then the axial force may not be dominating compared to radial and circumferential [19–21].

It can be noticed (From Fig. 4) by visual inspection that part had initially radius of curvature, and then, its thickness was uniform throughout the length. Initial radius zone of the part is of no use, and hence, it should be cut off from final product. Estimation of the wastage material can be done by measuring the axial length of the radius of curvature. From the graph of radial force vs. time, axial length of wastage material could be estimated for undeformed length of part. Wastage material axial length could be calculated from the total time taken for forming of 50 mm length. For finding wastage material axial length, time required to stabilize the radial force should be noted. Axial wastage material length was proportional to the time required to stabilize the radial force. Given formula was valid assuming machine that gives constant feed rate.

$$AL = \frac{T_S \times F_L}{F_T} \tag{2}$$

where AL = axial length (mm); T_s = time required to stabilize the radial force; F_L is the total forming length (i.e., 50 mm); F_T is the total time taken for traveling the forming length.

Equation (2) gives the value of table travel (axial length) for which the initial curved part was formed. This proportional equation did not give the exact value of wastage material as the time obtained by the graph of radial load would be time required to travel that distance by carriage and not for the actual length of finished part as actual part had increased length than 50 mm of table travel according to the % reduction provided for particular experiment. Axial length calculated by formula (2) was compared to the actual value of axial length by measuring it in a tool maker's microscope. Part was placed on the tool maker's microscope, and axial length of the curved section was measured by accurate table travel of the microscope. Obtained value of the axial length is mentioned in Table 3, and it was compared with the calculated value from Eq. (2).

Resultant of forces was found by Eq. (1), and it was optimized by plotting signal to noise ratio (S/N) versus input parameter graph. In any forming process, main objective is to minimize the operating force so in the present work force was considered as "smaller-the-better" type quality characteristics. Signal to noise ratio for "smaller-the-better" type quality characteristics can be found by Eq. (3).

$$\eta_{SB} = -10\log_{10}\left(\frac{1}{n}\sum_{i=1}^{r} y_i^2\right) \tag{3}$$

where y_i is the response of the ith quality characteristic; n is the total number of repetition of a run. S/N ratio plot for resultant force was obtained by considering the S/N ratios from above Eq. (3). S/N ratio plot for resultant force is shown in Fig. 5.

Figure 5 represents the S/N ratio for resultant force. The optimum combination of parameter was A3-B1-C2 which was 420 rpm speed, 20% reduction, and 0.30 mm/rev feed. ANOVA for resultant force had been done to know the % contribution effects of input variables on response variable, and it is shown in Table 4. ANOVA had been done by considering S/N values for resultant force.

From the S/N ratio graph and ANOVA Table 4, it was noticed that % reduction had highest significance on the resultant force with p-value 0 and 66.9% contribution. As % reduction increases material ahead of roller increases, hence, more material will oppose the axial travel of roller that leads to higher axial force with the higher % reduction value. Axial force was main constituent for resultant force so this condition leads to higher resultant force. The R^2 value of this ANOVA was 90.81%.

From analysis, it was noticed that % reduction had highest impact, and as % reduction increases resultant force increase. Speed and feed does not have significant impact on the resultant force as observed from the ANOVA; as these terms had higher p-value than 0.05. It was reported by Taguchi et al. [7] that criteria for any factor to be significant is its p-value should be less than 0.05 for the 95% confidence interval. It is always important to verify the results through confirmation experiments. Prediction of responses is done at optimum setting of parameter as suggested by Taguchi et al.

Table 3 Comparison of axial length by equation and experiment

Exp. No.	Axial length		Deviation	% error
	By equation	Experiment		
1	2.472	2.567	0.095	3.844
2	1.336	1.380	0.044	3.330
3	1.868	1.825	−0.043	−2.327
4	6.673	6.233	−0.440	−6.601
5	7.216	7.040	−0.176	−2.441
6	7.131	7.178	0.047	0.661
7	6.293	6.483	0.190	3.022
8	2.319	2.518	0.199	8.594
9	1.988	2.187	0.199	9.985
10	5.357	4.865	−0.492	−9.180
11	7.848	7.278	−0.570	−7.259
12	3.927	4.237	0.310	7.894
13	6.375	6.751	0.376	5.893
14	1.627	1.470	−0.157	−9.673
15	6.187	5.893	−0.294	−4.753
16	2.354	2.193	−0.161	−6.837
17	5.907	6.355	0.448	7.589
18	3.863	3.825	−0.038	−0.973
19	6.697	6.930	0.233	3.477
20	3.039	3.185	0.146	4.799
21	5.428	4.888	−0.540	−9.946
22	1.808	1.753	−0.055	−3.037
23	5.855	5.350	−0.505	−8.621
24	2.156	2.350	0.194	8.973
25	4.398	4.522	0.124	2.821
26	5.382	4.860	−0.522	−9.698
27	3.938	4.054	0.116	2.938

[7]. The estimated mean can be calculated using formula given below.

$$\eta_{\text{predicted}} = T + \sum_{i=1}^{n} (\eta_i - T) \tag{4}$$

where T is overall mean of response parameter; η is average SN value of significant parameters. Confidence intervals can be calculated from this formula:

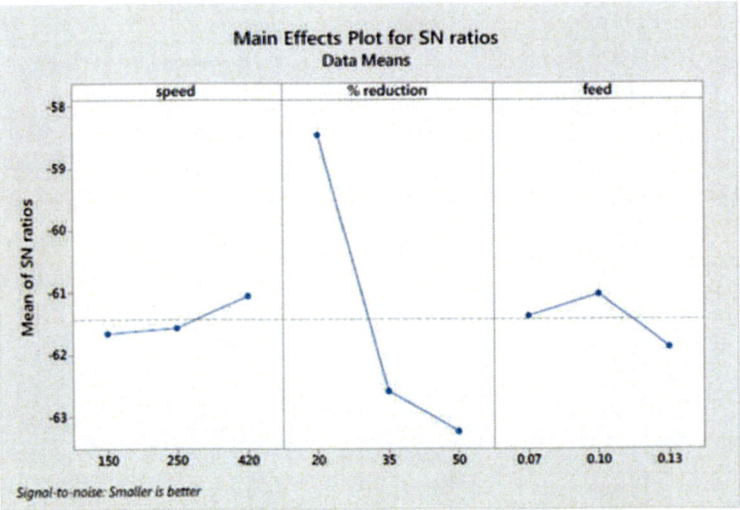

Fig. 5 S/N ratio plot for the resultant force

Table 4 ANOVA for resultant force

Factors	DF	Adj SS	Adj MS	*F*-value	*P*-value	Contribution (%)
Speed (A)	2	1.941	0.97	0.47	0.642	1.08
% reduction (B)	2	120.65	60.3	29.1	*0.000*	*66.9*
Feed (C)	2	3.34	1.66	0.81	0.480	1.85
A * B	4	4.05	1.01	0.49	0.744	2.25
A * C	4	10.1	2.52	1.22	0.374	5.61
B * C	4	23.6	5.91	2.86	0.096	13.13
Error	8	16.5	2.07	–	–	9.19
Total	26	180.32	–	–	–	100

$$CI_{CE} = \sqrt{F_{\alpha}(1, f_e)v_e\left[\frac{1}{\eta_{\text{eff}}} + \frac{1}{R}\right]} \tag{5}$$

where $F_{\alpha}(1, f_e)$ is the F ratio at confidence level of $(1 - \alpha)$ against DOF. $(1, f_e)$ is error degree of freedom v_e is error variance which is the ratio of the total variation of factors to the total degrees of freedom of its factors. R shows number of repetitions, and η_{eff} is given by:

$$\eta_{\text{eff}} = \frac{N}{1 + DF_{\text{total}}} \tag{6}$$

Table 5 Confirmation run output for resultant force

Trial no.	Parameter at their optimum level			Response parameter	
	Speed	%reduction	Feed	Resultant force (N)	S/N ratio
1	420	20%	0.1	732.89	- 57.300
2	420	20%	0.1	712.34	- 57.053

where N shows total number of experimental runs; DF_{total} is the summation of total degree of freedom of significant factors. The results of experiment performed at optimum combination of parameters should fall in 95% confidence interval (CI) range as:

$$\eta_{\text{predicted}} - CI_{CE} < \eta_{-\text{(TH)}} < \eta_{\text{predicted}} + CI_{CE} \tag{7}$$

By considering above equations, $\eta_{\text{predicted}} = -58.461$ and $\text{CI}_{\text{CE}} = \pm 2.5939$. So S/N ratio for resultant force trial results should fall in $-61.0549 < \eta < -55.8671$ region. This confirmation runs were performed at A3-B1-C2 optimum combination. In Table 5, output of confirmation test is shown.

Based on Table 5, it was observed that trial run results fall in calculated levels of confirmation test. Further, the experimental investigations it is found that flow forming is mainly affected by three force components, i.e., axial, radial, and circumferential. Among three, axial force is found to be highest, radial force is intermediate, and circumferential force is lease. These force analysis can be utilized for tooling design for various geometrical and material conditions.

4 Conclusions

Present work deals with analysis of resultant force in flow forming process, and some important conclusions can be drawn which are as follows.

- Force in flow forming had mainly three components, i.e., circumferential, axial, and radial force. It was found that axial force was main component of resultant force. The resultant force is having major impact of axial force among three force components because the magnitude of the axial force is highest compared to other two. Therefore, controlling of axial force is more important during the single roller backward flow forming. Also, working at low to moderate speed can reduce the axial force.
- From force observation data, it was noticed that radial and axial force had compressive nature while circumferential force had tensile nature which is inline with the published literature.
- ANOVA of resultant force shows that % reduction was most significant parameter affecting the resultant, force and as % reduction increases resultant force increases. It means, higher the reduction percentage, higher the resultant force.

Hence, lower to moderate reduction is preferred to reduce the resultant and other force components.

- Initial radius of curvature was identified on the flow forming part, and main cause for curvature was due to deflection of roller assembly and depth of forming.
- Optimum combination of input variable that minimizes the resultant force was speed 420 rpm, 20% reduction, and 0.30 mm/rev feed.

References

1. Parsa MH, Pazooki AMA, Ahmadabadi MN (2009) Flow-forming and flow formability simulation. Int J Adv Manuf Technol 42(5–6):463–473
2. Mohebbi MS, Akbarzadeh A (2010) Experimental study and FEM analysis of redundant strains in flow forming of tubes. J Mater Process Technol 210(2):389–395
3. Kim N, Kim H, Jin K (2013a) Minimizing the axial force and the material build-up in the tube flow forming process. Int J Pre Eng Manuf 14(2):259–266
4. Xiao G, Xia Q, Cheng X, Zhou Y (2014) Metal flow model of cylindrical parts by counter-roller spinning. Proc Eng 81:2397–2402
5. Bhatt RJ, Raval HK (2016) Influence of operating variables during flow forming process. Proc CIRP 55:146–151
6. Bhatt RJ, Raval HK (2017) Investigation of effect of material properties on forces during flow forming process. Proc Eng 173:1587–1594
7. Taguchi G, Chowdhury S, Wu Y (2005) Taguchi's quality engineering handbook, 2nd edn. Wiley, New York
8. Cao Z, Wang F, Wan Q, Zhang Z, Jin L, Dong J (2015) Microstructure and mechanical properties of AZ80 magnesium alloy tube fabricated by hot flow forming. Mater Des 67:64–71
9. Srinivasulu M, Komaraiah M, Rao CKP (2013) Prediction of the surface roughness of AA6082 flow-formed tubes by design of experiments. J Mech Sci Technol 27(6):1835–1842
10. Kim N, Kim H, Jin K (2013b) Minimizing the axial force and the material build-up in the tube flow forming process. Int J Prec Eng Manuf 14(2):259–266
11. Song X, Fong KS, Oon SR, Tiong WR, Li PF, Korsunsky AM, Danno A (2014) Diametrical growth in the forward flow forming process: simulation, validation, and prediction. Int J Adv Manuf Technol 71(1–4):207–217
12. Abedini A, Ahmadi SR, Doniavi A (2014) Roughness optimization of flow-formed tubes using the Taguchi method. Int J Adv Manuf Technol 72(5–8):1009–1019
13. Srinivasulu M, Komaraiah M, Rao CKP (2012) Experimental studies on the characteristics of AA6082 flow formed tubes. J Mech Eng Res 4(6):192–198
14. Molladavoudi HR, Djavanroodi F (2011) Experimental study of thickness reduction effects on mechanical properties and spinning accuracy of aluminum 7075-O, during flow forming. Int J Adv Manuf Technol 52(9–12):949–957
15. Wang L, Long H (2011a) A study of effects of roller path profiles on tool forces and part wall thickness variation in conventional metal spinning. J Mater Proc Technol 211(12):2140–2151
16. Wang L, Long H (2011b) Investigation of material deformation in multi-pass conventional metal spinning. Mater Des 32(5):2891–2899
17. Arai H (2003) Robotic metal spinning-shear spinning using force feedback control. Proc Robot Autom 3:3977–3983
18. Hayama M (1981) Study on the spinnability of aluminium and its alloys. Bulletin Faculty Eng 30:63–72
19. Shinde H, Mahajan P, Singh A, Singh R, Narasimhan K (2016) Process modeling and optimization of the staggered backward flow forming process of maraging steel via finite element simulations. Int J Adv Manuf Technol 87(5–8):1851–1864

20. Jolly SS (2010) Analysis of power and forces in the making of long tubes in hard-to-work materials. Proc World Congress Eng 2:2–6
21. Xia QX, Cheng XQ, Hu Y, Ruan F (2006) Finite element simulation and experimental investigation on the forming forces of 3D non-axisymmetrical tubes spinning. Int J Mech Sci 48(7):726–735

Analysis of TIG-Welded Aluminum Alloys During Single Point Incremental Forming at Different Wall Angles

M. M. Ghadmode, Arpit R. Patil, B. U. Sonawane, and Amrut Mulay

1 Introduction

Nowadays, in automobile sectors, more emphasis is given on to reduce the weight of automotive vehicles. Tailor-welded blanks (TWBs) became popular since late 90s to achieve such reduction in weight of the vehicles. Lot of problems was discussed in past by many researchers regarding the forming of tailor-welded blanks. The main problem which may usually occur during forming of TWBs is "formability deterioration." Most of the researchers have suggested many options to overcome such type of problem. Initially, some researchers have concentrated on TWBs made from steel and its high-strength alloys. Laser welding became popular to manufacture TWBs from high-strength steels. However, recently, the use of aluminum alloys has attracted automotive field, and hence, in last decade, more research work is going on to study the forming behavior of Aluminum alloys. As material is changed, its joining methods are changed and joining methods like TIG welding, and Friction Stir Welding (FSW) came into focus to prepare TWBs from different Aluminum alloys. Different grades of aluminum, whose weldability is good, have proven themselves suitable for many applications in ship building, aerospace industries, and automobile industries. A lot of research has been conducted in the area of single point incremental forming, and various process parameters have been studied for their

M. M. Ghadmode (✉) · B. U. Sonawane
Department of Production Engineering & Industrial Management, College of Engineering Pune, Pune 411005, India
e-mail: manojghadmode@gmail.com

A. R. Patil
Department of Metallurgy and Materials Science, College of Engineering Pune, Pune 411005, India

A. Mulay
Department of Mechanical Engineering, S V National Institute of Technology, Surat 395007, India

H. K. Dave and D. Nedelcu (eds.), *Advances in Manufacturing Processes*, Lecture Notes in Mechanical Engineering, https://doi.org/10.1007/978-981-15-9117-4_15

Fig. 1 Fixture of SPIF to clamp the blank

effect on surface roughness, forming limit diagram, sheet thinning, etc. However, the little attention has given in area of SPIF of TWBs. Silva et al. [1] evaluated the SPIF process on TWBs prepared by friction stir welding. Authors confirm the use of SPIF to form TWBs is a good combination, and TWBs can be formed with good forming depth. The study promotes the use of SPIF with existing machines like VMC, CNC milling machine. Authors have also pointed out the necessity of forming behavior of TWBs prepared from dissimilar materials. In this study, the AA6061-T6 and AA5052-H32 (both having 1.5 mm thickness) materials have joined by TIG welding. Bhattacharya et al. [2] concluded that the surface roughness of SPIF-formed component increases with increase in vertical step depth and tool diameter. However, surface roughness is observed decreasing for certain value of wall angle. Malwad and Nandedkar [3] conducted tests for various angles, and they observed that uniform thinning can be achieved for wall angles less than 65°. Alinaghian et al. [4] have studied the formability of TWBs prepared by FSW and then formed by SPIF in terms of bowl height. They optimized FSW parameters by response surface methodology for greater formability in their prescribed ranges of parameters for aluminum alloy AA6061 with 2 mm thick sheet. They studied the effect of welding direction on bowl height. Ebrahimzadeh et al. [5] have carried out comparison study between SPIF and Two Point Incremental Forming (TPIF), and they observed that TPIF gives 40% improvement than SPIF to form friction stir welded blanks made of AA5083. Gatea et al. [6] have investigated the effect of SPIF parameters like step depth, tool diameter, and feed rate on fracture occurrence for pure tungsten material (grade 1 and grade 2); wherein, they concluded that by keeping these parameters at their low level, higher wall angle can be achieved. Yogesh Kumar and Santosh Kumar have developed an innovative set up for incremental forming [7].

In this research work, the focus was kept on TWBs prepared by joining aluminum alloy AA5052-H32 and AA6061-T6 with TIG welding. These TWBs are then formed by SPIF, and formability of formed components is measured in terms of thickness distribution, forming strains, and forming depth. The Slow Strain Rate Test (SSRT) is also carried out to investigate the causes of failure occurred.

2 Experimental Procedure

Figure 1 consists of a fixture which was fixed on a three-axis vertical milling machine— PVM 40. A spherical end HSS tool of diameter of 10 mm was used to perform SPIF, and the lubricant 10W30 was used to avoid more friction between moving tool and sheet. To clamp the sheet during incremental forming process, a fixture with one backing plate of 170 mm diameter hole, a clamping plate, clamping screws, and four rods are used. The 6 mm filet radius is also provided at the hole of backing plate to facilitate the required material flow. ARTCAM software was used to generate tool path for conical shape with 130 mm bottom diameter. Forming depth could achieve its maximum extent till fracture during SPIF operation. Formability can be measured with the forming angle, strain, and forming depth till fracture.

2.1 Blank Material

Aluminum alloys AA5052-H32 and AA6061-T6 of 1.5 mm thickness were used for this study. These alloys are used in automobile and marine applications due to their weight, strength, and anti-corrosive properties. The mechanical properties are depicted in Table 1, and chemical compositions are listed in Table 2; the remaining percentage is of pure aluminum.

The wall angles were varied from 40° to 70° with an increment of 5° for parent material AA5052-H32, AA6061-T6. The same sets of experiments are performed for TWBs as well. The process parameters such as spindle step size (Δz), speed (n), feed (f) are kept constant throughout the experimental run. (Table 3).

Table 1 Mechanical properties of AA6061-T6 and AA5052-H32

Property	AA6061-T6	AA5052-H32
Tensile strength (MPa)	368	240
Yield strength (MPa)	279	193
Total elongation (%)	10	10.4

Table 2 Chemical compositions (%) of AA5052-H32 and AA6061-T6

Element	Fe	Mn	Cu	Si
AA5052-H32	0.300	0.095	0.005	0.120
AA6061-T6	0.34	1.038	0.25	0.64
Element	Cr	Ti	Zn	Mg
AA5052-H32	0.191	0.015	0.001	2.463
AA6061-T6	0.193	0.023	0.004	1.03

Table 3 Constant parameters during SPIF

Factors	Step size (mm)	Speed (rpm)	Feed (mm/min)
Value	0.2	1000	1000

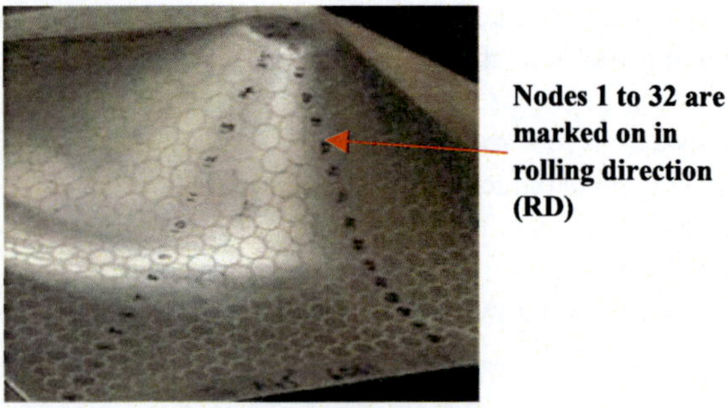

Fig. 2 Circular grids on one side of sheet

The blank size was kept to 176 mm × 176 mm for SPIF experiments carried out for parent materials. TWBs were obtained by joining two pieces of size 88 mm × 176 mm (each piece of AA5052-H32 and AA6061-T6). All blanks are then marked by laser engraving as shown in Fig. 2. The diameter of a circle during laser engraving was kept 5 mm for economical point of view. All circles, touching each other make a grid of total 32 circles × 32 circles are marked over the back side of the sheet which covers a surface area of 160 mm × 160 mm. 100 W capacity laser engraver was used to engrave hairline thick circles. After SPIF process, these circles get elongated or sometimes contracted depending upon the direction and intensity of forming force. The conical shape was chosen for the experiment with bottom diameter of 130 mm. The spherical tool of 10 mm diameter is used during experimentation as shown in Fig. 3.

2.2 *Measurement of Forming Strains Along Node Points*

Usually, stretching action takes place during SPIF, and all the circles, which come under the forming area, get elongated due to this stretching effect. Mylar tape, as suggested by Kumar et al. [8], was used to measure the dimensions of these elongated circles along the center line (as marked in Fig. 2). A 3 × zoom magnifying glass was used for the correctness of the measurement. The major strains, minor strains, and thickness strains were calculated based on these measured values. This procedure was followed for each experiment performed for both materials AA5052-H32, AA6061-T6, and their TWBs. The result of the same is discussed in Sect. 3.1.

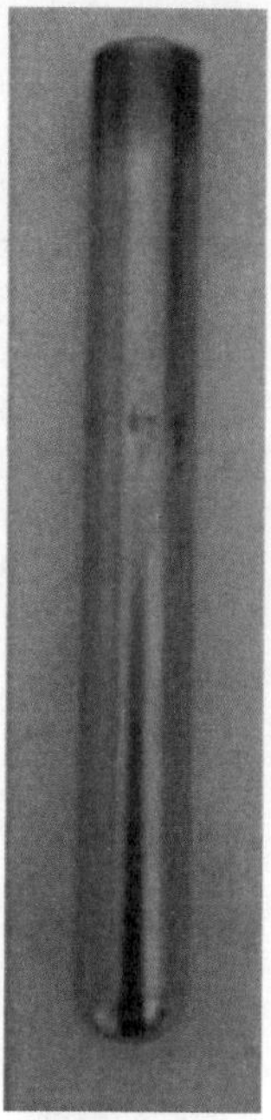

Fig. 3 Spherical tool (ϕ10 mm)

2.3 *Slow Strain Rate Test*

Slow Strain Rate Test (SSRT) is a test carried out on special purpose test machine, and it is similar to a normal tensile test carried out on general purpose universal testing machine. The only difference is the strain rate value. In slow strain rate test machine, the value of strain rate can be kept to very low value, i.e., up to 0.0017 mm/min (2.8×10^{-8} m/s). The SSRT is usually carried out to see the effect of environmental corrosive action on a component which is always undergo a very small amount of loading. This test is usually carried out in different environments, i.e., in air environment, different solution environment like sea water, different chemicals, etc.

Any component used in automotive part will be under continuous loading including TWBs. This loading may deform the weld at a very low strain rate. SSRT gives the information about how material responds to very small amount of straining. Most of the previous researcher concentrated on study of parent metal for slow strain rate testing. Some researchers had evaluated the welded joint for slow rate testing but they had used only those tensile specimens for the Slow Strain Rate Tests (SSRT), which were prepared with the tensile direction perpendicular to the welding direction. Thus, there is scarcity of research in the area of SSRT of welded aluminum alloy. Therefore, it is required to analyze the behavior of such joints under application of very small strain rate. The slow strain rate testing behavior of Tungsten Inert Gas Welded (TIG) AA5052-H32 and AA6061-T6 is studied. The dumbbell shape tensile specimens as per ASTM E8 standards (sub-size specimen) were prepared by wire cut process for the slow strain rate tests as shown in Fig. 4.

These specimens were prepared such a way that in specimen (a) and (b) loading direction of tensile load is kept parallel to the welding such a that the 6 mm width of specimen contains 3 mm AA5052-H32 with 3 mm weld, and another tensile specimen consisting of 3 mm weld with AA6061-T6 alloy. Specimen (c) is of transverse loading, where the tensile direction is kept perpendicular to the welding. Specimen (d) is of longitudinal loading with the 6 mm width of weld alone. The parent materials AA5052-H32 and AA6061-T6 are also wire cut in the above shape. This type of specimen testing will cover the overall direction study of slow strain rate testing on the Tailor-Welded Blanks (TWBs).

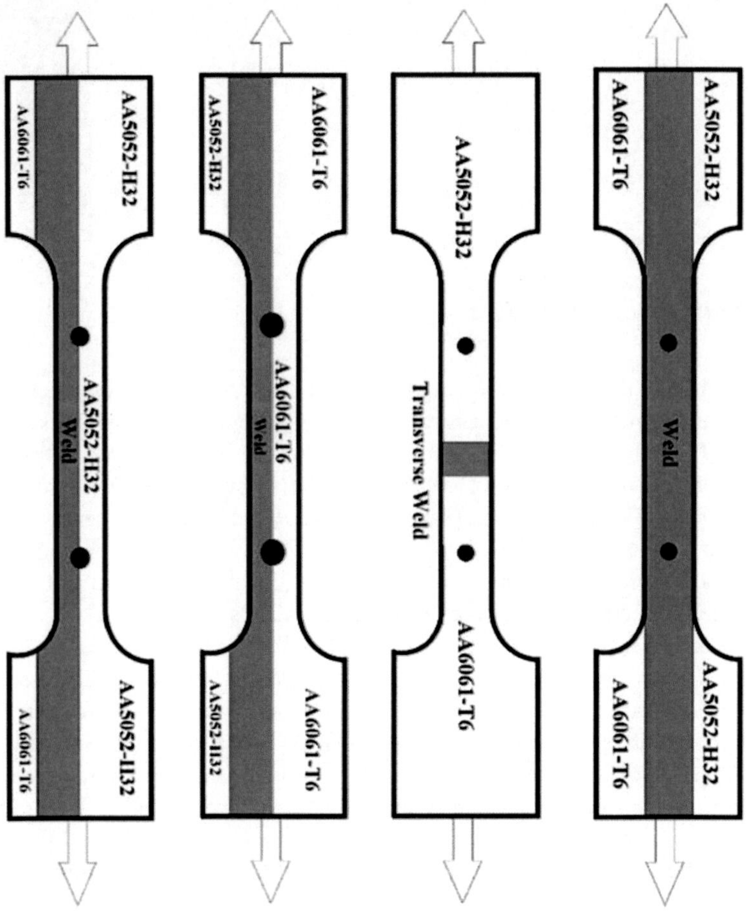

(a) (b) (c) (d)

Fig. 4 Tensile specimens of TWBs for SSRTs

The in situ slow strain rate tests were performed in air environment with strain rate of 0.0017 mm/min (i.e., 2.8×10^{-8} m/s) on a slow strain rate testing machine (the lowest possible strain rate). The results of the same are discussed in Sect. 3.4.

3 Results and Discussion

Results and discussion section is divided into various sub-sections wherein formability of formed component is discussed in terms of forming strains (Sect. 3.1), thickness reduction (Sect. 3.2), and forming depth achieved (Sect. 3.3). Results of SSRT are also discussed in Sect. 3.4. The effect of heat treatment on forming depth is discussed in Sect. 3.5 of this article.

3.1 *Formability Measured in Terms Forming Strains:*

The conical-shaped components formed by SPIF process are shown in Fig. 5a–d. Total seven experiments were performed for individual material, with wall angles 40°, 45°, 50°, 55°, 60°, 65°, and 70°. Figure 5a shows that parent material AA5052-H32 has been successfully formed up to 65°, and cracks are not observed. A horizontal crack is developed at initial bending region in the sample formed with wall angle 70° made of AA5052-H32 (Fig. 5c).

Figure 5b shows the red-marked circles, where cracks are developed in AA6061-T6 in samples having wall angles 50° and above. An enlarged view (Fig. 5c) of crack is clearly visible in failed components 70° (AA5052-H32) and 50° (AA6061-T6).

Table 4 shows the values of maximum major strains developed after forming the parent and TWB sheets using SPIF process. It can be observed that parent materials AA6061-T6 and AA5052-H32 are formed successfully, and no cracks observed for wall angles 40° and 45° (Table 4). However, in parent material AA6061-T6, cracks started developing for 50° at major strain equals to 0.52, for 55° at major strain equals to 0.44, and for 60° and for 65° cone angle it was observed to be 0.40. Whereas, for parent material AA5052-H32, these values are noted 0.46 for 50°, 0.60 for 55°, 0.74 for 60°, and 0.98 for 65°, and cracks were not observed. In other words, with increase of wall angle, more straining of the component takes place but material AA6061-T6 fails early than material AA5052-H32 (even though the value of major strain is more as compared with that of AA6061-T6). Figure 6 shows the formed components of TWBs.

The enlarged views of failure portion reveal that the failure occurred in the weld zone are of "brittle in nature." The strain values at these locations are mentioned in Table 4. From these values of strains of TWBs, it can be observed that all TWBs have failed with similar values of that of AA6061-T6. In other words, even though there are two materials available in TWBs, weld follows the forming behavior of stronger material, i.e., AA6061-T6.

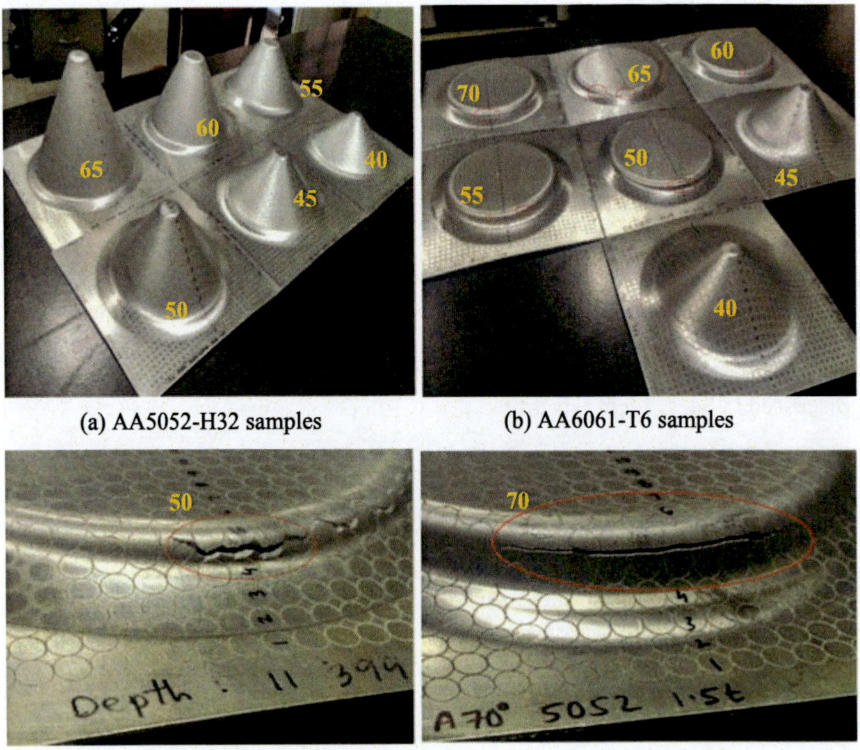

(a) AA5052-H32 samples (b) AA6061-T6 samples

(c) Failed component 50^0 (AA6061-T6) & 70^0 (AA5052-H32)

Fig.5 a–c Parent materials formed by SPIF (wall angle 40°–70°)

Table 4 Major strain values (max.) for both materials and TWBs after SPIF

Wall angle	AA5052-H32	AA6061-T6	TWB
40°	0.24 (no fracture)	0.26 (no fracture)	0.26 (fractured)
45°	0.34 (no fracture)	0.36 (no fracture)	0.24 (fractured)
50°	0.46 (no fracture)	0.52 (fractured)	0.38 (fractured)
55°	0.60 (no fracture)	0.44 (fractured)	0.52 (fractured)
60°	0.74 (no fracture)	0.40 (fractured)	0.32 (fractured)
65°	0.98 (no fracture)	0.40 (fractured)	0.24 (fractured)
70°	0.82 (fractured)	0.42 (fractured)	0.32 (fractured)

3.2 Thickness Distribution of Formed Components

During incremental forming process of conical shape, material gets stretched uniformly inwards in radial direction, and only the region in contact with tool radius is stretched more. Due to this, thinning takes place and failure occurs in this region

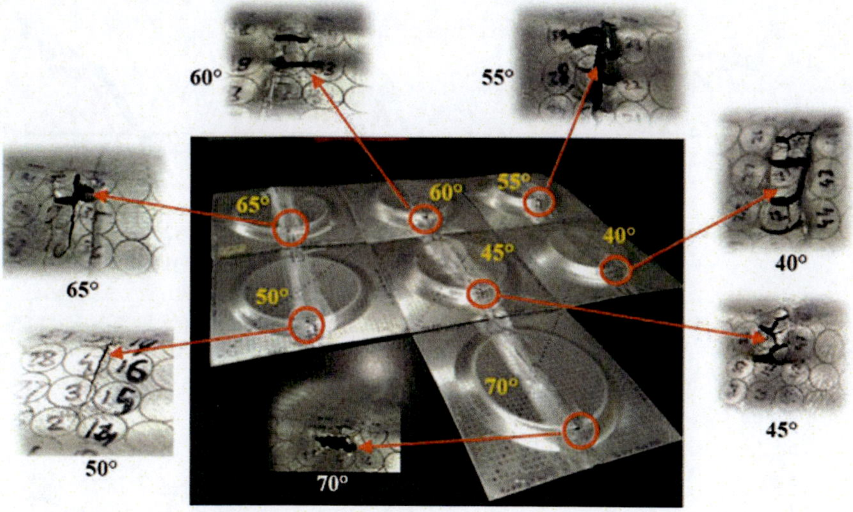

Fig. 6 TWBs formed by SPIF (wall angle ranging from 40° to 70°)

only (red circles marked in Fig. 5c and Fig. 6). Hence, it is necessary to plot the thickness distribution of the sheets along the direction of stretching. Thickness of the formed component can be calculated from the values of major strains and minor strains which were obtained using mylar tape for each node along the center line shown in Fig. 2.

Figure 7a–f shows comparison between thickness distribution of formed component with the desired thickness reduction (suggested by sine law [9]). Figure 7a, b shows the uniform thickness distribution for wall angle 40° and 45°, and thinning is observed near to desired value suggested by sine law of thinning. At center node points, i.e., at node point 17 (for 40°) and node points 16, 17, and 18 (for 45°), tool has not reached to that point, and hence, forming is not done here, and hence, no thickness reduction is observed. In Fig. 7c–g, at node points 6 and 28, local thinning is occurred in parent material AA6061-T6 and it is fractured. Whereas, AA5052-H32 material has shown uniform thickness distribution in all region for all wall angles (except it got failed at 70° wall angle). Likewise AA6061-T6, local thinning is also observed in all TWBs for wall angle 40°–70°. This excessive local straining is caused because material in the nearby region is restricting itself to get deformed, and hence, thinning at very small region is observed, and if the material in this region is brittle in nature, then it crack suddenly. It is necessary to check the tensile behavior of weld portion. Hence, slow strain rate tests are also carried out, and the results of the same are discussed in Sect. 3.4.

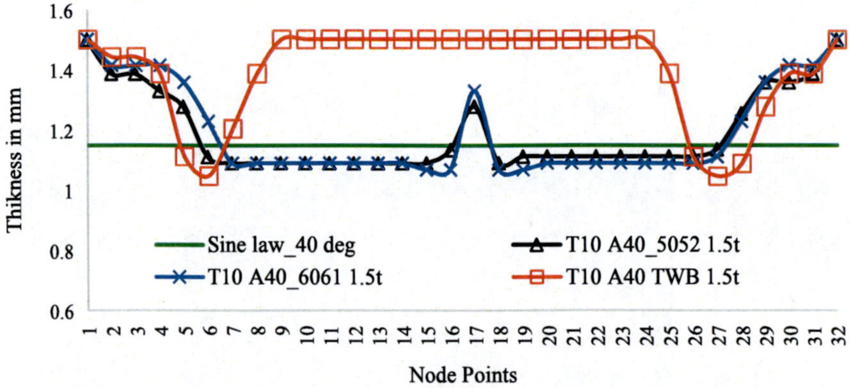

(a) Wall angle 40°

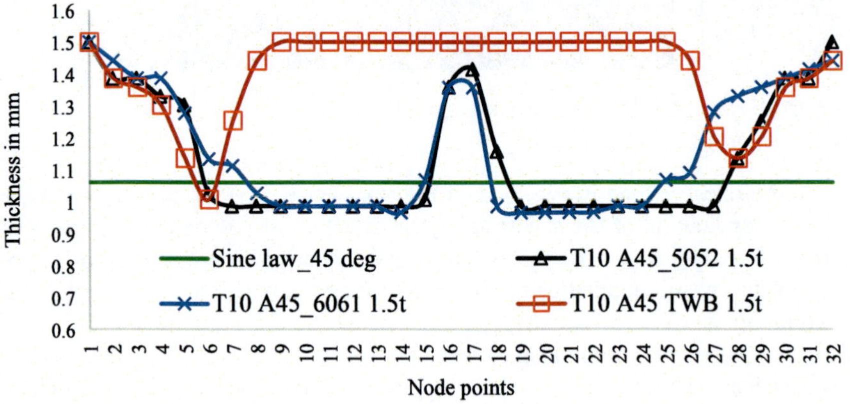

(b) Wall angle 45°

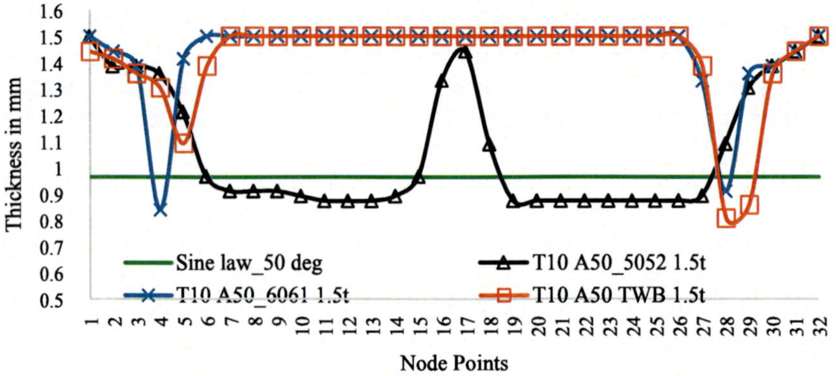

(c) Wall angle 50°

Fig. 7 a–g Thickness distribution along center line of formed blanks

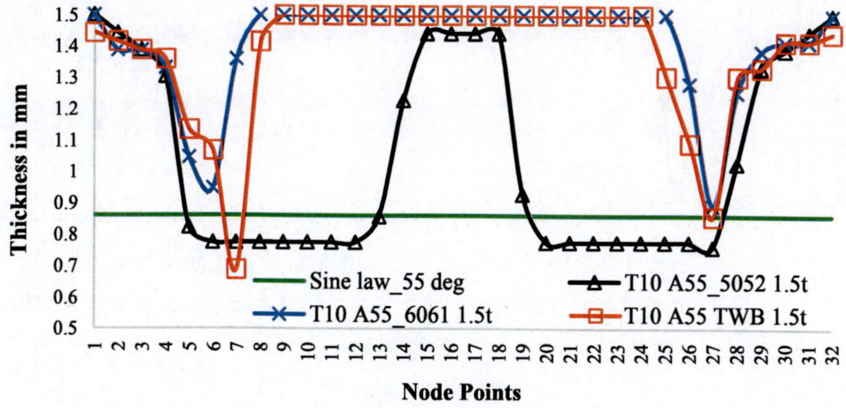

(d) Wall angle 55°

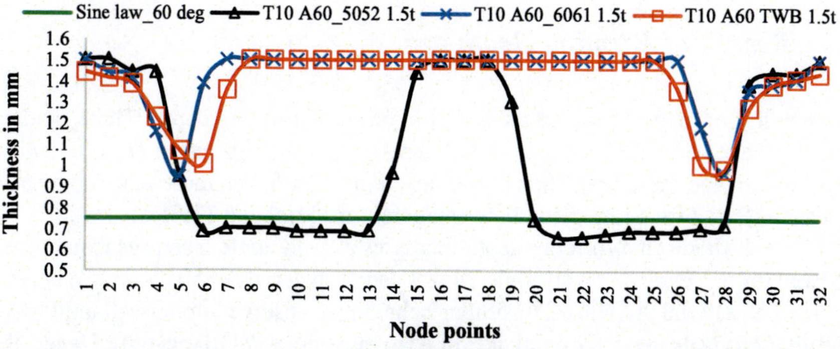

(e) Wall angle 60°

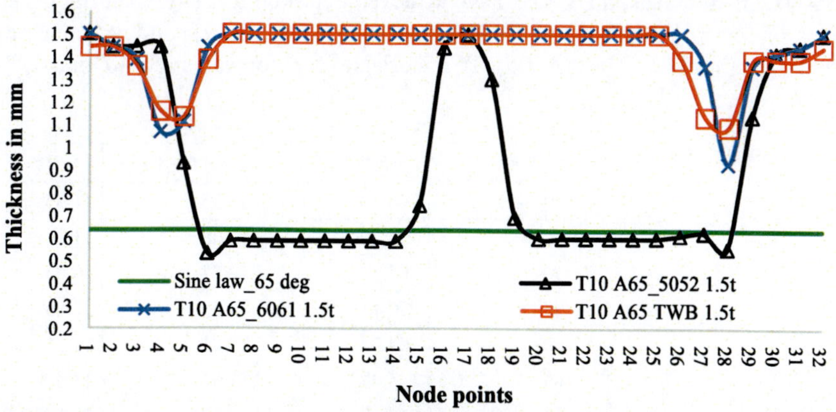

(f) Wall angle 65°

Fig. 7 (continued)

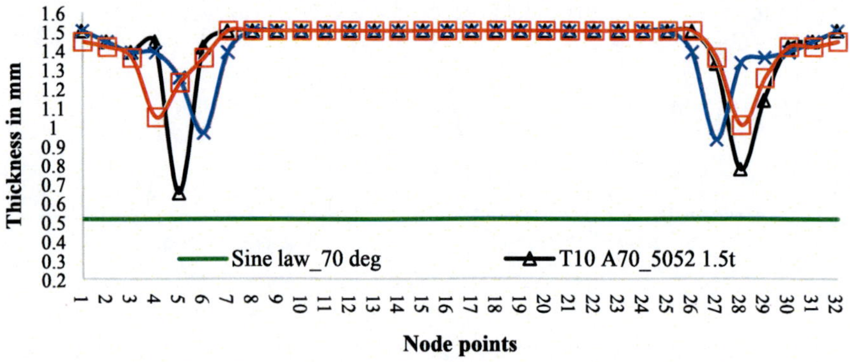

(g) Wall angle 70°

Fig. 7 (continued)

3.3 Results of Forming Depth

Formability of material can be measured in terms of forming depth. Three random readings have been taken for each component by digital height gage LH-600E (Mitutoyo Make), and the average heights are listed in Table 5, and these reading are also plotted in Fig. 8 for comparison between parent material and TWBs.

Figure 8 shows that forming depth increases as wall angle increases from 40° to 65° and starts reducing at wall angle 70° for parent material AA5052-H32. Whereas, for parent material AA6061-T6, similar behavior is observed up to wall angle 45°, and after this wall angle, the depth is limited to maximum 12–15 mm only. Hence, the formability of AA5052-H32, in terms of forming depth, is more than that of material AA6061-T6. Whereas, all TWBs have achieved depth up to 11–12 mm only. The values of forming depth of all TWBs are similar to the corresponding values of forming depth achieved by AA6061-T6 material (except for wall angle 40° and 45°).

Table 5 Average forming depth in mm

Wall angle	AA5052-H32	AA6061-T6	TWBs
40°	48.72	49.55	11.93
45°	54.79	52.08	11.01
50°	64.23	12.77	10.95
55°	70.43	15.85	12.85
60°	84.34	12.81	11.85
65°	115.47	11.54	11.50
70°	16.02	12.95	11.59

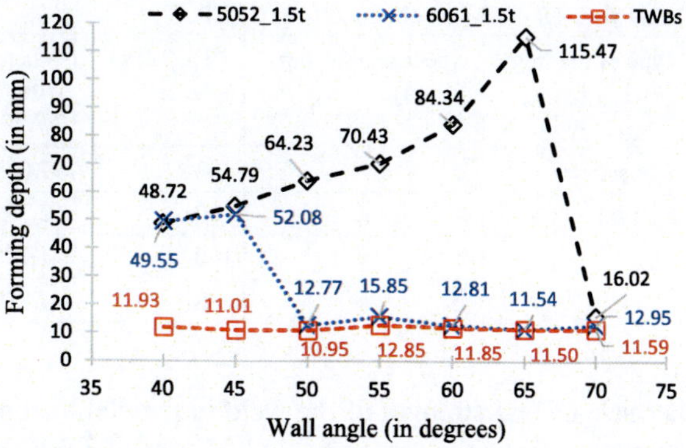

Fig. 8 Forming depth versus wall angle

3.4 *Results of Slow Strain Rate Tests (SSRTs)*

The dumbbell shape tensile specimens (sub-size specimens), as per ASTM E8 standards, were prepared by wire cut process for the slow strain rate tests as shown in Fig. 4, and the results in terms of stress–strain diagram are shown in Fig. 9. It can be clearly seen from Fig. 9 that the stress–strain curve of parent materials (without welding) have shown ductile failure, and their yield tensile strength (YTS) is noted up to 223 MPa for AA5052-H32 and 338 MPa for AA6061-T6. However, for all TWBs (both transverse and parallel loading), the stress–strain curves are showing brittle failure, and their respective YTS values are mentioned in Table 6. Table 6 shows that when weld undergoes transverse loading which exhibits good strength than that of when it is loaded in parallel directions. The study reveals that if material is getting stretched perpendicular to weld direction, then it will have good strength. But, if material is getting stretched parallel to weld direction, then it may fail early

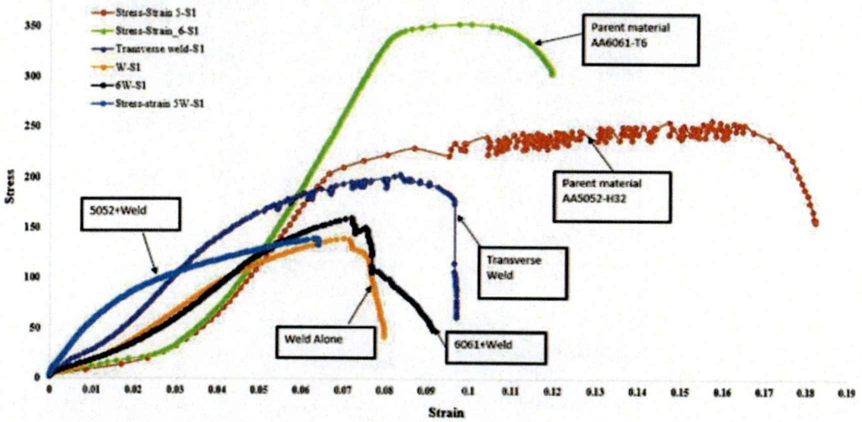

Fig. 9 Stress–strain diagrams of TWBs obtained from SSRTs

Table 6 YTS of TWBs obtained from SSRT

Type of loading	Specimen (as per Fig. 4)	YTS (MPa)	When compared with YTS of AA5052-H32	When compared with YTS of AA6061-T6 (%)
Transverse	c	203	91.03%	60.06
Parallel	a	140	62.78%	41.42
	b	160	71.75%	47.34
	d	138	61.88%	40.83

because of less strength of the weld in parallel loading. This deterioration in weld strength may be the reason behind failures of TWBs during SPIF process. In SPIF process, when conical shapes are formed, material gets elongated in radial direction, and if we try to form TWBs, then weld portion of TWBs gets elongated similarly, which means weld is getting stretched in parallel direction, and as the strength of weld in parallel direction is weak, it fails. Brittle nature of the weld portion may be the reason behind this weak strength. But it is not getting clear idea from stress–strain curve that really the weld portion exhibits brittle failure or not.

Further, the portions of all failed specimens are observed under Rapid-I Vision Measuring System with 67 × zooming capacity. Figure 10 shows the SSRT failed samples and their respective microstructures. It can be clearly seen from the

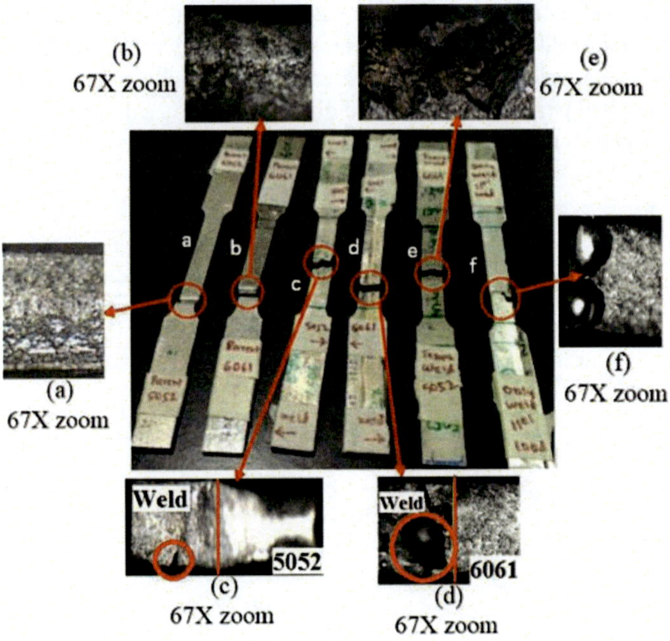

Fig. 10 Failed tensile samples of SSRTs and their respective microstructures

microstructures of weld samples (c), (d), (e), and (f) that the failure occurs due to porosity of the weld. The weld portion responds like brittle material and parent material shows ductile failure (please see red-marked circle in Fig. 10c, d). The researchers Singh et al. [11] and Wang et al. [12] have also quoted that porosity of weld is the reason behind failure of TIG-welded joints. According to them, excessive porosity reduces static strength of the weld. Hence, there is further scope of more microstructure study to find out the affecting element which is contributing in the brittle fracture of the weld. As per author opinion, the silica content in AA6061-T6 (i.e., 0.64%) may cause brittle type of failure in weld.

As porosity is observed in weld, it was decided to carry out heat treatment on TIG-welded blank, and then, this heat-treated blank is formed by SPIF process by keeping wall angle of 45°. The results of the same are discussed in Sect. 3.5 of this paper.

3.5 *Effect of Heat Treatment on Formability of TWBs*

Heat treatment is carried out on TIG-welded TWB by heating it at 480 °C, and holding time was kept for 45 min. Then, it is allowed to cool in furnace itself. This heat-treated blank is formed by SPIF process by keeping wall angle at 45°, tool diameter 10 mm, tool speed 1000 rpm, feed rate 1000 mm/min, and step depth was kept 0.2 mm. Figure 11 shows the formed component after heat treatment.

Forming depth of heat-treated TWB is measured by digital height gage, and the same is observed up to 22.63 mm, i.e., almost twice to the forming depth achieved for 45° wall angle non-heat-treated TWB. This clearly indicates that the formability of welded blank can be improved after heat treatment. This post-weld heat treatment may have distributed the microstructure homogeneously in the weld, and hence, improvement in formability in terms of forming depth might have achieved. Fadaei-fard et al. [13] had also suggested promising improvement in mechanical properties of weld zone after post-weld heat treatment of AA6061-T6.

Fig. 11 Heat-treated TWB after SPIF

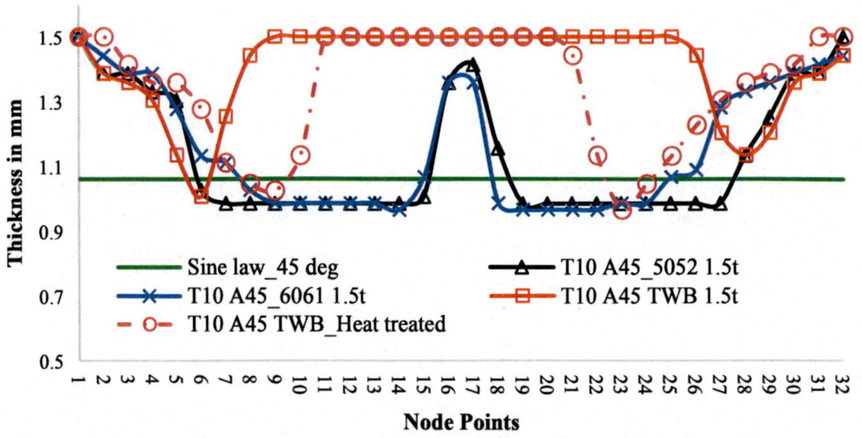

Fig. 12 Thickness distribution heat-treated TWB versus non-heat-treated blanks

It is also observed that the crack is initiated when the above forming depth is achieved. This crack was developed along the weld line, whereas in all non-heat-treated TWBs (Fig. 6) cracks are developed in transverse direction to weld. It means that the weld strength may be reduced after heat treatment, and due to forming forces weld got separated at forming depth of 22.63 mm.

Figure 12 shows comparison of thickness reduction between heat-treated TWB with non-heat-treated TWB and parent materials which were formed at 45° wall angle. Heat-treated TWB shows good improvement in thickness distribution, and it follows the pattern of AA6061-T6 parent material. Due to heat treatment the original mechanical properties of the parent material might have changed. In future, there is much scope in the study related to microstructural changes and material behaviors after heat treatment.

4 Conclusions

Formability of parent materials AA5052-H32 and AA6061-T6 and their TWBs prepared by TIG welding were studied by varying wall angle from 40° to 70°, and from the results and discussion above, it can be concluded that,

1. The parent material AA5052-H32 flows easily when compared with parent material AA6061-T6. Whereas, formability of TWBs depends upon stronger material.
2. Maximum forming depth up to 115.47 mm can be achieved for parent material AA5052-H32 with a wall angle of 65°, and the same is achieved up to 52.08 mm with a wall angle of 45° for parent material AA6061-T6. Whereas, maximum forming depth 12 mm was achieved for all TWBs.

3. Further investigation is done by performing slow strain rate testing and microstructure study (67x). It reveals that all TWBs have failed early because the weld zone behaves like brittle material. Silica content in weld may be the reason behind the failure of weld. More study is required to study the effect of silica on the formability behavior of AA6061-T6 and weld.
4. Heat treatment on tailor-welded blanks may improve the ductility of weld, and hence, it may improve the formability in terms of forming depth of TWBs.

Acknowledgements We are thankful to the Department of Production Engineering and Industrial Management, College of Engineering Pune, for their technical and financial support to carry out this research work. We are also thankful to Department of Metallurgy and Material Science, College of Engineering Pune, for their continual technical support.

References

1. Silva MB, Skjoedt M, Vilac P, Bay N, Martins PAF (2009) Single point incremental forming of tailored blanks produced by friction stir welding. J Mater Process Technol 209:811–820
2. Bhattacharya A, Maneesh K (2011) Formability and surface finish studies in single point incremental forming. J Manuf Sci Eng 133:061020-1 to 061020-8
3. Malwad DS, Nandedkar VM (2014) Deformation mechanism analysis of single point incremental sheet metal forming. Proc Mater Sci 6:1505–1510
4. Alinaghian I, Ranjbar H, Beheshtizad MA Forming limit investigation of AA6061 friction stir welded blank in a single point incremental forming process: RSM approach. Trans Indian Inst Met. https://doi.org/10.1007/s12666-017-1093-y
5. Ebrahimzadeh P, Baseri H, Mirnia MJ (2018) Formability of aluminum 5083 friction stir welded blank in two-point incremental forming process. Proc IMechE Part E J Process Mech Eng 232(3):267–280. https://doi.org/10.1177/0954408917692370
6. Gatea S, Bin Lu, Chen J, Hengan Ou, McCartney G (2018) Investigation of the effect of forming parameters in incremental sheet forming using a micromechanics based damage model. IntJ Mater Form. https://doi.org/10.1007/s12289-018-1434-3
7. Kumar Y, Kumar S (2018) Design and development of single point incremental sheet forming machine. AIMTDR 2014, 94-1 to 94-4
8. Kumar R (2013) Analysis of major strains and minor strains in sheet metal forming. Int J Appl Innov Eng Manage 2:194–198
9. Li Y, William JT, Silva DMB, Liu Z, Haibo Lu, Meehan PA (2015) Deformation mechanics and efficient force prediction in single point incremental forming. J Mater Process Technol 221:100–111
10. Oleksik V (2014) Influence of geometrical parameters, wall angle and part shape on thickness reduction of single point incremental forming. Proc Eng 81:2280–2285
11. Singh G, Kang AS, Singh K, Singh J (2017) Experimental comparison of friction stir welding process and TIG welding process for 6082–T6 aluminium alloy. Mater Today Proc 4:3590–3600
12. Wang X, Wang K, Shen Y, Kai Hu (2008) Comparison of fatigue property between friction stir and TIG welds. J Univ Sci Technol Beijing 15(3):280–284
13. Fadaeifard F, Matori KA, Garavi F, Al-Falahi M, Sarrigani GV (2016) Effect of post weld heat treatment on microstructure and mechanical properties of gas tungsten arc welded AA6061-T6 alloy. Trans Nonferrous Met Soc China 26:3102–3114

Autogenous TIG Welding of Al-5083-H111 Butt Joint

Keyurkumar D. Tandel and Jyoti V. Menghani

1 Introduction

Aluminum alloys are extensively used in structures where high strength to weight ratio, good corrosion resistance and high low temperature ductility with good formability and weldability is required such as automotive components, aerospace equipments and marine structures [1]. Fusion welding processes are widely used in fabrication of aluminum alloy structures. Fusion welding of aluminum alloys is difficult due to the fact that they are having high thermal conductivity, high thermal expansion coefficient, high solidification shrinkage, relatively wide solidification temperature ranges, tendency of refractory aluminum oxide (Al_2O_3) formation and low stiffness [2]. Arc welding of aluminum alloys associated with porosity due to high solubility of hydrogen when in the molten state, lack of wetting, lack of fusion, hot cracking, tensile residual stresses, strength reduction and poor joint efficiency [3, 4]. Distortion is one of the major fabrication challenges in welding as it causes dimensional inaccuracies of the structure and component [5, 6]. However, increased application of aluminum alloys in almost all industries is a driving force for the researchers to develop feasible and efficient technologies to join aluminum alloys without much adverse effect on their mechanical, chemical and metallurgical performances desired for longer life.

Arc welding processes such as TIG commonly known as GTAW (Gas Tungsten Arc Welding) and MIG is commonly known as GMAW (Gas Metal Arc Welding) were very popular to weld aluminum alloys during early phases of welding development. Usage of inert gas for shielding liquid weld pool produced high quality and high strength weld without corrosive fluxes compared to SMAW (Shielded Metal

K. D. Tandel (✉) · J. V. Menghani
Department of Mechanical Engineering, Sardar Vallabhbhai National Institute of Technology, Ichchhanath, Surat, Gujarat 395007, India
e-mail: kdtandel@gecv.ac.in

H. K. Dave and D. Nedelcu (eds.), *Advances in Manufacturing Processes*, Lecture Notes in Mechanical Engineering, https://doi.org/10.1007/978-981-15-9117-4_16

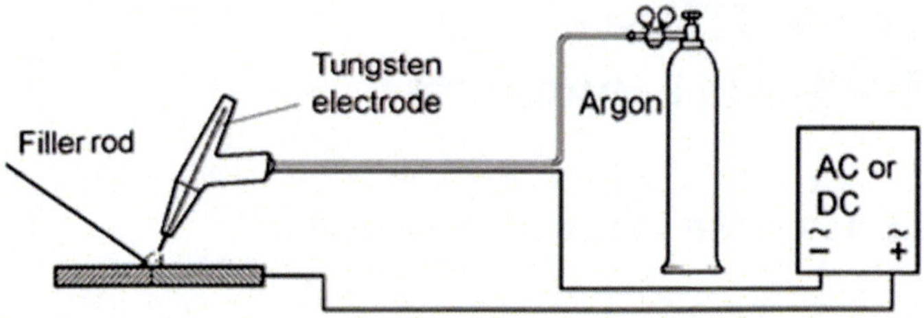

Fig. 1 Welding circuit of GTAW

Arc Welding). GTAW was developed earlier than GMAW process for welding of aluminum and was used for all thicknesses and joint types [7]. GMAW was mostly used to weld thicker section and also due to the fact that welding can be performed at higher speed. However, GTAW still has an important place in the aluminum welding industry. GTAW with AC current and pure argon shielding gas is now most often used to weld thinner sections of aluminum and also for applications where aesthetics is most important [8]. However, in recent era welding processes such as Laser Beam Welding (LBW) [9], Friction Stir Welding (FSW) [10], Pulsed arc GTAW [11] and Pulsed arc GMAW [12] has been developed to achieve better quality weld joint and better productivity.

In conventional GTAW process, non-consumable tungsten electrode is used to generate arc between base metal and electrode. Filler metal is separately added into arc which melts along with the base metal in common weld pool and upon solidified produces a good quality weld joint. High purity (99.95%) inert gases like argon, helium is used for shielding of liquid weld pool. Welding circuit of GTAW process is shown in Fig. 1. Weld quality depends on variables like selection of filler material, shielding gas, gas flow rate, welding current, arc voltage, welding travel speed and weld polarity. Change in one parameter requires other parameters to set accordingly to get sound quality weld joint [5]. Welding parameters associated with higher heat input (high current, low travel speed) leads to greater welding temperature and lower post-weld cooling rate, which in turn results in thicker HAZ. Grain coarsening due to higher heat input is the main cause of lower mechanical properties [13]. Porosity is one of the common issues in fusion welding and it is very difficult to get weld joint of aluminum which is free from porosity. Porosity generates in the weld metal due to entrapment of gases in solidified weld metal. Main reason behind presence of porosity in aluminum weld metal is hydrogen gas, which is having very high solubility in molten weld metal but very low solubility in solid. Amount of porosity is minimum in autogenous welds as chances of contamination is minimum due to no usage of filler metal but when filler metal is used porosity tends to increase due to contamination from the wire [4]. Aluminum alloy 5xxx series containing Mg in the range of 1–2.5% may be prone to hot cracking if they are welded autogenously or with the filler metal having similar chemical composition. Filler metal having higher Mg content (Mg > 3.5%) is used to avoid hot cracking during welding of these materials [4]. Lincoln Electric Aluminum Welding Guide for GMAW states that heat--treatable

Table 1 Chemical composition of as received Al-5083 showing weight % of elements

Mg	Mn	Si	Fe	Cu	Cr	Zn	Al
5	0.57	0.15	0.35	0.05	0.05	0.02	Rest

aluminum alloys (2xxx, 6xxx, 7xxx series) are more prone to hot cracking during autogenous weld. It is observed that Base metal having higher magnesium content (AA-5083) is less prone to hot cracking during autogenous welding. Qiang et al. [14] has successfully executed double-sided coaxial GTA flat-overhead welding on aluminum 5083 alloy 8 mm thick plate. Full penetration autogenous welding was performed without any kind of imperfections in the weld. Ancona et al. [15] executed autogenous butt weld of 5083 alloy having 3 mm thickness using Laser welding process showing acceptable quality of weld joints in their experimental work.

In the present investigation, 6 mm thick Al-5083 is welded using autogenous TIG welding process. The welds are characterised in terms of macro observation, optical microstructure, microhardness and tensile strength. SEM fractographic studies are carried out to investigate the nature of weld failure. Possible advantage in terms of distortion control has also been derived through present work.

2 Materials and Methods

2.1 Base Metal

Base material used in this experimental work is aluminum alloy 5083-H111, 6 mm thick plate. Aluminum alloy 5xxx series is Al–Mg alloy, possesses maximum strength among non-heat-treatable alloys. Aluminum alloy 5083 is non-age hardenable Al–Mg based alloy, having excellent corrosion resistance, superplasticity and high strength [16]. Alloy 5083 performs exceptionally well in extreme environment. It is highly resistant to industrial chemical and seawater attack. It is having good weldability and not prone to hot cracking. The chemical composition of 5083-H111 as per ASTM B209/B928M is shown in Table 1. The common physical and mechanical properties of 5083-H111 are enlisted in Table 2.

2.2 Autogenous TIG Welding

The plate was cut into pieces of size 150 mm × 75 mm by using cutting wheel grinder to make the butt joint. Plate faying surfaces are flat grounded to make zero root gap square butt joint configuration. Weld joint set-up is as shown in Fig. 2. Square butt joint of AA 5083-H111, 6 mm thick plate was performed by TIG welding process without addition of filler wire. Lincoln Electric make Aspect 300 TIG Welding

Table 2 Common physical and mechanical properties of AA5083

Property	Value
Density	2.65 g/cm^3
Melting point	570 °C
Thermal expansion	25×10^{-6}/K
Thermal conductivity	121 W/mK
Electrical resistivity	0.058×10^{-6} Ω m
Modulus of elasticity	72 GPa
Proof stress	125 MPa
Tensile strength	275–350 MPa
Brinell hardness	75 HN

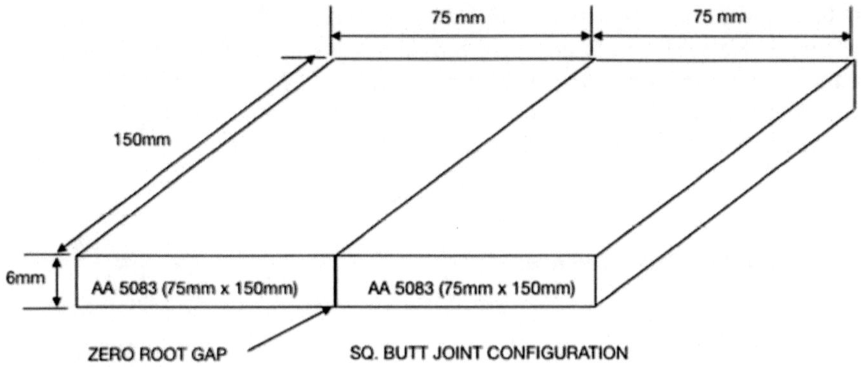

Fig. 2 Square butt joint configuration

machine was used for TIG welding trials. Welding was performed in machine mode to precisely control welding travel speed, heat input and to have repeatability in the quality of weld joint produced. A fixture to hold the TIG torch was mounted on Milling machine as shown in Fig. 3. Faying surfaces and top–bottom face up to 10 mm from edge were thoroughly cleaned by coarse grade emery paper (400 Grit Size) to remove any traces of aluminum oxide before clamping them on the table of milling machine. Plates to be welded were clamped on table of milling machine which can be moved along with table at constant travel speed against stationary held TIG torch. Single-pass per side technique was employed and welding was performed from both sides to achieve full penetration joint. Arrangement was made to protect opposite side of welding from oxidation. Industrial grade pure argon was used as shielding gas and Thoriated Tungsten electrode of 3.0 mm diameter was used for all trials of TIG welding. Electrode tip was ground in blunt conical shape. Pointed conical shape tungsten electrode tip deteriorates rapidly and arc wandered during AC welding which leads to creating a weld defect. Using blunt electrode tip, arc becomes stable and it forms round ball at the tip naturally during welding.

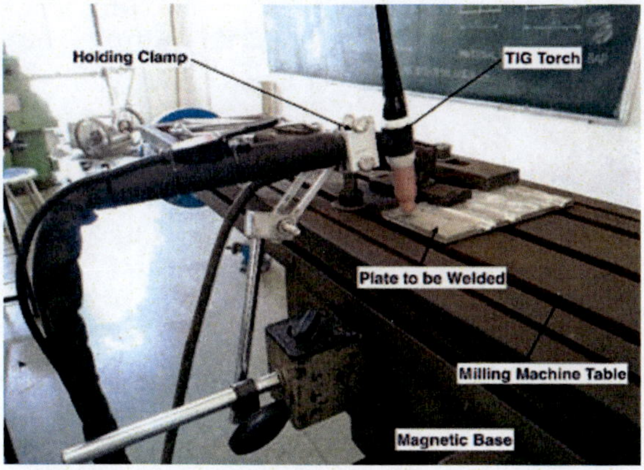

Fig. 3 TIG torch holding fixture

Initial trials, Bead-on-Plate were carried out at different current levels keeping constant travel speed (150 mm/min) by autogenous TIG welding, to check penetration in the base metal. Welding parameters for bead-on-plate trials are as shown in Table 3. Welding parameters, which will give at least 3 mm penetration are considered for further welding trails of square-butt joint of 6 mm thick Al-5083-H111 plate by autogenous TIG welding process. Plate on which bead-on-plate trials were performed was cut across the weld. Cut section of the plate was etched to reveal macrograph of weld penetration. Refer Fig. 4. Welding current 150 Amp and 170 Amp at 150 mm/min shows more than 3 mm penetration in the base metal. Therefore, any travel speed slower than 150 mm/min will ensure more than 3 mm penetration for both welding current levels of 150 and 170 amp. Hence, Autogenous TIG welding trials on square-butt joint are carried out at two current levels (150, 170 amp) with combination of three travel speeds (80, 110, 140 mm/min) to evaluate the effect of travel speed and welding heat input on mechanical property of weld joint.

Welding voltage was 15–16 V and shielding gas flow rate was kept in the range of 13–15 lpm for all welding trails [17]. Details of welding parameters for Autogenous TIG welding are shown in Table 4.

Table 3 Welding parameters of autogenous TIG welding for bead-on-plate trials

Trial No.	Current (amp)	Voltage (V)	Travel speed (mm/min)	Shielding gas flow rate (lpm)	Penetration (mm)
1	100	15–16	150	13–15	1.5
2	120	15–16	150	13–15	2.0
3	150	15–16	150	13–15	3.5
4	170	15–16	150	13–15	4.0

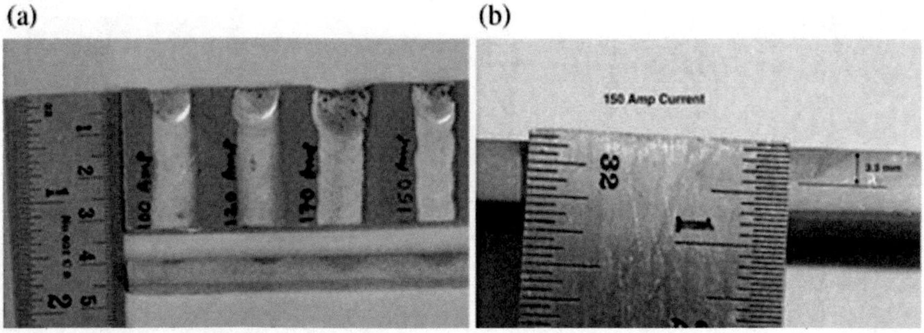

Fig. 4 Bead-on-plate trials to reveal welding penetration: a weld penetration at different current levels, b Weld penetration for 150 amp current

Table 4 Welding parameters of autogenous TIG welding of alloy 5083-H111

Trial No	Sample ID	Current (A)	Travel speed (mm/min)
1	T1	150	70–80
2	T2	150	100–110
3	T6	150	140–150
4	T3	170	70–80
5	T4	170	100–110
6	T5	170	140–150

2.3 Characterisation of Weld Joint

Macro Observation. Weld joint produced by Autogenous-TIG welding process was observed with naked eye in as-welded condition to check presence of macro defects like pin hole, under cut, etc. and weld bead finish. Amount of angular distortion was measured after plates were cooled to room temperature post welding. Weld joint was sectioned at several places and the sectioned surface was rough polished using emery paper. Kellers etch (Distilled Water—192 ml, Nitric acid—5 ml, Hydrochloric acid—3 ml, Hydrofluoric acid—2 ml) was applied for 20–25 s on the polished surface to perform macro observation of weld area. Depth of penetration was also assessed.

Microstructural Characterisation. Microstructural characterisation was performed using optical microscope with CCD camera attachment (Carl Zeiss, Jena Make, Model—EPY TYP –II) and scanning electron microscope (SEM) (Hitachi Make, Model—S-3400 N). For metallographic examination, specimens were cut perpendicular to the weld axis in such a way that base metal, HAZ and weld metal can be analysed in single specimen. Specimens were then etched using Kellers etch for 10–15 s. Specimens were then observed under optical microscope at 100 and 400× magnification. Fracture surface of tensile test specimen was observed under SEM at different magnification from 250 to 2000×. Fracture surface morphology of TIG-welded specimen was analysed from the captured image.

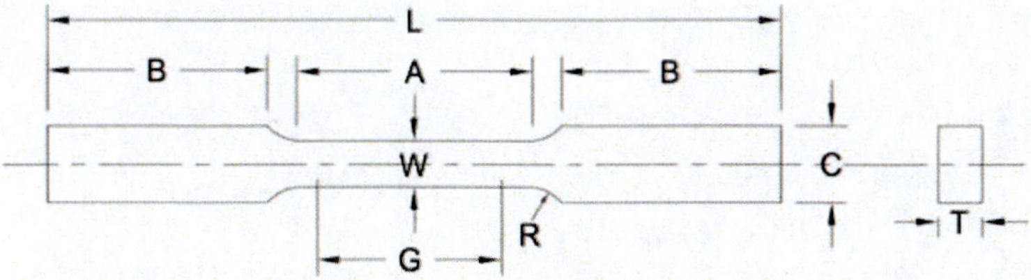

Nomenclature	Dimension in mm
Gauge length, G	25.00 ± 0.10
Width, W	6.01 ± 0.05
Thickness, T	Thickness of material (6mm)
Radius of fillet, R	6
Overall Length, L	100
Length of reduced section, A	32
Length of grip section, B	30
Width of Grip Section, C	10

Fig. 5 Subsize flat tensile specimen dimensions as per ASTM B557

Microhardness Measurement. Microhardness measurement was performed on digital microhardness tester (Clemex Make, Model—HWMMT-X). A precision diamond indenter was impressed on each sample at a load of 200 gm for a dwell time of 20 s. The indentation was measured microscopically at 100× magnification.

Tensile Test. Sub sized flat tensile test specimens were prepared from the welded plate and unwelded base metal as per ASTM B557. Dimensions of sub size flat tensile specimen are shown in Fig. 5. Specimens were cut perpendicular to the weld axis. Specimens were tested on Kudale Instruments (P) Ltd. make computerised tensometer, having 2 T loading capacity. Tensile test was performed at test speed of 10 mm/min. Tensile strength of autogenous TIG-welded plate is compared with that of base metal.

3 Result and Discussion

3.1 Macro and Microstructural Analysis

Weld joint produced by Autogenous TIG welding process showed excellent bead finish in as-welded condition and requires no post-processing. Weld joint was free from macro surface defects like pinholes and undercut. Figure 6a shows Autogenous TIG-welded plate in as-welded condition. Since welding was performed in machine

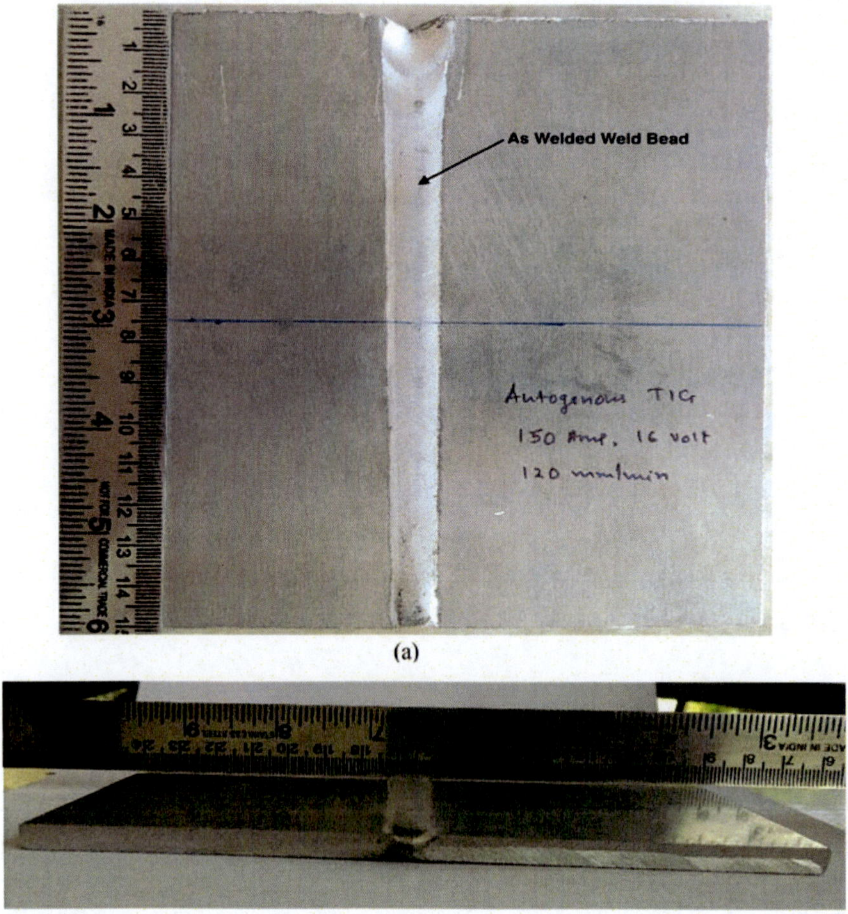

(a)

(b)

Fig. 6 **a** Autogenous TIG weld bead finish in as-welded condition, **b** no measurable angular distortion in autogenous TIG-welded plate

mode at constant travel speed, a uniform weld bead was achieved without any serration. After completion of welding, plate was allowed to cool to room temperature in still air and angular distortion was measured. No measurable angular distortion was observed in Autogenous TIG-welded plate in as-welded condition (refer Fig. 6b). Square-butt joint configuration is the prime reason for absence of angular distortion. Cumulative welding heat input is very less in square-butt joint type due to the fact that total number of passes to complete the joint is only two (one pass each on both sides) compared to V groove joint types in which multipass welding is to be performed to complete the joint. It is evident that more welding involves more heat input leads to higher distortion. Here welding is performed from both sides facilitates balanced welding, which in turn helps in controlling distortion. Also, in square-butt

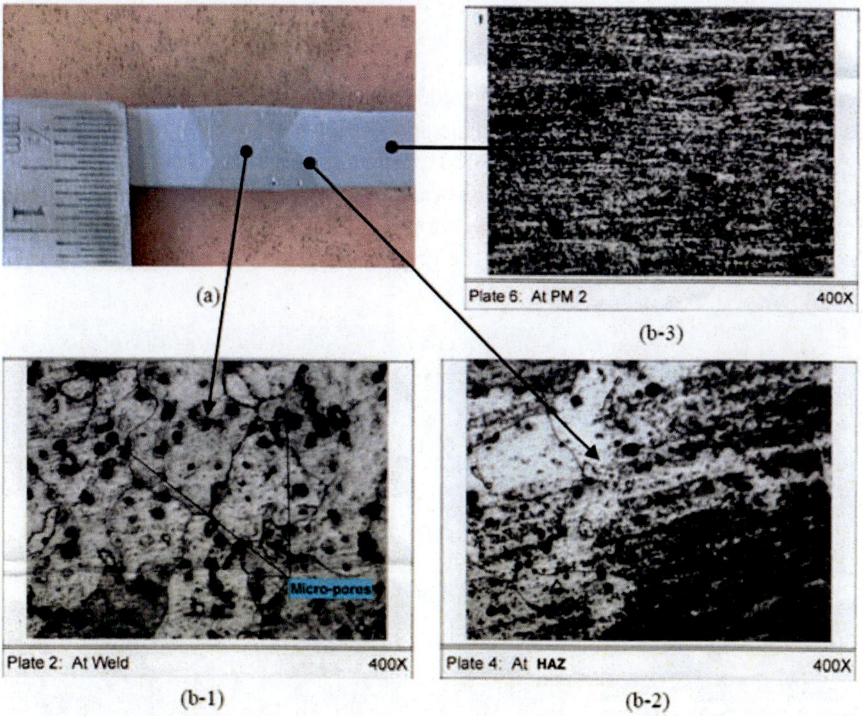

Fig. 7 **a** Macrograph of Autogenous TIG-Welded Specimen T6; **b** microstructure of Specimen T6 at 1—Weld, 2—HAZ and 3—unaffected Base Metal.

joint type with zero root gap gives minimum scope for angular distortion in welding compared to V-groove joint types as absence of groove restricts flexibility of base metal to distort during welding. As shown in Fig. 7a, macrograph of autogenous TIG weld cross-section corresponds to maximum UTS value (Sample ID—T6) reveals complete overlap of both passes. This ensures capability of welding parameters to achieve full penetration weld joint in square-butt joint type. Figure 7b–1, 2 and 3 show weld, HAZ and base metal microstructure of autogenous TIG-welded specimen T6 corresponds to maximum UTS value at 400× magnification respectively.

Microstructural observation reveals that weld and surrounding region is free from defects like voids, cracks and lack of fusion. Microstructure of unaffected base metal shows elongated un-recrystallized grain in aluminum solid solution. Microstructure at weld shows fine columnar grain in aluminum solid solution due to sudden heating and cooling associated with welding. However, columnar grain structure is absent in HAZ region. Microstructure of weld at 400× magnification shows micro-pores that are not observed at 100× magnification. HAZ microstructure shows coarse recrystallized grain in aluminum solid solution. Welding heat input is responsible for grain coarsening in HAZ region.

3.2 Mechanical Properties

Tensile strength of Autogenous TIG-welded specimens at different process parameters was tested. For comparison, tensile strength of base metal without weld was also carried out. Photographs of fractured tensile test specimens are shown in Fig. 8. Fracture location of all Autogenous TIG-welded specimens is within the weld region. Tensile test results of base metal and Autogenous TIG weld are shown in Table 5. As well as comparison of test results are graphically illustrated in Fig. 9.

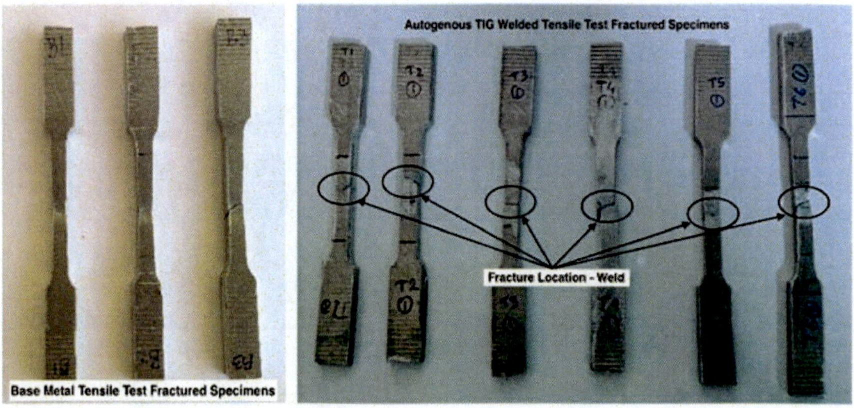

(a) (b)

Fig. 8 Fractured tensile specimen: **a** base metal; **b** autogenous TIG Weld

Table 5 Results of tensile testing of autogenous TIG-welded joints and base metal

Trial No.	Sample ID	Welding current (amp)	Travel speed (mm/min)	UTS (MPa)	% Elongation	Fracture location
Base metal						
–	B1	–	–	298.27	20	–
–	B2	–	–	299.67	20	–
–	B3	–	–	295.81	20	–
Autogenous TIG weld						
1	T1	150	70–80	265.3	16	Weld
2	T2	150	100–110	278.5	20	Weld
3	T6	150	130–140	284.7	22	Weld
4	T3	170	70–80	200.6	10	Weld
5	T4	170	100–110	249	14	Weld
6	T5	170	130–140	258	16	Weld

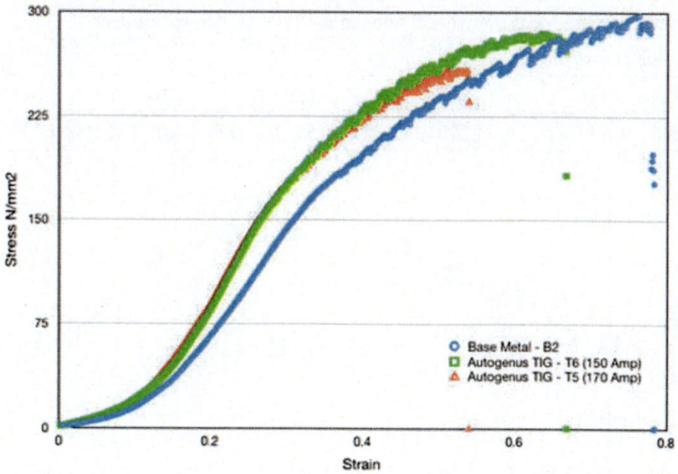

Fig. 9 Comparison of UTS of base metal (B2) and autogenous TIG-welded specimens (T6, T5)

Three specimens (B1, B2 and B3) are tested to determine base metal tensile strength, which shows UTS of base metal value in the range of 295–298 MPa and 20% elongation. Tensile test results of Autogenous TIG-welded specimen shows decreasing UTS trend with increase in welding current and reduction in welding travel speed. Refer tensile test results of sample ID T1, T2 and T6 corresponding to 150 amp welding current and sample ID T3, T4 and T5 corresponding to 170 amp welding current as shown in Table 5. This is due to the fact that with increase in welding current and decreasing welding travel speed, welding heat input will increase which will tend to promote coarser dendritic structure in the weld and HAZ region. Faster travel speed leads to faster cooling of weld, resulting in finer grain structure formation in the weld which is responsible for increased strength of the weld joint [18]. Welding parameters corresponding to specimen no. T6 shows best value of UTS among all specimens which is 284.7 MPa, which is lower than measured base metal UTS (avg. 298 MPa) by approximately 5%. Presence of micro-pores in the weld is the reason for reduced tensile strength compared to base metal. Increase in % elongation, which is a measure of ductility, with increase in strength was observed from the tensile test results. Homogenous distribution of particles can increase both strength and ductility. Compared to elongated, inhomogeneous cast microstructure, weld metal shows refined, homogenised structure, which increases both strength and ductility of the weld joint.

The microhardness profiles of Autogenous TIG-welded joints are measured at the centre thickness of the cross section of the weld as enlisted in Table 6. Microhardness measurement is carried out in weld, HAZ and base metal region. Micrphardness variation trend across the weld is depicted in Fig. 10. Weld region shows lower hardness compared to base metal and HAZ. Slightly increased hardness value is reported in HAZ region near to fusion boundary compared to base metal. Re-precipitation of the β-phase Mg_2Al_3 may be the reason behind this [19]. Al-5083 is non-heat treatable

Table 6 Microhardness measurement of autogenous TIG-welded specimen

Hardness value across the Weld (Hv)				
Base metal	HAZ	Weld	HAZ	Base metal
178, 184, 215	219, 233, 254	155, 162, 174	261, 258, 243	207, 184, 171

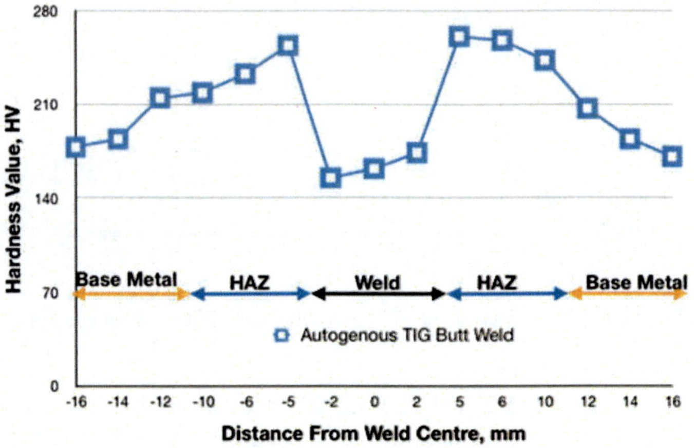

Fig. 10 Microhardness profile across the autogenous TIG weld

alloy hence reduction in hardness due to weld heat input is not anticipated, as plastic deformation is the prime hardening mechanism [20].

3.3 Fractography Study

Fractured tensile test specimen T6 of Autogenous TIG-welded plate in which fracture location is within weld was taken for SEM scan. Figure 11 shows SEM images of fractured tensile test specimen surfaces confirm the presence of micro porosities which is responsible for reduction in tensile strength of the weld joint compared to base metal. Furthermore, fracture surface of Autogenous TIG-welded specimen shows dimple type topography which confirms completely ductile nature of the weld joint.

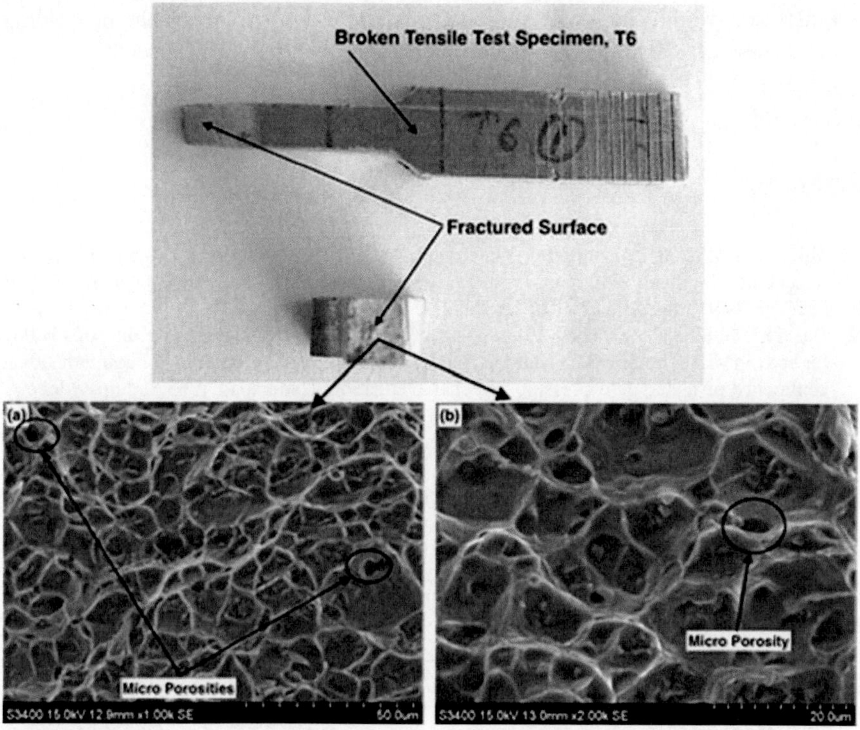

Fig. 11 SEM images of fractured surface of tensile test specimen (T6) of Autogenous TIG weld: **a** 1000×, **b** 2000×

4 Conclusion

Following conclusions are drawn from the test results of the experimental work.

Tensile strength of autogenous TIG-welded specimen is lowered by around 5% compared to unwelded base metal due to presence of micropores and formation of dendritic structure in the weld.

Microhardness in the weld region is lower than base metal but slight increase in hardness near fusion boundary is observed due to re-precipitation of β-phase.

Autogenous TIG-welded plate having square-butt joint configuration shows no angular distortion. Absence of weld groove reduces total amount of weld metal deposited, leads to reduction in cumulative heat input is advantageous in controlling distortion. Balance welding due to both side welding is also contributing to controlling distortion.

Microstructure of weld region shows fine columnar structure in aluminum solid solution whereas in HAZ region grain coarsening is observed due to heat of welding.

Full fusion welding joint with optimum tensile strength is achieved having welding parameters of 150 amp welding current and 140 mm/min travel speed.

References

1. Miller WS, Zhuang L, Bottema J, Wittebrood A, De Smet P, Haszler A, Vieregge A (2000) Recent development in aluminium alloys for the automotive industry. Mater Sci Eng A 280(1):37–49. https://doi.org/10.1016/S0921-5093(99)00653-X
2. Doshi SJ, Gohil AV, Mehta ND, Vaghasiya SR (2018) Challenges in fusion welding of Al alloy for body in white. In: Singh SK, Purohit R (eds) Materials today: proceedings, 7th international conference of materials processing and characterization, Hyderabad, p 6370. https://doi.org/10.1016/j.matpr.2017.12.247
3. Zhou W (1999) Problems in welding of high strength aluminium alloys. Singapore Welding Soc Newsl 1
4. Mathers G (2002) The welding of aluminium and its alloys, 1st ed. Woodhead Publishing Limited, Cambridge, pp 10–50
5. Weman K (2011) Welding processes handbook, 2nd ed. Elsevier, Amsterdam, pp 63–70
6. Masubuchi K, Blodgett OW, Matsui S, Ruud CO, Tsai CL (2001) Welding science and technology. In: Cynthia L. Jenney, Annette O'Brien (eds) Welding handbook, vol 1, 9th ed. American Welding Society, Miami, pp 297–354
7. Anderson T (2010) Welding aluminum—questions and answers, 2nd ed. American Welding Society, Miami, pp 13–17
8. KumarSingh S, Tiwari RM, Kumar S, Kumar S (2018) Mechanical properties and micrstructure of Al-5083 by TIG. In: Islam M, Ming TC, Arici-Bogner E, Mani VN, Malla SG (eds) Materials today: proceedings, international conference on processing of materials, minerals and energy, Ongole, Andhra Pradesh, p 819. https://doi.org/10.1016/j.matpr.2017.11.152
9. Leong KH, Sabo KR, Sanders PG, Spawr WJ (1997) Laser welding of aluminum alloys. In: Leonard R. Migliore, Ronald D. Schaeffer (eds) International conference on photonics west, lasers as tools for manufacturing II, proceedings of SPIE, San José, CA, USA, vol 2993, p 37. https://doi.org/10.1117/12.270039
10. Thomas WM (1998) Friction stir welding and related friction process characteristics. Paper presented at 7th international conference on joints in aluminium (INALCO'98), Abington, Cambridge, 15–17 April 1998
11. Kumar A, Sundarrajan S (2009) Optimization of pulsed TIG welding process parameters on mechanical properties of AA 5456 aluminum alloy weldments. Mater Des 30(4):1288–1297. https://doi.org/10.1016/j.matdes.2008.06.055
12. Subramaniam S, White DR, Jones JE, Lyons DW (1999) Experimental approach to selection of pulsing parameters in pulsed GMAW. Welding J New York 78:166–172
13. Meseguer-Valdenebro JL, Martinez-Conesa EJ, Serna J, Portoles A (2016) Influence of the welding parameters on the heat affected zone for aluminium welding. Thermal Sci 20(2):643–653. https://doi.org/10.2298/TSCI140503106M
14. Qiang W, Wang K (2019) Double-sided coaxial GTA flat overhead welding of 5083 aluminum alloy. J Mater Process Technol 272:9–16. https://doi.org/10.1016/j.jmatprotec.2019.04.042
15. Ancona A, Sibillano T, Tricarico L, Spina R, Lugara PM, Basile G, Schiavone S (2005) Comparison of two different nozzles for laser beam welding of AA5083 aluminium alloy. J Mater Process Technol 164:971–977. https://doi.org/10.1016/j.jmatprotec.2005.02.048
16. Zhou C, Yang X, Luan G (2005) Fatigue properties of friction stir welds in Al 5083 alloy. Scripta Mater 53(10):1187–1191. https://doi.org/10.1016/j.scriptamat.2005.07.016
17. Peasura P, Watanapa A (2012) Influence of shielding gas on aluminum alloy 5083 in gas tungsten arc welding. Procedia Eng 29:2465–2469. https://doi.org/10.1016/j.proeng.2012.01.333

18. Baghel PK, Nagesh DS (2018) Mechanical properties and microstructural characterization of automated pulse TIG welding of dissimilar aluminum alloy. Indian J Eng Mater Sci 25:147–154
19. Jiang Z, Hua X, Huang L, Wu D, Li F (2017) Effect of multiple thermal cycles on metallurgical and mechanical properties during multi-pass gas metal arc welding of Al 5083 alloy. Int J Adv Manuf Technol 93(9–12):3799–3811. https://doi.org/10.1007/s00170-017-0771-6
20. Borrego LP, Costa JD, Jesus JS, Loureiro AR, Ferreira JM (2014) Fatigue life improvement by friction stir processing of 5083 aluminium alloy MIG butt welds. Theoret Appl Fract Mech 70:68–74. https://doi.org/10.1016/j.tafmec.2014.02.002

Review on Friction Stir Welding of Polymer to Aluminium Alloys: Process and Properties

Rutvik Ghiya and Vishvesh J. Badheka

1 Introduction

Global developments are driving the transport industry to make lighter, faster, greener and eventually cheaper vehicles [1]. In an effort to increase the efficiency of automotive and aerospace vehicles, a significant development in the field of high performance lighter metals like aluminium and magnesium has been observed. Plastics have also become increasingly popular due to their lightweight and good mechanical properties. In an emerging class of vehicle, hybrid structures are being designed which require the permanent joining of dissimilar materials like aluminium and polymers [2]. Aluminium offers properties like low weight, high strength and excellent thermal and electrical conductivity. Polymers, like high density polyethylene (HDPE), polypropylene (PP) and polycarbonate (PC), offer properties like high strength to weight ratio and corrosion resistance. The first example of the successful application of this technical breakthrough was recorded at the end of 1996 when the front end of the Audi A6 (made by Ecia, Audincourt/France) was developed as a hybrid framework incorporating steel sheet with elastomer-modified polyamide PA6-GF30 [3]. Carbon/epoxy sections in the luxury-sports car, Lamborghini Murcièlago, are bonded to lightweight alloys and steel (bumpers, fenders, hoods) [4]. In addition to the above-mentioned usage, the polymer-metal hybrids are commonly used in the instrument panel and bumper cross-beams, door systems and tailgate applications as well as in non-automotive applications, ranging from refrigerator housings to bicycle frames [3]. In almost two decades, the use of plastic, in a family car, has increased

R. Ghiya (✉) · V. J. Badheka
Department of Mechanical Engineering, School of Technology, Pandit Deendayal Petroleum University, Raisan, Gandhinagar, Gujarat 382330, India
e-mail: rutvikghiya@gmail.com

V. J. Badheka
e-mail: Vishvesh.Badheka@spt.pdpu.ac.in

H. K. Dave and D. Nedelcu (eds.), *Advances in Manufacturing Processes*, Lecture Notes in Mechanical Engineering, https://doi.org/10.1007/978-981-15-9117-4_17

from 6 to 15% [5]. A new concept vehicle is being developed by Mercedes-Benz, named F-125, expected by 2025, which will be consisting of metal-polymer hybrid to reduce emissions. The new-generation aeroplanes like Airbus A350 XWB and Boeing 787 Dreamliner are some large structural systems, which are comprised of more than 50% CFRP (Carbon Fibre Reinforced Polymer) composites combined with metal alloys like aluminium and titanium [6].

Amancio-Filho et al. [7] and Kah [8] et al. have published extensive studies on hybrid metal-polymer joints. The joining techniques for metal and polymer can be divided into three categories- adhesive bonding, mechanical fastening and welding. Adhesive bonding requires surface preparation and special handling for solvents and chemicals. These joints cannot be disassembled without damaging parent material and may degrade due to environmental factors like temperature, moisture and humidity. These joints are unpredictable over a prolonged period of time and present challenges for performing NDT (Non-Destructive Testing). On the other hand, mechanical fastening is usually highly susceptible to hole stress concentration, which induces strength degradation. Some other concerns with mechanical fastening are- increased component weight, loosening of fastener due to moisture, stress relaxation and creep, increase in residual stresses due to difference in thermal expansion coefficients of metal and polymer.

For these dissimilar materials- laser, ultrasonic and friction stir welding has been proposed and implemented in some recent work. Available weld geometries and weldable materials, appear to be the limitations to laser welding. In ultrasonic welding, there are limitations to sample size, noise concerns and the process is usually compatible with a thermoplastic. In addition to this, both, laser and ultrasonic welding require expensive equipment [7].

Such difficulties can be countered by the solid-state welding process, friction stir welding (FSW). In recent years, a lot of work has been done on variations of friction-based solid-state welding such as friction lap welding (FLW) [9–14], friction stir spot welding (FSSW) [6, 15, 16] and friction spot joining (FSpJ) [17, 18]. But limited work has been recorded on butt welds formed using FSW. This article comprehensively reviews work covering several significant aspects of dissimilar metal-polymer butt weld by FSW, such as joint efficiency, microstructural characterization, hardness and thermal evolution.

2 Friction Stir Welding

Friction Stir Welding is amongst the processes which have offered capabilities of welding the dissimilar metals and polymers. In FSW, the pin of a non-consumable tool is inserted in the material, at the edge of the butt joint to be created. The shoulder of the tool is allowed to come into contact with the sample surface. The tool is given rotational speed and it traverses into the materials as shown in Fig. 1. Due to the heat generated by the interaction of tool and workpiece, the material gets softened locally and plastic deformation of the workpiece material is observed. The rotational and the

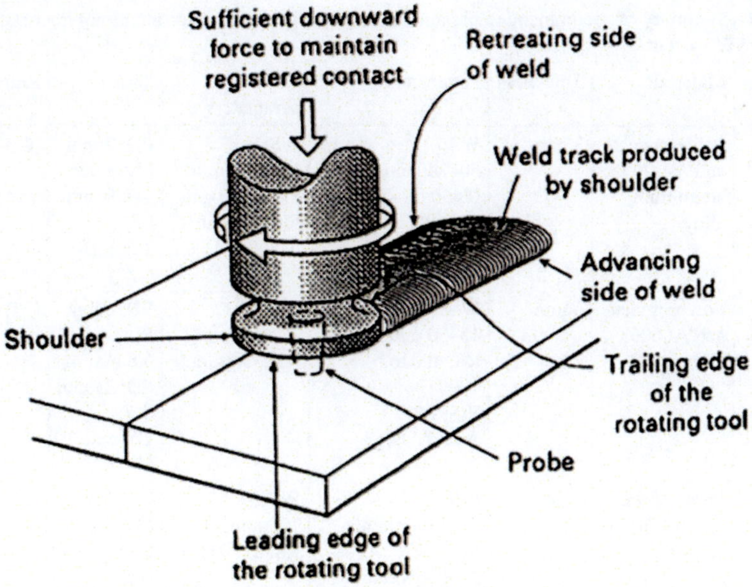

Fig. 1 Schematic representation of the FSW process [21]

linear motion of the tool forces the material from front to back of the pin of the tool and the joint is formed [19]. The process has found its great application in welding of aluminium alloys and realizing its potential, it has also been used for welding of steel, copper, titanium, metal matrix composites, magnesium, thermoplastics and various other combinations (Table 1) [20].

Sahu et al. [23] formed joint using a slot type design by inserting 3.5 mm thick PP in an Al cavity as shown in Fig. 2.

3 Mechanical Properties of Dissimilar Al-Polymer Welds

3.1 Tensile Strength and Bend Strength

In most studies, the joint strength has been found to be much lower than the parent material. Rahmat et al. [22] found the maximum tensile load to be 283.38 N, corresponding to the D/S ratio (Depth of shoulder penetration and traverse speed ratio) of 0.2%. The strength is less than 10% of PC material. The failure in test samples was concluded because of formation of kissing bond defect and due to a combination of other factors- reduction in strength and the residual stresses in HAZ (Heat affected zone).

A maximum tensile load of 75.79 N was sustained by PP and Al 6063 joint and this capacity was increased to 120.79 N with the joint formed using the slot as observed

Table 1 Summary of recommended parameters and tool geometries for metal-polymer butt joint using FSW system

S. No.	Materials	Thickness	Parameters	Result parameters	Tool	Reference
1	Polycarbonate and AA 7075 aluminium alloy	3 mm	WS: 50–200 mm min^{-1} Pld: 0.05–0.40 mm RtS: PC AvS: Al 7075	WS: 150 mm min^{-1} RS: 3250 rpm Pld: 0.1 mm	PD: 2 mm PL: 3 mm SD: 9 mm PSp: Tapered (15°)	[22]
2	Polypropylene and Al 6063 alloy	6 mm	WS: 15–50 mm min^{-1} RS: 400 to 1000 rpm RtS: PP AvS: Al 6063	WS: 25 mm min^{-1} RS: 750 rpm	PD: 6 mm PL: 5.6 mm SD: 25 mm SSp: Concave 6°	[23]
3	Polycarbonate and AA 7075 aluminium alloy	3 mm	WS: 50–150 mm min^{-1} RS: 3000–3500 rpm RtS: PC AvS: Al 7075	WS: 100 mm min^{-1} RS: 3250 rpm	PD: 1 mm SD: 9 mm PSp: Taper (25°)	[24]
4	High-density polyethylene and AA 5059 aluminium alloy	4 mm	WS: 30–200 mm min^{-1} RS: 400–2000 rpm RtS: HDPE AvS: Al 7075 TPO: 1.4 mm on Al	WS: 63 mm min^{-1} RS: 710 rpm	SD: 16 mm PSp: Cylindrical	[2]
5	High-density polyethylene and AA 5059 aluminium alloy	4 mm	Pld: 0.2 mm TTA: 2.5° RtS: HDPE AvS: Al 5059 TPO: 1.5 mm on Al	WS: 63 mm min^{-1} RS: 710 rpm	PD: 5 mm PL: 3.7 mm SD: 16 mm	[25]
6	Polypropylene and aluminium	6 mm	WS: 25 mm min^{-1} TTA: 2° RS: 600 rpm RtS: PP AvS: Al	WS: 25 mm min^{-1} RS: 600 rpm	SD: 20 mm PD: 7 mm PL: 5 mm PSp: Threaded Cylindrical	[26]

WS Welding speed, *RS* rotational speed, *RtS* retreating side, *AvS* advancing side, *TTA* tool tilt angle, *TPO* tool pin offset, *PL* pin length, *PD* pin diameter, *PSp* pin surface profile, *SSp* shoulder surface profile, *Pld* plunge depth

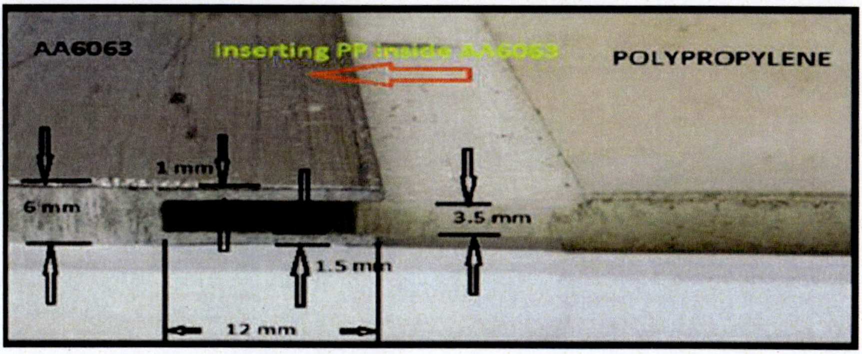

Fig. 2 Slot type design of base weld joint [23]

in Fig. 3 [23]. Due to improper mixing of PP and aluminium the welding efficiency of less than 10% w.r.t. PP was noted.

Moshwan et al. [24] reported tensile strength for PC and AA 7075 butt weld joint. Maximum tensile load of 586 N was acquired with 100 mm min^{-1} and 3250 rpm. Like Sahu et al. [23], less than 10% ultimate tensile strength of PC is secured in the weld. This can be attributed to mechanical interlocking and no formation of second phase particles.

Khodabakhshi et al. [2] evaluated joint strength for HDPE and AA 5059 and discovered it to be 50% w.r.t. HDPE (Fig. 4), which are the finest results observed in polymer and metal butt welds. The relatively low strength was credited to (1) inferior interfacial bonding at material interface, (2) shrinkage micro-voids and (3) Stress concentration in aluminium fragments.

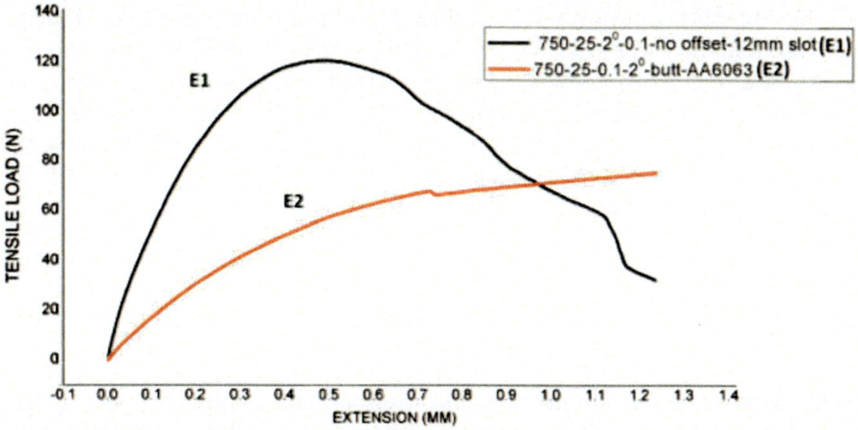

Fig. 3 Load versus extension during the tensile test [23]

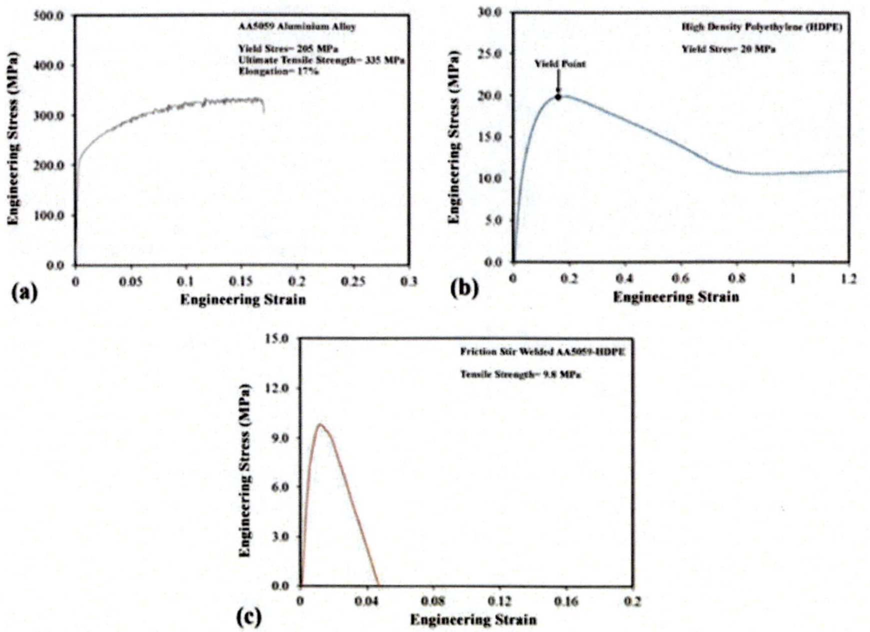

Fig. 4 Engineering stress–strain curves for **a** AA5059, **b** HDPE, and **c** friction stir welded Al–HDPE [2]

Khodabakhshi et al. [25] have suggested two ways to improve transverse tensile strength—(1) Pre-heating Al sheets and (2) Using chemical or plasma etching to activate polymer.

Rout et al. [26] formed leak-proof joints with a UTS (Ultimate Tensile Strength) of 1.221 MPa (Fig. 5a), which was less than UTS of parent material PP. SEM (Scanning Electron Microscopy) analysis of these fractured joints indicated three different zones- metal-rich, polymer-rich and mixed morphology. Most important being the mixed region, where hook-like morphology (Fig. 5c–e) can be observed resisting the crack propagation. The authors observed the same resistance to fracture in the analysis of bend test, which resulted in 30 MPa of flexural stress and 0.5 GPa of modulus of elasticity (Fig. 5b).

3.2 *Hardness*

Khodabakhshi et al. [2, 25] performed load–displacement indentation testing over Al to HDPE joint, to understand hardness characteristics of different regions of the weld (Fig. 6). The hardness of AA5059 was found to be 86.7 HV, which was reduced in HAZ (Heat Affected Zone) and TMAZ (Thermo-Mechanically Affected Zone) regions to 55.8 HV. Due to the softening of cold-worked material, there was a loss

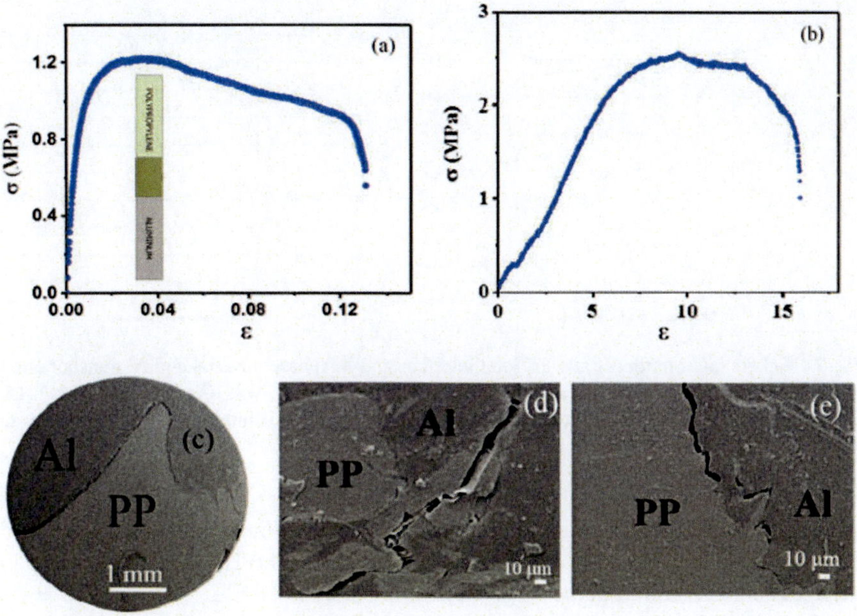

Fig. 5 **a** Stress–strain curve, **b** bend test curve, **c** SEM micrograph showing crack propagation through the interface and becoming blunt in the hooked region, **d** high-magnification image shows crack propagation through the PP suggesting good Al PP bonding, **e** crack dissipation is blunt towards the end region, indicating strong interlocking between Al and PP [26]

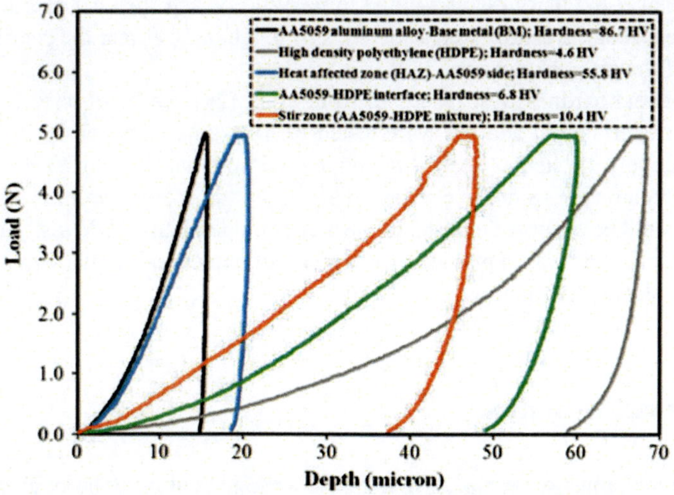

Fig. 6 Indentation hardness load–depth graphs for AA5059, HDPE, HAZ and SZ [2]

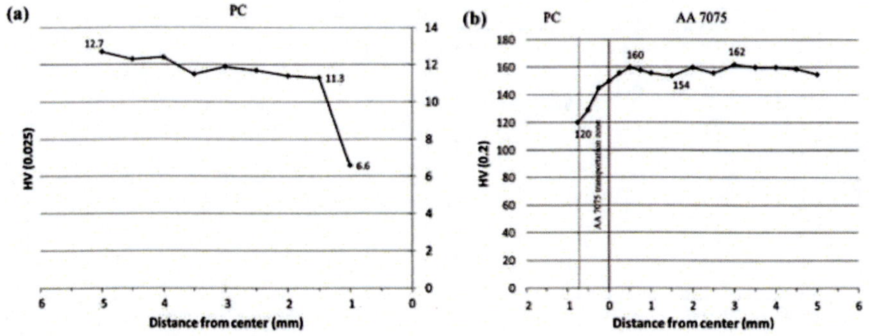

Fig. 7 Vickers microhardness at **a** PC joint interface and **b** Al base material and Al transportation interface [24]. From the International Journal of Materials Research, Vol. 106, Issue 3, March 2015, Dissimilar friction stir welding between polycarbonate and AA 7075 aluminium alloy, by Moshwan et al. © Carl Hanser Verlag GmbH & Co.KG, Muenchen

in strength. Al-HDPE interface and SZ regions exhibited a hardness of 6.8 HV and 10.4 HV, respectively. The hardness of HDPE was just 4.6 HV and the increase in strength can be credited to the Al particles embedded in HDPE.

Moshwan et al. [24] performed Vickers hardness test on PC interface and Al interface (Fig. 7). In the PC interface, at a distance of 1 mm from the centre, the lowest hardness was reported—6.6 HV, which is explained by the thermal degradation of PC. Testing at distances from centreline revealed a sudden increase in hardness to 11.3 HV and then gradually increase to 12.7 HV. At 0.75 mm from the centreline, in Al interface, the hardness was noted to be 120 HV and then a surge to 160 HV at distances from the centre was observed. The low hardness near the centre in Al is because of the grain elongation.

The Vickers hardness test results of Rout et al. [26] can be observed in Fig. 8. The base materials, Al and PP, were tested to have a hardness of 46 HV and 5.6 HV respectively. In the interlocked region, a minimum hardness of 20 HV could be observed even on PP's side. In contrast to previous studies, the hardness of Al in the interface could be observed to be 65 HV, which is higher than 45 HV, tested on parent material itself. This was explained by the grain refinement of Al that was caused by the severe plastic deformation.

4 Thermal Evolution

Four thermocouples were placed in a k-type arrangement by Sahu et al. [23], two (T1 and T2) in Al and two (T3 and T4) in PP (Fig. 9). T1 reads 120°C as the welding begins because of high initial heat generation during plunging. As tool traverse, T2 reads its maximum temperature of 60°C and then reduces. There was no change in temperature in the thermocouples placed in PP, due to the extremely low thermal

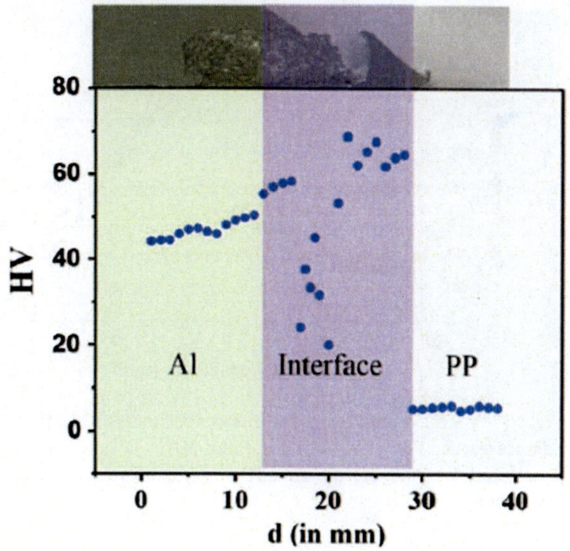

Fig. 8 Vickers microhardness test across cross-section [26]

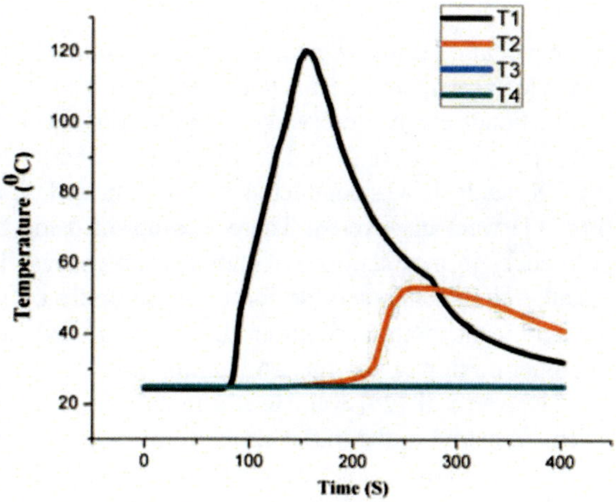

Fig. 9 Thermal cycle at various distance from weld centreline [23]

conductivity. The heat generation is also very low because of the low yield strength of PP.

Moshwan et al. [24] used ten thermocouples, five on each side of joint, in a K-type arrangement. The highest temperature of 164.4 °C was reported in AA 7075 and 66.5 °C in PC, as shown in Fig. 10. It was noted that at peak temperature, the highest volume of Al was transported to PC. As the spindle speed increases the amount of heat generation also increases but this can cause flash. At high traverse speeds, the temperature achieved is at the lower ends because of the short duration for heat generation.

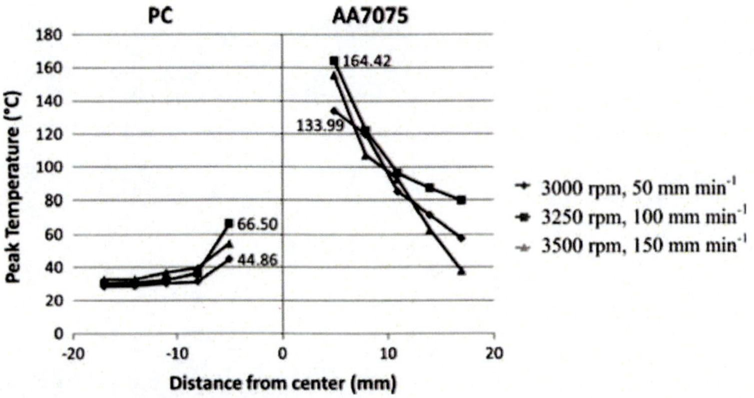

Fig. 10 Process temperatures for three experiments [24]. From the International Journal of Materials Research, Vol. 106, Issue 3, March 2015, Dissimilar friction stir welding between polycarbonate and AA 7075 aluminium alloy, by Moshwan et al. © Carl Hanser Verlag GmbH & Co.KG, Muenchen

5 Microstructural Characterization

Rahmat et al. [22] investigated the cross-section of joints and found AA 7075 transported to PC at upper side due to tool rotation. There was no onion ring formation and kissing bonds can be observed in cross-sections. Due to different flow and mechanical properties, kissing bond was expected. Energy-dispersive X-ray spectroscopy (EDXS) analysis was conducted for the joint and profiles for Al, O, C, Cu and Mg (Fig. 11) were discovered. There was no diffusion between materials because of differences in mechanical and chemical properties. There were no traces of intermetallic compounds and the joint is based only on mechanical interlocking, with no significant mixture. Sahu et al. [23] also attributed the low tensile strength to improper mixing of polymer and aluminium.

Moshwan et al. [24] evaluated the microstructures to conclude elongation in grains when compared to the initial structure of parent materials (Fig. 12). Due to differences in properties, a strong mechanical interlocking bond can be achieved with good intermixing, which would increase the interfacial contact area. But SEM (Scanning Electron Microscopy)'s cross-section revealed no such particle scattering. Like previous research, there was no onion ring formation, rather kissing bond and voids could be observed. On magnification, between SZ (Stir Zone) and AA 7075, a thin layer of PC was observed. Using EDXS (Energy-dispersive X-ray spectroscopy) analysis, Al, O, C, Cu and Mg, profiles were formed. Al was found to be scattered in SZ. There were no diffraction lines of ceramic type products in the X-ray diffraction patterns for the dissimilar joint as shown in Fig. 13. Due to plastic deformation, Al is transported into PC. Using the analysis and observations from the cross-section, it can be concluded that the joint between Al and PC was because of mechanical interlocking, instead of molecular bonding.

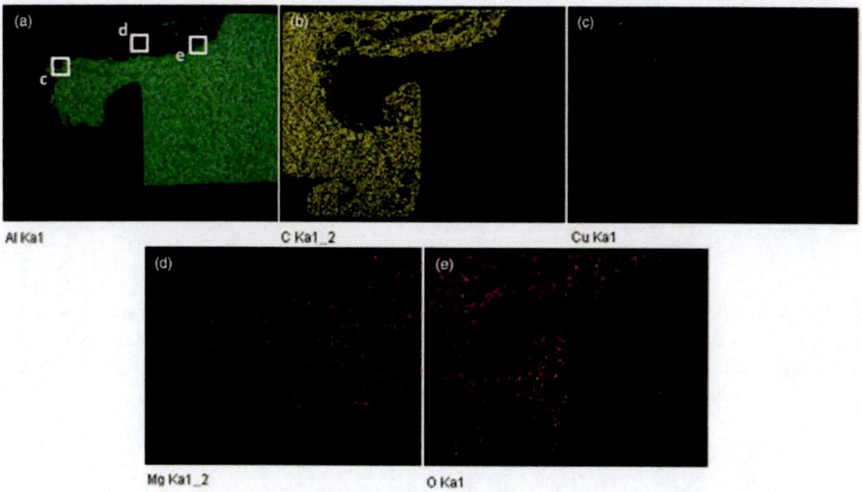

Fig. 11 EDXS map analysis of the joint interface between PC and AA7075. **a** Al, **b** C, **c** Cu, **d** Mg and **e** O [22]

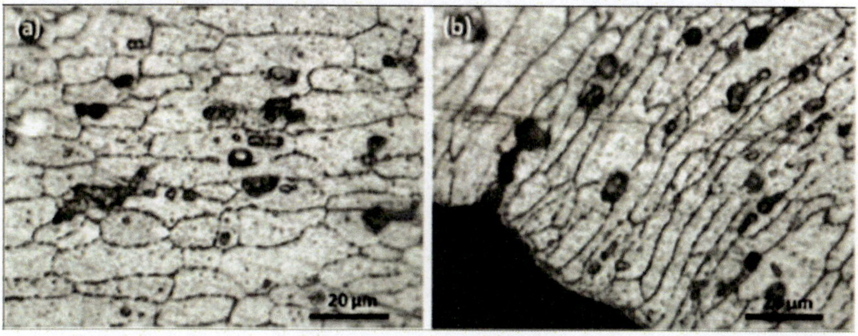

Fig. 12 Grain boundary images taken at **a** AA7075 base material and **b** AA7075 transportation [24]. (From the International Journal of Materials Research, Vol. 106, Issue 3, March 2015, Dissimilar friction stir welding between polycarbonate and AA 7075 aluminium alloy, by Moshwan et al. © Carl Hanser Verlag GmbH & Co.KG, Muenchen)

In a 2014 study conducted by Khodabakhshi et al. [2], observations similar to previous studies were drawn- (1) non-uniform grain elongation due to non-uniform temperature distribution and (2) aluminium scattered particles in SZ. In addition to bright grey contrast of AA 5059 and dark grey contrast of solidified HDPE, some mixed grey-white contrast could be seen in SEM images (Fig. 14). EDXS analysis was performed on this region, to identify elements present. Peaks of Al, Mg and C were found in spectra, though this indicated the presence of elements, there was no evidence of any molecular bond formation. To understand the nature of bond better,

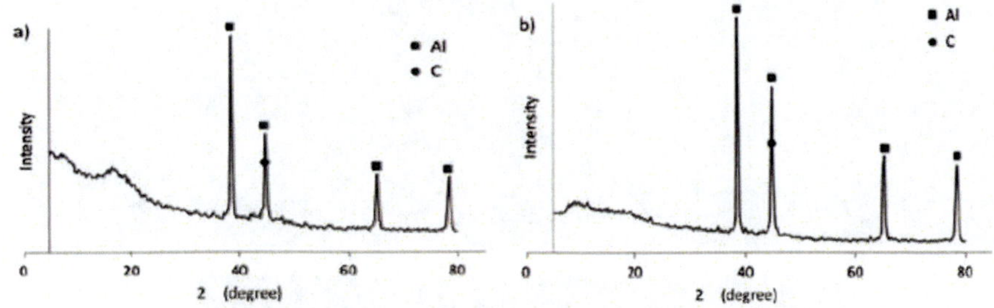

Fig. 13 X-ray diffraction analysis results: **a** PC side of welded joint and **b** AA7075 side of welded joint [24]. From the International Journal of Materials Research, Vol. 106, Issue 3, March 2015, Dissimilar friction stir welding between polycarbonate and AA 7075 aluminium alloy, by Moshwan et al. © Carl Hanser Verlag GmbH & Co.KG, Muenchen

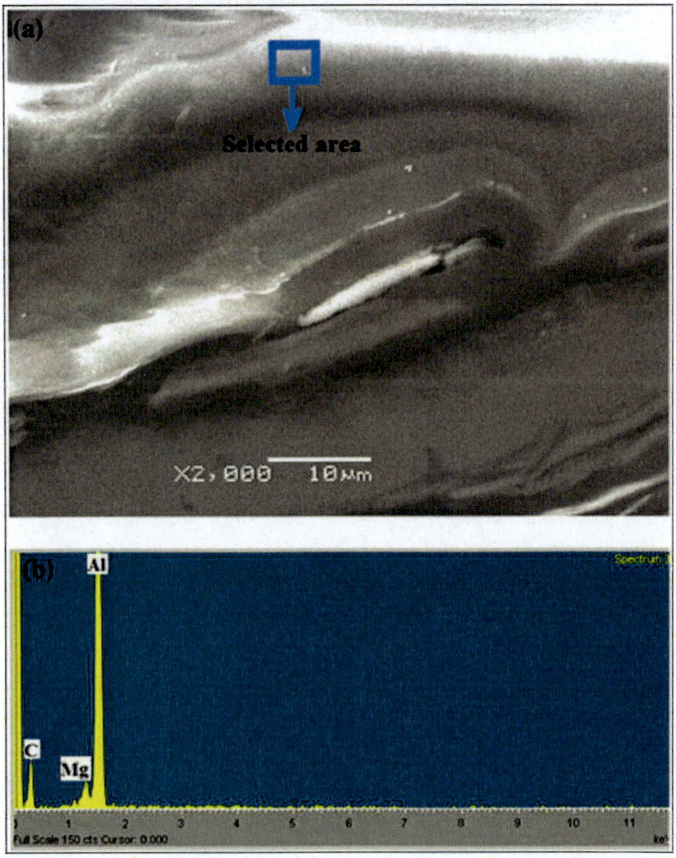

Fig. 14 Al–HDPE interface; **a** SEM image and **b** EDXS area elemental analysis spectrum [2]

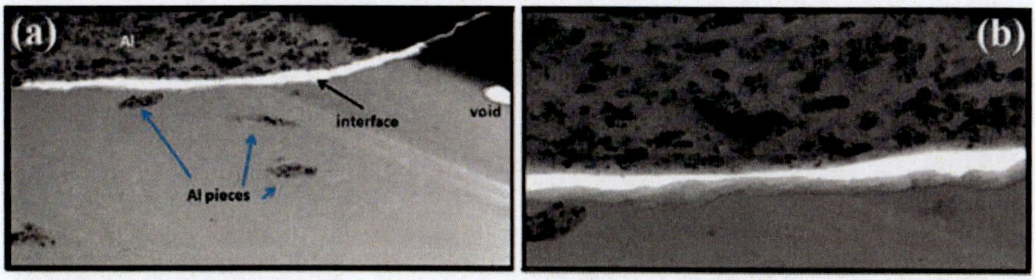

Fig. 15 **a, b** STEM-BF images from the Al/polymer interfaces and the trapped Al pieces in the polymer side in the mixing zone [25]

TEM (Transmission Electron Microscopy) and XRD tests were performed. XRD (X-ray Powder Diffraction) of SZ indicated the presence of Al, Mg and $C_{46}H_{94}$ but this does not confirm the presence of any new phases. TEM analysis shows the presence of a 30 nm interface layer between Al and HDPE. The calculated Fast Fourier Transform (FFT) diffraction pattern implies that the interfacial region doesn't contain Al_2OC [27], Al_4O_4C [2] and $Al_3O_{3.5}C_{0.5}$ [28]. However, the spacing and angles match with Al_2O_3 (corundum). So, there might be a chance of aluminium oxide's presence in the interface layer.

Khodabakhshi et al. [25] prepared a detailed study on the nature of bonding in Al and polymer using structural characterization analysis like Scanning Transmission Electron Microscopy (STEM), EDXS, High-Resolution Transmission Electron Microscopy (HR-TEM), Fast Fourier Transform (FFT) and Optical Microscope. Presence of aluminium fragments in SZ and their entrapment in the polymer was verified by optical microscopy. Small rod-shaped impurities were detected, with diameter and length of 75 nm and 250 nm, respectively. Grain was refined to a size of less than 100 nm. In the vicinity of Al-HDPE interface, bubbles and gaps were formed. It was observed that as the heat input decreases the void formation also decreases. At high temperatures, there was the formation of a large number of bubbles. Researchers focused on trapped Al particles in HDPE, to understand bonding nature and new phase development (Fig. 15). An interface layer of ~30 nm was realized, where chemical interaction might have been possible. As HDPE has no functional group, there might be a possibility of Van der Waals bonding in the form of aluminium oxides. Similar to Khodabakhshi et al. [2], the authors concluded the chances of formation of corundum. The nature of bonding was found to be mechanical interlocking and surface bonding. By decreasing traverse speed or increasing rotational speed, reduction in Al fragments can be brought. But this can lead to more void formation and non-uniform mixing.

Rout et al. [26] observed mixing and mechanical interlocking between Al and PP from macro images. From optical microscopy, it was noted that the weld was free from defects like voids, cracks and porosities. Presence of metal particles in the polymer matrix was confirmed from SEM and TEM analysis. Such presence is attributed to the partially molten phase of polymer and plasticized Al at the same temperature during welding. Using Bright Field HR-TEM, the interface of Al-polymer was observed,

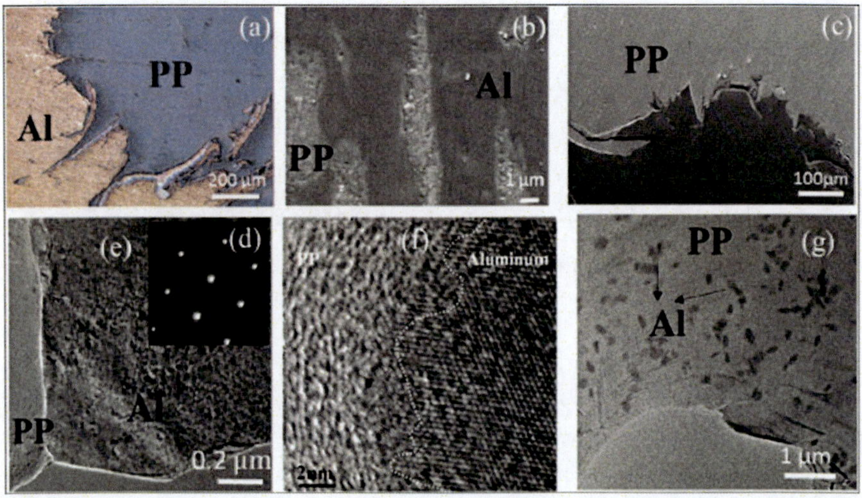

Fig. 16 **a** The interlocking of Al and PP observed under the optical microscope, **b** SEM micrograph of polished samples with bright parts indicating polymer, **c** SEM micrograph showing the hooked region, **d** SAED pattern corresponding a plane of Al, **e** TEM micrograph showing the continuous interface of Al and PP, **f** HRTEM image revealing Al, PP, and interface, **g** Entrapment of Al in the PP matrix is observed [26]

which indicated hook-like morphologies. To understand the type of bond formation, Fourier Transform Infrared Spectroscopy (FTIR) was performed, but no new peaks were observed, confirming that the joint formation is only due to mechanical interlocking. (Fig. 16). Like Khodabakhshi [25], no conclusive evidence of the presence of aluminium oxide was found in XRD and FTIR.

Rout et al. [26] used molecular simulations to better understand the molecular interaction and verify experimental results. The conditions of the experiment are mimicked using heating, cooling, annealing and other protocol processes. Hook like formations and absence of any bonds between C and Al were some of the observations drawn from simulations, which are in line with the experimental data. Using molecular dynamics calculations, a theoretical stress–strain curve (Fig. 17) was plotted, which showed the failure during the tensile test was due to complete failure in Al-PP interface and detachment. Although the exact values of experimental and theoretical don't match, the material interaction and the mode of failure were similar and could be used for a better understanding at the molecular level.

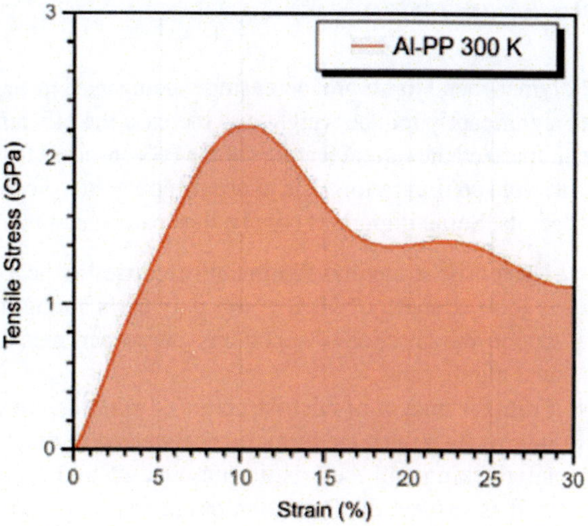

Fig. 17 The stress–strain curve plotted using simulations, for a tensile procedure in the Al-PP structure at 300 K [26]

6 Future Scope and Conclusion

6.1 Research Gap

Friction Stir Welding is an effective method for joint formation for dissimilar materials, polymer-metal, due to its solid-state characteristic. Although, some studies have been conducted on butt weld formation in polymer-metal and some primary results have been achieved, but still there seems to be scope for further study in this prospect, to better understand the bond formation and expanding its application to new fields of engineering. Some areas that need to be addressed in future work:

- Effect of parameters of the system: No pattern or conclusions have been drawn on the effect of parameters (such as tool tilt angle, tool-tip profile, tool offsets, feed rate and spindle speed) on mechanical and microstructural properties of the weld.
- Effect of tool design and geometry: Elements of the tool, like shoulder diameter, pin geometry, including its size and shape, and the nature of tool surface, affects the force and torque variations, heat input and plasticized material flow in FSW system.
- Thermal and material flow behaviour: Due to a large difference in mechanical, chemical and thermal properties of these dissimilar materials, understanding the thermal and material behaviour during weld could be complex. Modelling it could help control excessive heat and could also help decide the correct offset.
- Different metals: Most of the research work is limited to aluminium alloys and similar work could be considered to be expanded to other metals which have a relatively low melting point. But the selection of metal not only relies on its properties but also the application of joint.

6.2 Conclusion

Polymer-metal joints are increasingly being used in automobile and aerospace sector, to significantly reduce weight and increase the efficiency of vehicles. This work has summarized the salient features of the FSW process for polymer-metal and focuses on butt weld configuration. The paper demonstrates the methodology and the properties of joints. Some important conclusions can be summarized as follows:

- The tensile strength is significantly reduced at the joint and the best result obtained is joint strength of 50% of the polymer's strength. The reduction in strength is due to the absence of secondary phase particles and defects like kissing bonds and micro-voids.
- From the images of microstructure of joints and interface of Al-polymer, it can be inferred that the bond formation between dissimilar materials is due to mechanical interlocking and there is no presence of molecular interaction. Although some studies suspect the formation of secondary phase particles, there is no evidence of the presence of such bonds.
- There was a loss in hardness in HAZ and TMAZ and an increase was observed at the joint in case of HDPE and a decrease in case of PC. The increase in hardness is attributed to the presence of Al particles in the polymer matrix. Loss of hardness in Al is because of grain elongation.
- Higher temperatures were observed at distance in Al because of its high thermal conductivity. At peak temperature maximum volume of Al is transported to the polymer.
- There is further scope of study on the effect of parameters, microstructure studies, thermal and material behaviour and extending the process to other metals.

Acknowledgements The authors would like to thank the Student Research Program of Office of Research & Sponsored Program, Pandit Deendayal Petroleum University (PDPU), Gandhinagar, India for providing necessary support and infrastructure to carry out this work under project No.: ORSP/R&D/SRP/2019-20/1454/61.

References

1. Cho JH, Jae Kim W, Gil Lee C (2014) Texture and microstructure evolution and mechanical properties during friction stir welding of extruded aluminum billets. Mater Sci Eng A 597:314–323. https://doi.org/10.1016/j.msea.2013.12.087
2. Khodabakhshi F, Haghshenas M, Sahraeinejad S, Chen J, Shalchi B, Li J, Gerlich AP (2014) Microstructure-property characterization of a friction-stir welded joint between AA5059 aluminum alloy and high density polyethylene. Mater Charact 98:73–82. https://doi.org/10.1016/j.matchar.2014.10.013
3. Grujicic M, Sellappan V, Omar MA, Seyr N, Obieglo A, Erdmann M, Holzleitner J (2008) An overview of the polymer-to-metal direct-adhesion hybrid technologies for load-bearing

automotive components. J Mater Process Technol 197:363–373. https://doi.org/10.1016/j.jmatprotec.2007.06.058
4. Feraboli P, Masini A (2004) Development of carbon/epoxy structural components for a high performance vehicle. Compos Part B Eng 35:323–330. https://doi.org/10.1016/j.compositesb.2003.11.010
5. Amanat N, James NL, McKenzie DR (2010) Welding methods for joining thermoplastic polymers for the hermetic enclosure of medical devices. Med Eng Phys 32:690–699. https://doi.org/10.1016/j.medengphy.2010.04.011
6. Goushegir SM (2015) Friction spot joining of metal-composite hybrid structures. 123
7. Amancio-Filho ST, Dos Santos JF (2009) Joining of polymers and polymer-metal hybrid structures: Recent developments and trends. Polym Eng Sci 49:1461–1476. https://doi.org/10.1002/pen.21424
8. Kah P, Suoranta R, Martikainen J, Magnus C (2014) Techniques for joining dissimilar materials: metals and polymers. Rev Adv Mater Sci 36:152–164
9. Costa J (2016) friction stir welding in dissimilar joints
10. Derazkola HA, Elyasi M (2018) The influence of process parameters in friction stir welding of Al-Mg alloy and polycarbonate. J Manuf Process 35:88–98. https://doi.org/10.1016/j.jmapro.2018.07.021
11. Derazkola HA, Kashiry Fard R, Khodabakhshi F (2017) Effects of processing parameters on the characteristics of dissimilar friction-stir-welded joints between AA5058 aluminum alloy and PMMA polymer. Weld World 62:117–130. https://doi.org/10.1007/s40194-017-0517-y
12. Liu FC, Liao J, Nakata K (2014) Joining of metal to plastic using friction lap welding. Mater Des 54:236–244. https://doi.org/10.1016/j.matdes.2013.08.056
13. Ratanathavorn W, Melander A (2015) Dissimilar joining between aluminium alloy (AA 6111) and thermoplastics using friction stir welding. Sci Technol Weld Join 20:222–228. https://doi.org/10.1179/1362171814Y.0000000276
14. Shahmiri H, Movahedi M, Kokabi AH (2017) Friction stir lap joining of aluminium alloy to polypropylene sheets. Sci Technol Weld Join 22:120–126. https://doi.org/10.1080/13621718.2016.1204171
15. André NM, Goushegir SM, Dos Santos JF, Canto LB, Amancio-Filho ST (2016) Friction Spot Joining of aluminum alloy 2024–T3 and carbon-fiber-reinforced poly(phenylene sulfide) laminate with additional PPS film interlayer: Microstructure, mechanical strength and failure mechanisms. Compos Part B Eng 94:197–208. https://doi.org/10.1016/j.compositesb.2016.03.011
16. Esteves JV, Goushegir SM, dos Santos JF, Canto LB, Hage E, Amancio-Filho ST (2015) Friction spot joining of aluminum AA6181-T4 and carbon fiber-reinforced poly(phenylene sulfide): Effects of process parameters on the microstructure and mechanical strength. Mater Des 66:437–445. https://doi.org/10.1016/j.matdes.2014.06.070
17. Yusof F, Miyashita Y, Seo N, Mutoh Y, Moshwan R (2012) Utilising friction spot joining for dissimilar joint between aluminium alloy (A5052) and polyethylene terephthalate. Sci Technol Weld Join 17:544–549. https://doi.org/10.1179/136217112x13408696326530
18. Yusof F, bin Muhamad MR, Moshwan R, bin Jamaludin MF, Miyashita Y (2016) Effect of surface states on joining mechanisms and mechanical properties of aluminum alloy (A5052) and Polyethylene Terephthalate (PET) by dissimilar friction spot welding. Metals (Basel) 6. https://doi.org/10.3390/met6050101
19. Ghosh M, Kumar K, Kailas SV, Ray AK (2010) Optimization of friction stir welding parameters for dissimilar aluminum alloys. Mater Des 31:3033–3037. https://doi.org/10.1016/j.matdes.2010.01.028
20. Bozkurt Y (2012) The optimization of friction stir welding process parameters to achieve maximum tensile strength in polyethylene sheets. Mater Des 35:440–445. https://doi.org/10.1016/j.matdes.2011.09.008
21. Elangovan K, Balasubramanian V (2008) Influences of tool pin profile and tool shoulder diameter on the formation of friction stir processing zone in AA6061 aluminium alloy. Mater Des 29:362–373. https://doi.org/10.1016/j.matdes.2007.01.030

22. Rahmat SM, Hamdi M, Yusof F, Moshwan R (2014) Preliminary study on the feasibility of friction stir welding in 7075 aluminium alloy and polycarbonate sheet. Mater Res Innov 18:S6-515–S6-519. https://doi.org/10.1179/1432891714Z.0000000001035
23. Sahu SK, Panda M, Mahto RP, Pal SK, Pal K, Dash P (2017) Experimental investigation to join Al 6063 alloy to polypropylene using friction stir welding. In: 10th international conference on precision, meso, micro nano engineering (COPEN 10), pp 727–731
24. Moshwan R, Rahmat SM, Yusof F, Hassan MA, Hamdi M, Fadzil M (2015) Dissimilar friction stir welding between polycarbonate and AA 7075 aluminum alloy. Int J Mater Res 106:258–266. https://doi.org/10.3139/146.111172
25. Khodabakhshi F, Haghshenas M, Chen J, Shalchi Amirkhiz B, Li J, Gerlich AP (2017) Bonding mechanism and interface characterisation during dissimilar friction stir welding of an aluminium/polymer bi-material joint. Sci Technol Weld Join 22:182–190. https://doi.org/10.1080/13621718.2016.1211583
26. Rout A, Pandey P, Oliveira EF, da Silva Autreto PA, Gumaste A, Singh A, Galvão DS, Arora A, Tiwary CS (2019) Atomically locked interfaces of metal (aluminum) and polymer (polypropylene) using mechanical friction. Polymer (Guildf) 169:148–153. https://doi.org/10.1016/j.polymer.2019.02.049
27. Grass VE, Ryabkov YI, Goldin BA, Sitnikov PA (2004) X-Ray diffraction study of aluminum monoxycarbide abstract. J Struct Chem 45:100–106
28. Asaka T, Kotani R, Kudo T, Yoshida H, Fukuda K (2013) Synthesis and disordered crystal structure of $Al_3O_{3.5}C_{0.5}$. Inorg Chem 52:2648–2653. https://doi.org/10.1021/ic302674g

Effect of Process Parameters on Tensile Strength in FSW of Aluminium and Stainless Steel

Niraj Kumar, Dhrupal Kotadiya, Vishvesh J. Badheka, and Vijay S. Gadakh

1 Introduction

Presently, different types of metals, ceramics and compounds can be joined together but still there are many unsolved issues in the joining due to their different thermal, mechanical and structural properties. On the contrary, there is an increasing demand towards the use of dissimilar joints in shipbuilding, military vehicles, aerospace and automobile industries. The industry acceptable sound joint is a major concern where the joint quality has given more priority than other concerns. Honda Accord used dissimilar joint of Aluminum and steel where they achieved weight reduction with increased fuel efficiency [1]. To date, many works are related to join dissimilar materials using different joining processes but the joint is a big challenge due to the brittle intermetallics formation [2]. As mentioned joining of Aluminium and steel is complex due to different thermal properties, dissimilar thermal expansion, heat capacity and thermal conductivity, lattice transformation, large difference between the melting points (660 °C for Al alloy and 1497 °C for steel) and nearly zero solid

N. Kumar (✉) · D. Kotadiya
Department of Mechanical Engineering, Silver Oak College of Engineering and Technology, Ahmedabad, Gujarat 382481, India
e-mail: nk365624@gmail.com

D. Kotadiya
e-mail: dhrupalkotadiya.me@socet.edu.in

V. J. Badheka
Department of Mechanical Engineering, Pandit Deendayal Petroleum University, Gandhinagar, Gujarat 382330, India
e-mail: vishvesh.badheka@spt.pdpu.ac.in

V. S. Gadakh
Department of Production Engineering, Amrutvahini College of Engineering, Ahmednagar, SPPU, Pune, Maharashtra 422608, India
e-mail: vijay.gadakh@avcoe.org

H. K. Dave and D. Nedelcu (eds.), *Advances in Manufacturing Processes*, Lecture Notes in Mechanical Engineering, https://doi.org/10.1007/978-981-15-9117-4_18

solubility of iron in aluminium are creating the large discrepancy between the metals causing distortion, formation of the cavities and cracks, leading to the reduction of the mechanical properties after the joining processes [3]. Laser roll bonding [4], friction welding [5], FSW [6–10], laser brazing/welding [11], and laser pulse welding [12] are the processes employed till date to join different grades of the steels to the aluminum alloys.

The objective of the present work is to find process window to get defect-free welds and to study the effect of process parameters namely tool rotational speed (TRS) and weld speed (WS) on the FSWed joint strength of aluminium alloy (AA 6061) and austenitic stainless steel (AISI 304). Literature on FSW of these material combinations was reported. However, Bang et al. [13] studied conventional FSW and GTAW assisted FSW by considering effect of TRS on mechanical properties and microstructure of the joints. Similarly, Harwani and Badheka [14] investigated the effect of tool shoulder size on the mechanical properties and microstructure of the joints. Ghosh et al. [15] joined these materials using different TRS. Bang and Bijoy [16] have done thermal studies of TIG assisted FSW using three-dimensional finite element analysis. Ogawa et al. [17] have studied residual stress measurements for these weld materials. The motivation behind to do FSW experiments on these materials is that reported literature lacks to make process windows to get defect-free welds by considering effect of individual process parameters on the mechanical properties.

2 Experimental Procedure

A 100 × 50 × 3 mm thick plate of AA 6061-T651 and AISI 304L was used as a workpiece material for FSW. The chemical composition of AA6061-T651 is 1.03% Mg, 0.56% Si and 0.12% Mn and AISI 304 is 18.78% Cr, 8.2% Ni and 1.69% Mn [14]. The tensile strength of 668 and 302 MPa was found for AISI 304 and Al 6061-T651, respectively. Al 6061-T651 temper is solution heat-treated artificially aged where the aluminium material is extruded and given 1–3% stretching, which relives the internal residual stresses. The experiments were conducted modified vertical milling machine shown in Fig. 1 at Welding Research Laboratory of School of Technology, PDPU, Gandhinagar. The tilt angle (TA) (2°), Tool offset (2 mm) towards Aluminium on advancing side (AS), plunging depth (0.1 mm), and dwell time of 30 s were kept constant.

The values of the FSW process parameters that have varied during the execution of experiments are shown in Table 1. Based on previous work [14] and pilot experimentation the working range was identified.

During experiments, a tool of an unthreaded tungsten carbide alloy tool (88% WC-12% Co by weight) is employed with flat conical shoulder having shoulder diameter of 18 mm, root and tip pin diameter as 5 mm and 3 mm, and pin length of 2.8 mm is shown in Fig. 2 [14]. The tool material was procured from Sinter Sud Pvt. Ltd., Italy and was given cryogenic heat treatment after machining for enhancing

Fig. 1 Modified vertical milling machine employed for FSW

Table 1 Weld parameters and their levels

Factors	Level 1	Level 2	Level 3
TRS (rpm)	545	765	1070
WS (mm/rev)	20	31.5	50
TA (°)	1	2	3

Fig. 2 FSW tool

the hardness. The hardness of the tool was 92.8 HRC, and the torsional resistance was about 2100 N/mm^2. A stainless steel fixture (having dimensions 230 × 195 × 25 mm) was specially designed for the FSW of Al plates [14], as shown in Fig. 3a.

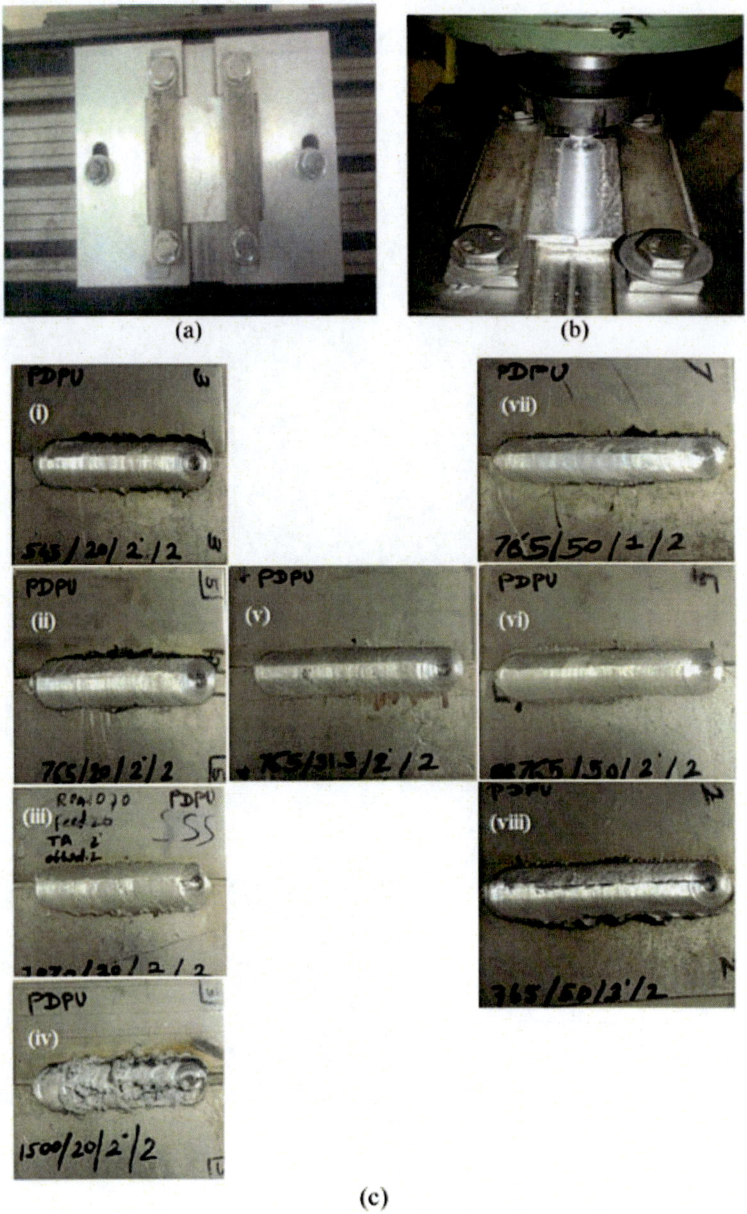

Fig. 3 **a** FSW Fixture, **b** experimentation during welding and **c** welded samples

Figure 3b, c shows the experimental setup for joining Aluminum and steel alongwith experimentation during welding and FSWed joints at different process conditions. To measure the temperature during the process, two holes were drilled 2 mm away from the nugget zone and two K-type contact thermocouples were inserted in the holes based on past experience, one on the advancing and one on the retreating side (RS). Three tensile specimens were prepared from a joint as per ASTM E8/E8M-11 standard and were tested at Electrical Research & Development Association (ERDA), Baroda with the help of a well-calibrated Universal Testing Machine.

3 Results and Discussion

In order to study the effect of individual process parameter on the joint tensile strength, bottom-up approach [18] was employed by exploring the Table 1. Firstly, TRS was varied keeping WS (20 mm/min) and TA (2°) constant (see Fig. 3ci–iv). Later WS was varied keeping TRS (765 rpm) and TA (2°) constant (see Fig. 3cv, vi). Lastly, TA was varied keeping TRS (765 rpm) and WS (50 mm/min) constant (see Fig. 3cvii, viii) A total eight number of experiments were performed, out of which three conditions showed defective joints (see Fig. 3civ, vii and viii) hence their mechanical properties were not evaluated and not shown in Table 2. The defective joints were produced due to high heat input. The details of process parameters and their corresponding tensile strength are shown in Table 2. All the joints were failed in the weld zone during tensile testing.

3.1 *Effect of Tool Rotational Speed on Tensile Strength*

The effect of tool rotational speed on tensile strength is shown in Fig. 4. The maximum joint tensile strength was observed at TRS of 765 rpm WS of 20 mm/min. This can be attributed due to the sufficient heat input in the weld region [13, 15, 19]. Furthermore, welding defects were observed at a higher value of TRS of 1070 rpm due to excessive

Table 2 Tensile strength and peak temperature of Al-Steel FSWed joint

S. No	TRS	WS	TS (MPa)			Avg. TS (MPa)	Peak temp. (°C)	
			Trial 1	Trial 2	Trial 3		RS (steel)	AS (Al)
1	545	20	22.514	16.785	44.647	27.982	328	194
2	765	20	35.13	31.752	56.151	41.011	345	197
3	765	31.5	30.178	42.315	34.115	35.536	298	170
4	765	50	24.457	11.391	9.716	15.188	244	188
5	1070	20	13.102	13.102	21.349	15.851	328	178

Fig. 4 Effect of tool rotational speed on FSWed Al-Steel joint

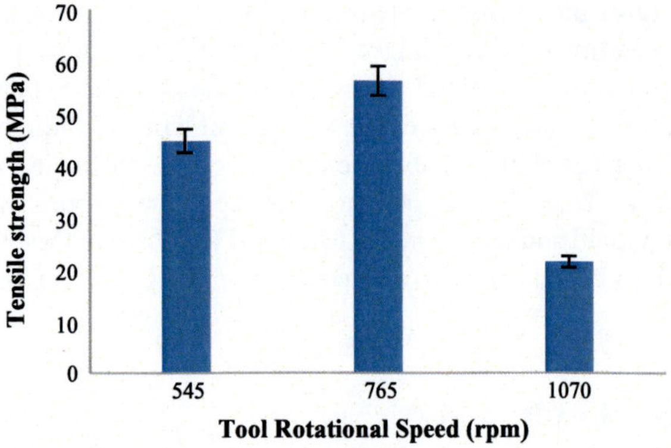

softening of the weld materials. These welding defects are observed due to improper materials flow and mixing [20].

3.2 *Effect of Weld Speed on Tensile Strength*

The weld speed has significant role in the mechanical properties of FSW of Al-steel joints. From Fig. 5, it has been found that increasing the weld speed decreases the tensile strength of the joint. Increasing the weld speed decreases the heat input in which insufficient intermixing of the material take place and thereby weakens the joint.

Fig. 5 Effect of weld speed on FSWed Al-Steel joint

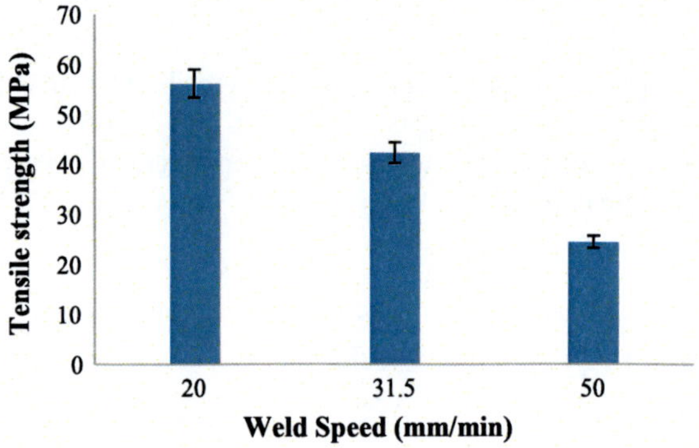

3.3 *Effect of Tool Rotational Speed and Weld Speed on Peak Temperature*

The effect of tool rotational speed and weld speed on peak temperature of Al-Steel FSWed joint is shown in Table 2. The maximum measured temperature for the aluminium alloy and steel are 345 °C and 197 °C, respectively at constant tool rotational speed and varying weld speed. A similar observation was reported by Chen and Kovacevic [21] in Al-steel joining. Also, increasing weld speed reduces the peak temperature on steel side, but on aluminium side first decrease and then increased. This may be attributed due to availability of time for heat conduction to the aluminium is reduced. It is estimated that the heat input into the steel is ~2 times that of the aluminium alloy at a tool rotational speed of 765 rpm and a weld speed of 20 mm/min by considering thermal properties of the materials. Similarly, results were obtained at constant weld speed and varying tool rotational speed. This may be attributed due to the mechanical mixing and deformation action of the tool pin and an increase in the frictional heat generated under the shoulder. Figure 6 shows the thermal history at TRS 765 rpm and WS 20 mm/min. It can be seen that time to attain maximum temperature above 200 °C is ~30 s whereas for cooling it is ~90 s. The temperature greater than 200 °C was considered due to Al typically recrystallizes between 200 and 300 °C range. The peak temperature and cooling rate, are the most essential parameters at any specific location in the joint during welding, which not only affect the microstructure but also the mechanical properties of welded joint [22].

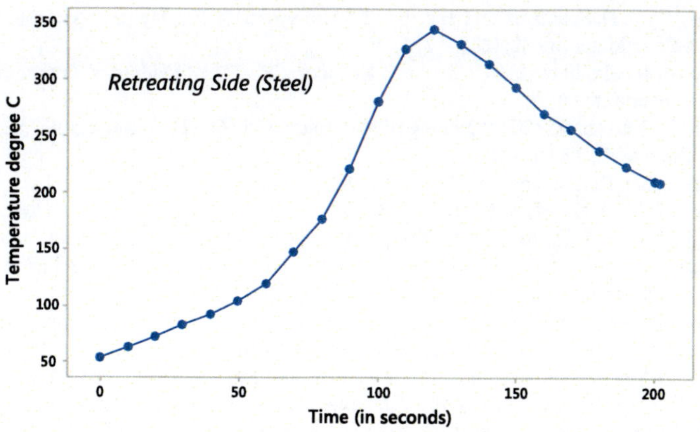

Fig. 6 Temperature history at TRS 765 rpm and WS 20 mm/min

4 Conclusion

In this work, FSW of dissimilar Aluminium (AA6061-T651) alloy and Stainless Steel (SS 304) material was successfully welded. The tool rotational speed and weld speed have a significant effect on tensile strength of the FSWed joint. From the obtained results, it is observed that at TRS of 765 rpm and 20 mm/min, highest tensile strength 41 MPa was obtained. Increasing the weld speed at constant tool rotational speed the tensile strength decreases. Whereas, at constant weld speed and varying tool rotational speed, the tensile strength increases from low to high tool rotational speeds then decreases. Tool rotational speed and weld speed are major FSW process parameters that decides the heat input and thereby the structure and properties of the joints. Microstructural characterization and their correlation with the mechanical properties of FSW of these joints will be carried out as future work.

Acknowledgements The authors would like to thank the authorities of Pandit Deendayal Petroleum University, Gandhinagar for providing the facilities to carry out this work.

References

1. Kusuda Y (2013) Honda develops robotized FSW technology to weld steel and aluminum and applied it to a mass-production vehicle. Ind Rob 40:208–212. https://doi.org/10.1108/01439911311309889
2. Agudo L, Eyidi D, Schmaranzer CH et al (2007) Intermetallic FexAly-phases in a steel/Al-alloy fusion weld. J Mater Sci 42:4205–4214
3. Atabaki MM, Nikodinovski M, Chenier P, Kovacevic R (2013) Welding of aluminum alloys to steels: an overview
4. Rathod MJ, Kutsuna M (2004) Joining of aluminum alloy 5052 and low-carbon steel by laser roll welding. Weld J 83:16–26
5. Shubhavardhan RN, Surendran S (2012) Friction welding to join stainless steel and aluminum materials. Int J Metall Mater Sci Eng 2:53–73
6. Ramachandran KK, Murugan N (2019) Influence of axial force on tensile strength and microstructural characteristics of friction stir buttwelded aluminum alloy/steel joints. Strength Mater 51:300–316. https://doi.org/10.1007/s11223-019-00076-7
7. Dehghani M, Mousavi SAAA, Amadeh A (2013) Effects of welding parameters and tool geometry on properties of 3003-H18 aluminum alloy to mild steel friction stir weld. Trans Nonferrous Met Soc China (English Ed 23). https://doi.org/10.1016/S1003-6326(13)62683-7
8. Tang J, Shen Y (2017) Effects of preheating treatment on temperature distribution and material flow of aluminum alloy and steel friction stir welds. J Manuf Process 29. https://doi.org/10.1016/j.jmapro.2017.07.005
9. Zheng Q, Feng X, Shen Y et al (2016) Dissimilar friction stir welding of 6061 Al to 316 stainless steel using Zn as a filler metal. J Alloys Compd 686:693–701. https://doi.org/10.1016/j.jallcom.2016.06.092
10. Sameer MD, Birru AK (2019) Investigations on microstructural evolutions and mechanical properties of dual-phase 600 steel and AA6082-T6 aluminum alloy dissimilar joints fabricated by friction stir welding. Trans Indian Inst Met 72:353–367. https://doi.org/10.1007/s12666-018-1487-5

11. Mathieu A, Shabadi R, Deschamps A et al (2007) Dissimilar material joining using laser (aluminum to steel using zinc-based filler wire). Opt Laser Technol 39:652–661. https://doi.org/10.1016/j.optlastec.2005.08.014
12. Torkamany MJ, Tahamtan S, Sabbaghzadeh J (2010) Dissimilar welding of carbon steel to 5754 aluminum alloy by Nd:YAG pulsed laser. Mater Des 31:458–465. https://doi.org/10.1016/j.matdes.2009.05.046
13. Bang H, Bang H, Jeon G et al (2012) Gas tungsten arc welding assisted hybrid friction stir welding of dissimilar materials Al6061-T6 aluminum alloy and STS304 stainless steel. Mater Des 37:48–55. https://doi.org/10.1016/j.matdes.2011.12.018
14. Harwani DM, Badheka VJ (2019) Effect of shoulder diameter on friction stir welding of Al 6061 to SS 304. In: Deb D, Balas VE, Dey R (eds) Innovations in infrastructure proceedings of ICIIF 2018, advances in intelligent systems and computing. Springer, Singapore, pp 355–366
15. Ghosh M, Gupta RK, Husain MM (2014) Friction stir welding of stainless steel to Al alloy: effect of thermal condition on weld nugget microstructure. Metall Mater Trans a 45A:854–863. https://doi.org/10.1007/s11661-013-2036-9
16. Bang H, Bijoy MS (2011) Temperature behavior in dissimilar butt joint during TIG assisted friction stir welding. J Korean Weld Join Soc 29:63–71. https://doi.org/10.5781/KWJS.2011.29.5.561
17. Ogawa D, Kakiuchi T, Hashiba K, Uematsu Y (2019) Residual stress measurement of Al/steel dissimilar friction stir weld. Sci Technol Weld Join 24:685–694. https://doi.org/10.1080/13621718.2019.1588521
18. Nadammal N, Kailas SV, Suwas S (2015) A bottom-up approach for optimization of friction stir processing parameters: a study on aluminium 2024-T3 alloy. Mater Des 65:127–138. https://doi.org/10.1016/j.matdes.2014.09.005
19. Mahto RP, Kumar R, Pal SK, Panda SK (2018) A comprehensive study on force, temperature, mechanical properties and micro-structural characterizations in friction stir lap welding of dissimilar materials (AA6061-T6 & AISI304). J Manuf Process 31:624–639. https://doi.org/10.1016/j.jmapro.2017.12.017
20. Ranjan R, Khan AR, Parikh C, et al (2016) Classification and identification of surface defects in friction stir welding: an image processing approach. J Manuf Process 22. https://doi.org/10.1016/j.jmapro.2016.03.009
21. Chen CM, Kovacevic RÃ (2004) Joining of Al 6061 alloy to AISI 1018 steel by combined effects of fusion and solid state welding. 44:1205–1214. https://doi.org/10.1016/j.ijmachtools.2004.03.011
22. Ramanjaneyulu K, Madhusudhan Reddy G, Venugopal Rao A, Markandeya R (2013) Structure-property correlation of AA2014 friction stir welds: role of tool pin profile. J Mater Eng Perform 22:2224–2240. https://doi.org/10.1007/s11665-013-0512-4

Study of Friction Stir Welding on Aerospace Grade ZE41AMg Alloy and Its Comparison with Laser Beam Welding on ZE41AMg Alloy

Adithyan Annamalai, T. R. Kishore Babu, S. Karthikeyan, N. Siddharth, and S. Muralidharan

1 Introduction

Magnesium, an element besides having copious presence on the surface of the Earth has unbounded supplies in the ocean beds. Magnesium and its alloys are approximately 40% less heavy than aluminum and 78% insubstantial than steel. Magnesium in its molten state can burn in air, hence difficult to cast and weld. Furthermore, in the case of magnesium, melting and solidification eventuating during classical welding processes instigates high porosity and contamination in the weld zone thereby limiting the joint performance. However, vast heat-affected zones (HAZ), fusion zone areas (FZ), high shrinkages, dissimilitude in microstructures, dissimilitude in mechanical properties, high residual stresses and distortion phenomenon in the arc-welded joints have attracted this undeniable attention towards high energy density laser welding and solid-state friction stir welding techniques [1].

Cao et al. [2] investigated the two principal types of industrial lasers, CO_2 and Nd: YAG, in influencing weldability of magnesium alloys. Laser welded joints with minimal cracks, very low porosity and enhanced surface quality were obtained for some magnesium alloys using appropriate laser processing parameters. Cao et al. [3] studied the effects of welding speeds on the weld soundness of ZE41A-T5 magnesium sand castings and concluded that the penetration depth, average width of the weld and weld zone area decreased on escalating the welding speed. Also, a notable grain refinement was observed in the FZ as a result of higher cooling rate. Singh et al. [4] suggested that a pin with 18 mm shoulder diameter, threaded cylindrical profile, 710 rpm rotational speed and 28 mm/min welding speed when employed for welding Mg alloys will give strong welds with high tensile strength. Silva et al. [5] investigated the effects of travel speed (200 mm/min) and rotation speed (1200 rpm)

A. Annamalai (✉) · T. R. K. Babu · S. Karthikeyan · N. Siddharth · S. Muralidharan
Thiagarajar College of Engineering, Madurai, Tamil Nadu 625015, India
e-mail: adithyanannamalai@gmail.com

H. K. Dave and D. Nedelcu (eds.), *Advances in Manufacturing Processes*, Lecture Notes in Mechanical Engineering, https://doi.org/10.1007/978-981-15-9117-4_19

on the residual stress and microstructure of friction stir welded ZK60-RE alloy. Asadi et al. [6] enumerated an alternative method to ameliorate the classical fusion welding technique. Using this technique, the metallurgical defects for magnesium and its alloys can be drastically reduced. Laser welded joints with minimal cracks, less porosity and better surface quality welds were obtained through laser welding. Carlone et al. [7] observed the microstructural characteristics in FSW and Tungsten Inert Gas (TIG) welding techniques. In TIG welding weld, small globular grains in fusion zone and coarse grains in the HAZ were observed. In FSW, Thermo Mechanically Affected Zone (TMAZ) and HAZ exhibited distorted grains. With T5 heat treatment, uniform hardness distribution was seen. Mishra et al. [8] addressed the recent progress in process modelling, microstructural characteristics, material related issues and the potential practical applications of friction stir welding. Mishra et al. [9] studied the microstructural and mechanical characteristics of friction stir welded ZE40 and concluded that the metallurgical aspects, bond strength and weld quality depend directly on mechanical stirring and penultimately the newly emerged grains. The grains formed were ascended by the rotational speed of the tool during welding. Zhao et al. [10] observed that a phased escalation in porosity in the fusion zone regardless of the prevailing conditions. The amalgamation and inflation of small pre-existing pores as a result of heating and curtailment of internal pressure contributed to the surge in porosity in the fusion zone. Weisheit et al. [11] delineated that the prime surface quality in combination with the deepest penetration depths could be accomplished by using helium inert gas. Welds with moderate porosity could be achieved by employing appropriate laser parameters. A speed of 2.5–3 m/min and 1.5 kW laser power seemed perfect conditions for welding thin plates (approximately 2.5 and 3 mm). A higher power (2–2.5 kW) and a lower speed (1–2 m/min) were quintessential for welding thicker plates.

In this research work, the effects of the various process parameters and tool geometry with special profile tools on the FSW of ZE41A Mg alloy are studied. Correlating the mechanical property of hardness with process parameters and tool geometry optimized weld parameters were concluded. A study on the laser beam welding on ZE41AMg alloy was done by [1]. A comparison on the micro-hardness results of Friction Stir Welded ZE41A Mg alloy with laser beam welded ZE41AMg alloy samples were made.The microstructural behaviour of laser beam welded specimen is analyzed through SEM analysis to have a deeper insight into the characterisation of various zones.

2 Experimental Proceedings

2.1 Selection of Material

Magnesium alloys possess superior impact resistance and unique damping capacity. They are labelled using the conventions prescribed by American Society for Testing and Materials (ASTM) and Society of Automotive Engineers (SAE). The former half comprises of the two centric alloying elements in the alloy and the latter half denotes their percentage. ZE41A is a magnesium alloy encompassing zirconium and zinc. The ZE41A is suitable for high casting applications as proffers favourable welding properties and is also pressure tight. It is manufactured by casting processes similar to other alloys using traditional techniques. The ZE41A alloy consists of 93.997% Mg, 4.2% Zn, 0.603% Zr and 1.2% Ce.

2.2 *Visual Examination and Metal Cutting*

The material was visually inspected with the naked eye to check for any flaw or crack in the surface of magnesium alloy. The required dimension (82 × 77 × 75) mm was set and it was using wire cut Electro Discharge Machining (EDM) and Bandsaw machine into the required number of pieces. Surface grinding was done to remove the surface defects. Finishing was done to the edges thereby removing the burr in the magnesium alloy. The specimen surface was cleared so that the undesired material was removed and the edges were finished off to remove any burr in it. Figures 1 and 2 shows the band saw cutting machine and wire cut EDM machine image which was captured while cutting ZE41A metal. The cut pieces were shown in Fig. 3.

Fig. 1 Band saw cutting machine

Fig. 2 EDM wire cutting

Fig. 3 ZE41A cut pieces

2.3 Selection Parameters

The primitive parameters considered for welding were welding traverse speed and tool rotational speed. Using this, four welds were carried out. Table 1 shows the different welding parameters like traverse speed and rotational speed.

Table 1 Parameters chosen for our experiment

Number of welds	Transverse speed (mm/min)	Rotational speed (rpm)
1	20	1100
2	20	1200
3	25	1100
4	25	1200

Other parameters like Tilt Angle—2°, Axial Force—8 kN, Shoulder Plunge Depth—0.2 mm, D(Shoulder width)/d(Diagonal dimension of the square portion) Ratio—18/6—3, Dwell Time—2 Min and Tool Material—H13 Tool Steel was considered as fixed parameters. Tool geometry—Square pin with concave shoulder.

2.4 Tool Design

Friction stir welding tool was modelled using CREO parametric software as depicted in Fig. 4. H13 tool steel was selected and machined to the required dimensions with pin profiles made using lathe which is shown in Fig. 5. Square pin geometry having four edges improves the stirring action.

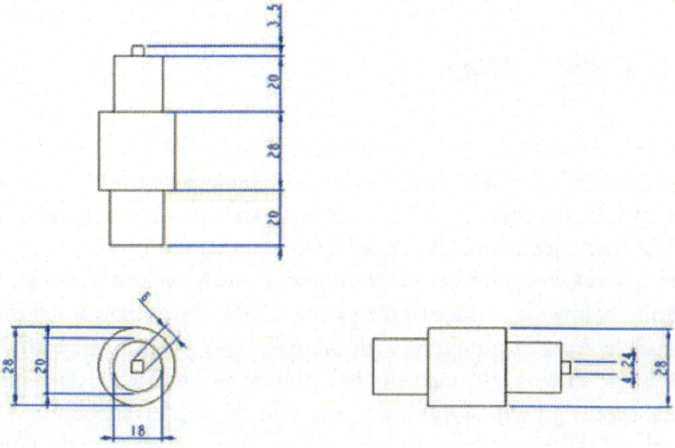

Fig. 4 Square tool drawing

Fig. 5 Square tool photo

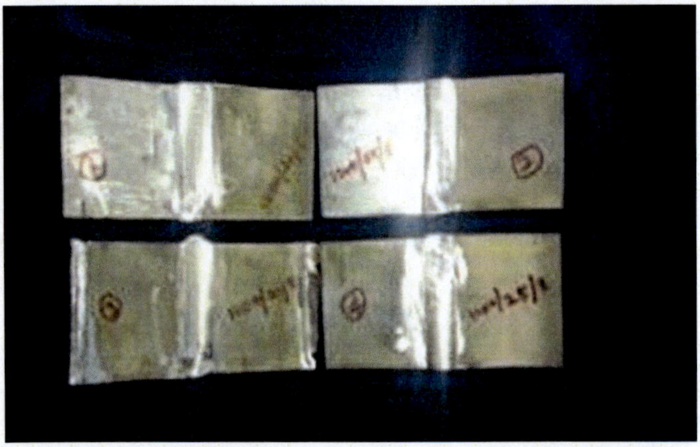

Fig. 6 Friction stir welded specimens

2.5 Friction Stir Welding

Friction stir welding of ZE41A plates was carried on LMW KODI-40 FSW machine to weld a butt joint of dimensions 150 × 75 mm. Two weld traverse speeds, 20 mm/min and 25 mm/min were coupled with tool rotation speeds of 1100 and 1200 rpm. The four specimens after welding are shown in Fig. 6.

The welding tools were plunged at the center line up to an instance where shoulder touched 0.2 mm below the surface of the plates. The tool was then advanced forward to execute the weld. After plunging with the FSW tool, a dwell time of 2 min was provided to enable stirring and ensuring fine flow of tool and material. This provides sufficient heat for the growth of fine sized grains in the welded specimen. The engendering butt joints (150 mm × 75 mm) were lacerated into strips with the help of EDM providing sufficient width normal to the welding line to perform hardness tests on the specimens.

2.6 Hardness Test

The welded sample was cut along the cross-section and etched to differentiate the various thermally affected zones within the weld. Those regions are Weld zone (WZ), Thermo Mechanical Affected Zone (TMAZ) and Unaffected Zone (or) the base metal. Vickers hardness test was performed to measure the hardness of these welded specimens. The hardness testing was done as per ASTM E384:2016 standard. Vickers hardness was measured in Commando lab—Coimbatore using a Hardness tester machine [BS 50(SPL)] bearing a capacity of 50 kg.

2.7 *Microstructural Analysis*

The molecular grain structure in the weld region plays an important role in analyzing the metallurgical properties of the weld. For determining these properties, we used a metallurgical microscope to characterize the weld zone, heat affected zone and the base metal for a particular welded specimen. The microstructure analysis was done for maximum power and minimum power at minimum and maximum speeds.

3 Experimental Proceedings

3.1 *Mechanical Analysis of Friction Stir Welding*

At 1100RPM rotational speed and 20 mm/min weld speed, the average hardness value at weld zone, TMAZ and base metal are 54HV0.5,68HV0.5 and 72HV0.5, respectively as shown in Fig. 7.

Similarly, at 1100 rpm and at 25 mm/min, the hardness variation in weld zone, TMAZ and base metal were observed to be 0.5–54, 0.5–70 and 0.5–76, respectively. This implies that when traverse speed is increased, the hardness value also increases proportionately as shown in Fig. 8.

At 1200 rpm and at 25 mm/min the hardness variation in weld zone, TMAZ, and base metal was 0.5–53, 0.5–63, and 0.5–68 respectively. This shows that as the rotational speed is increased, the hardness values decreased commensurately. To conclude, when rotational speed and traverse speeds are increased, the hardness

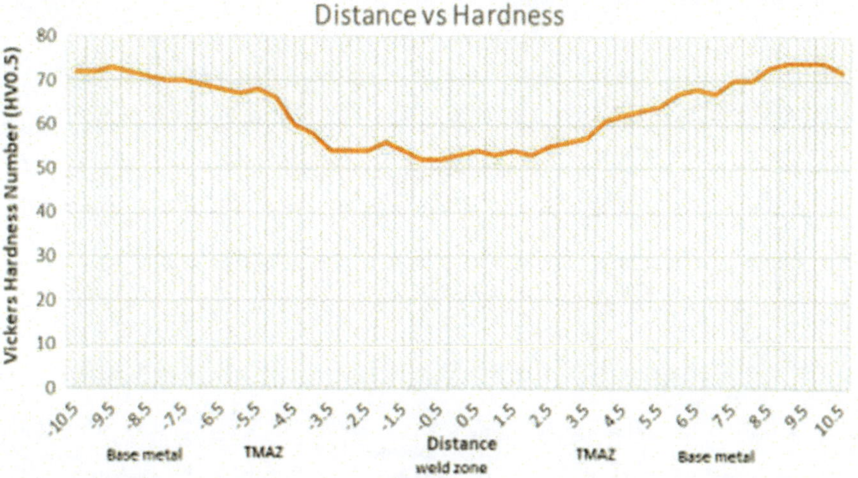

Fig. 7 Hardness plot for 1100 RPM and 20 mm/min

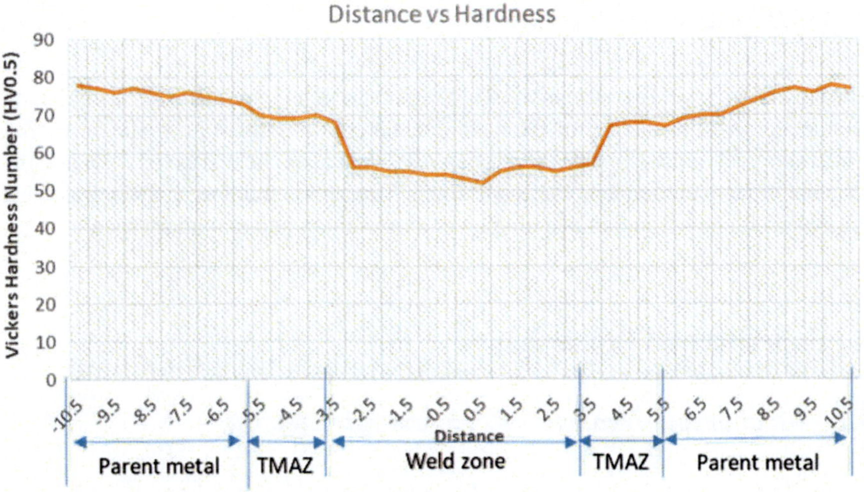

Fig. 8 Hardness plot for 1100 RPM and 25 mm/min

values get decreased in all zones revealing the fact that mechanical stirring plays an important role in determining the hardness. This phenomenon is depicted graphically in Fig. 9. A similar graphical trend was observed at 1200 RPM rotational speed and 20 mm/min traverse speed.

On the whole, it can be concluded that when a rotational speed of 1100 RPM and weld speed of 20 mm/min was employed, uniform hardness was seen throughout the weld zone thus proving it to be the optimized parameter. As the hardness is uniform in the weld region, the chances of fracture due to brittleness is very scarce.

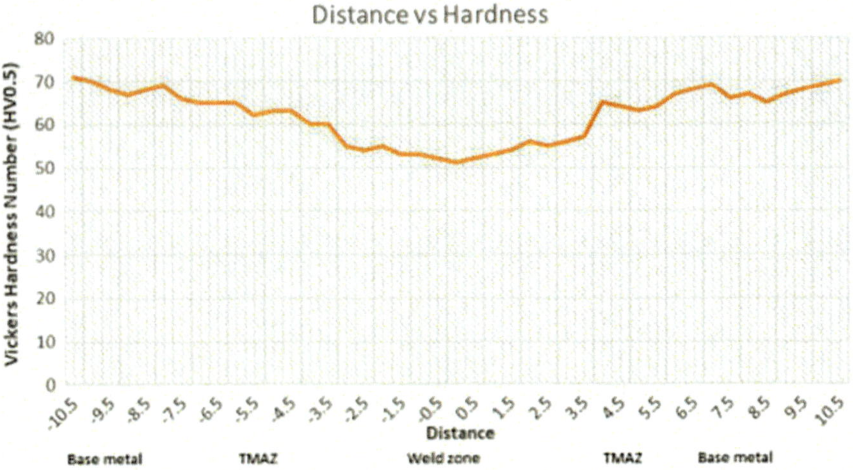

Fig. 9 Hardness plot for 1200 RPM and 25 mm/min

3.2 Laser Beam Welding Results

Vickers hardness test was taken to perceive the hardness of the welded specimen. The test was taken for all the samples and grouped in accordance to the welding power 1, 1.5 and 2 kW respectively shown in Table 2, and with each 0.1 mm distance from either side of the weld center and the graphs have been plotted as shown in Figs. 10, 11 and 12.

From the above graphs we can conclude that by altering the process parameters like welding power and welding speed, the variations in the weld zone can be seen

Table 2 Hardness variation plot

Power (kW)	Speed (mm/s)	Hardness (BHN)		
		Base metal	HAZ	Weld zone
1	1000	55.5	77.7	70.3
	1500	53.9	74.3	67.5
	2000	54.6	71.6	65.2
1.5	1000	53.9	82.8	73
	1500	55.5	80.2	71.6
	2000	53.9	78.5	69.2
2	1000	54.2	90.8	78.3
	1500	53.9	87.5	75.5
	2000	53.4	84.6	72.6

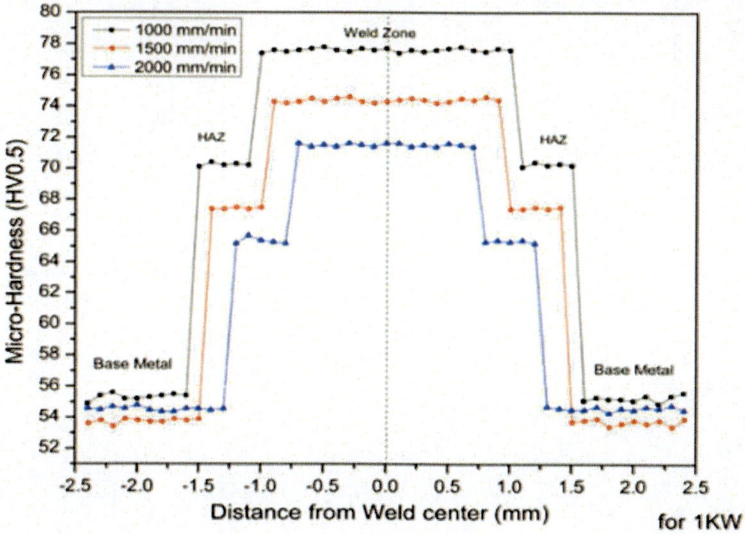

Fig. 10 Hardness variation for 1 kW

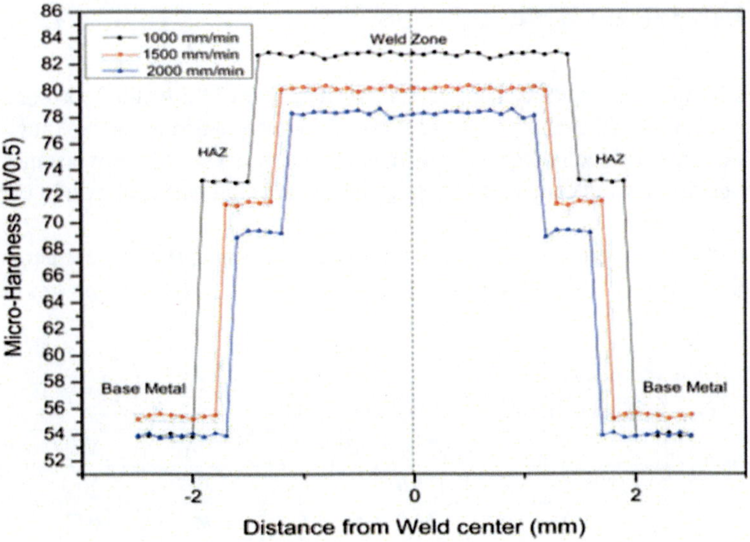

Fig. 11 Hardness variation for 1.5 kW

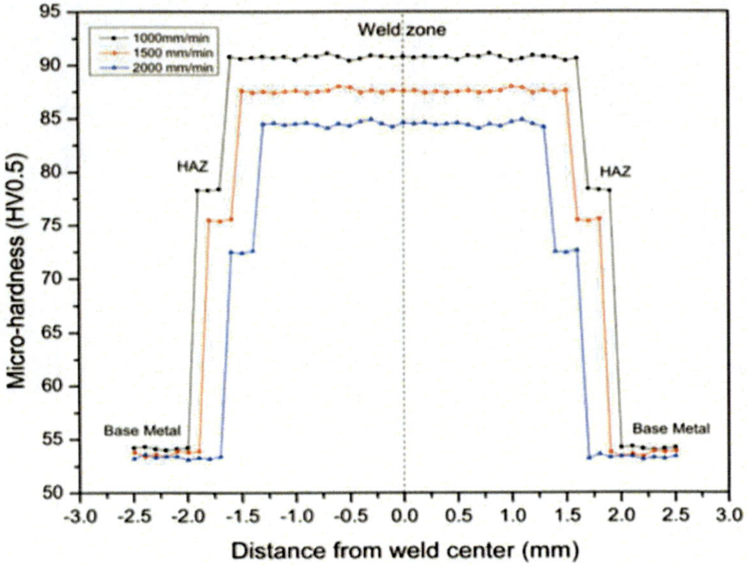

Fig. 12 Hardness variation for 2 kW

profoundly. When the welding power is increased, a wider weld zone and HAZ can be observed. But on the other hand, on increasing the welding speed, the weld zone and HAZ becomes narrower significantly. By comparing and analyzing, we can conclude that the weld with the parameters 1 kW, 1000 mm/min give us weld with minimum defects.

3.3 *Microstructural Analysis of Laser Beam Welding*

The sample corresponding to 1 kW, 1000 mm/min details clearly the shift from the weld zone to heat-affected zone (HAZ) and finally to the base metal. The weld zone possesses a very fine grain size owing to increased hardness resulting from the welding processes whereas the laser welding has a narrow HAZ made visible by certain sporadic distribution of grains on the surface of the base metal, i.e. the HAZ possess a mixture of both fine grain and coarse grain structures. The base metal is easily visible as they are very coarse grain in both the Figs. 13 and 14.

In specimen 2 with welding power of 2 kW and 2000 mm/min, a formation of pattern is seen in the HAZ. The weld region is made up of fine sized grains and the base metal is filled with coarse grains. But the HAZ forms fine grains in a root

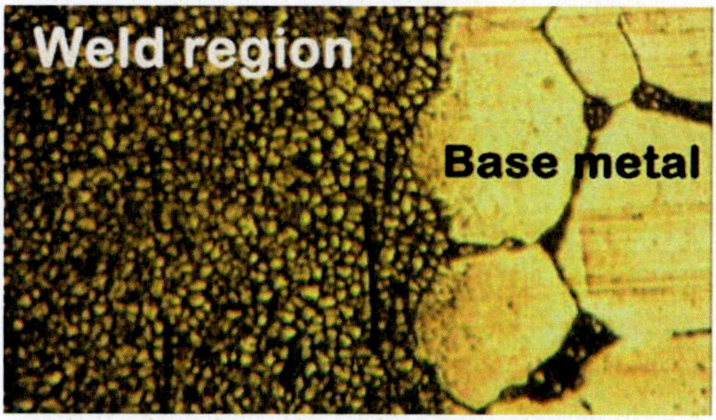

Fig. 13 Weld zone at 1 kW, 1000 mm/min in 100x magnification

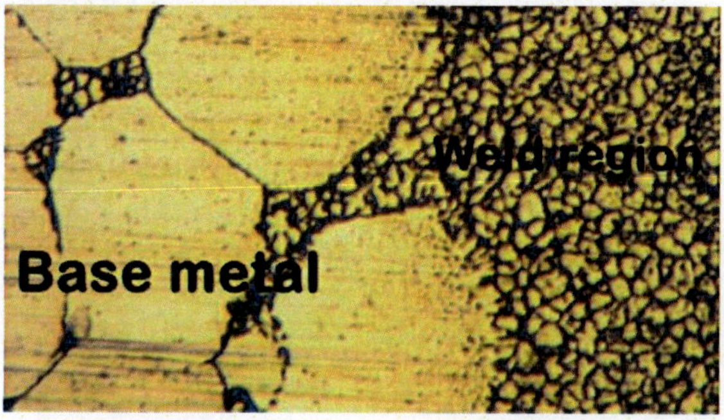

Fig. 14 Weld zone at 2 kW, 2000 mm/min in 100× magnification

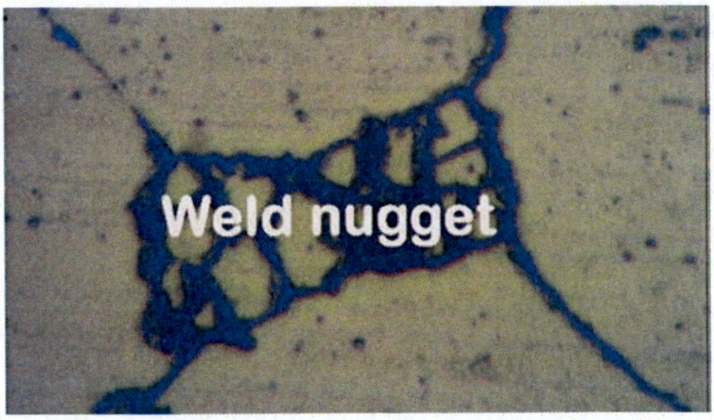

Fig. 15 Weld zone at 2 kW, 2000 mm/min in 400× magnification

shape orientation over the coarse grains. The high welding power of the laser makes a significant impact on the formation of the grain partition between the weld zone and the base metal.

In Fig. 15, a pocket of fine grains is trapped inside the coarse grain cluster. This is a continuation of the root shape orientation emerging from the weld zone to the base metal and is magnified 400 times to find out the methodology involved in this happening. This has duly occurred as a result of the solidification defect. This analysis shows us the dominance of welding power and speed in the formation of HAZ in a weld, i.e. when the welding power decreased, the HAZ is narrow and evenly formed avoiding the risk of hotspots near the weld that lead to a crack.

4 Conclusion

From the experiments performed and the figures presented above, the important inferences could be surmised as follows:

- Two different parameters, the tool rotational speed (1100 and 1200 rpm) and traverse speed (20 and 25 mm/min) that majorly have an effect on friction stir welding have been chosen to produce four different welded specimens.
- Sound welds were produced in all the FSW samples by fixing these parameters.
- The mechanical property—microhardness study has been done on all these specimens to study the hardness distribution and it is found that the hardness is significantly decreased at the weld zone.
- Average Hardness value in FSW follows a decreasing trend with escalation in rotational speed of the tool which reveals the fact that intense stirring action makes the material soft. This decrease in hardness is due to
- When both rotational speed and traverse speed are increased, the hardness value gets decreased.
- Uniform hardness distribution is observed in 1100 rpm at 20 mm/min which is the optimized parameter amidst all the other parameters chosen.

- It is found that in laser beam welding, by increasing the welding power, a wider weld zone and HAZ can be observed. Higher laser power substantially provides higher heat input, forming a wider weld zone. Also, the oxidation process occurring in the weld zone results in a higher hardness.
- By comparing and analyzing the results of hardness tests of laser beam welding, we can conclude that the weld with the parameters 1KW, 1000 mm/min give us a desirable weld with minimum defects.
- In both welding processes, hardness distribution is found to be uniform. But some defects were encountered in laser beam welded specimens at the weld zone due to the melting. On the contrary, in friction stir welding, the specimens are subjected to the plastic zone which makes it free from defects.
- By comparing both welding processes, the friction stir welding process is found to be defect-free and can be easily automated without providing much strain on the fixture design.

5 Experimental Proceedings

- Tensile test and fractography test can be taken to study the nature of fracturing, i.e. ductile or brittle.
- X-ray diffraction could be done for studying the precipitates present in the welded zone.
- Scanning electron microscope (SEM) can be performed to study the grain structure.
- Energy Dispersive X-Ray Spectroscopy (EDX) shall be performed to study the compositions present in the welded zones.
- Electro Back Scattered diffraction (EBSD) can be done to get deeper insight into the grain boundaries and grain orientation that eventually influence the mechanical properties.

References

1. Marimuthu P, Dinesh Babu P, Ram Prabhu T (2020) Laser welding of ZE41 Mg alloy: experimental investigations on the effect of parameters and nondestructive testing. Trans Indian Inst Met 73:1587–1593
2. Cao X, Xiao M, Jahazi M, Immarigeon JP (2005) Continuous wave ND: YAG laser welding of sand-cast ZE41A-T5 magnesium alloys. Mater Manuf Processes 20(6):987–1004
3. Cao X, Jahazi M, Xiao M, Medraj M (2005) Effect of welding speed on Nd:YAG laser weldability of ZE41A-T5 magnesium sand castings. Paper presented at 44th annual conference of metallurgists of CIM, Calgary, Canada, 21–24 Aug 2005
4. Singh K, Singh G, Singh H (2018) Review on friction stir welding of magnesium alloys. J Magnesium Alloys 6(4):399–416

5. Silva Da, Oliveira VB, Pereira VF, Maluf O, Buzolin RH, Pinto HC (2017) Microstructure and residual stresses in a friction stir welded butt joint of as-cast ZK60 alloy containing rare earths. Mater Res 20(3):775–779
6. Asadi, Kazemi-Choobi K, Elhami A (2012) Welding of magnesium alloys. New features on magnesium alloys, vol 6. InTech, Croatia, pp 121–159
7. Carlone P, Astarita A, Rubino F, Pasquino N (2016) Microstructural aspects in FSW and TIG welding of cast ZE41A magnesium alloy. Metallur Mater Trans B 47(2):1340–1346
8. Mishra RS, Ma ZY (2005) Friction stir welding and processing. Mater Sci Eng R Rep 50(1–2):1–78
9. Sameer MD, Manzoor Hussain (2014) Microstructure and Mechanical properties of friction stir welded Ze40 magnesium alloy. Paper presented at international conference on advance research and innovations in mechanical, material science, industrial engineering and management (ICARMMIEM-2014), KIT, 10–12 Jan 2014
10. Zhao H, Debroy T (2001) Pore formation during laser beam welding of die-cast magnesium alloy AM60B—mechanism and remedy. Weld J 80(8):204–210
11. Weisheit A, Galun R, Mordike BL (1998) CO_2 laser beam welding of magnesium-based alloys. Weld J 77(4):149–154

Influence of Layer Thickness, Infill Rate and Orientation on Thermal and Structural Loading of FDM Parts

Andrei-Danut Mazurchevici, Ramona-Iuliana Popa, Constantin Carausu, Simona-Nicoleta Mazurchevici, and Dumitru Nedelcu

1 Introduction

Additive manufacturing (AM) represents a modern manufacturing technology that allows realization of a physical prototype based on a three-dimensional CAD model, without involving other traditional operations such as turning and milling.

The most popular and widely used additive manufacturing technologies include Fused Deposition Modeling (FDM), selective laser sintering (SLS) and stereolithography (SLA), each of these having both strengths and weaknesses. However, all of these technologies are closely related to the initial setting of an increased number of process parameters (layer thickness, working temperature, filing speed and others) as opposed to conventional manufacturing technologies [1].

According to the technical literature, the basic mechanical characteristics (tensile strength, bending resistance and impact resistance) of the materials used in additive manufacturing, in the case of parts that use the same design and material, but are achieved by using process parameters that have different values, can have very large differences in mechanical responses, even by several hundred percent [1, 2].

A.-D. Mazurchevici · R.-I. Popa · C. Carausu · S.-N. Mazurchevici · D. Nedelcu (✉)
Department of Machine Manufacturing Technology, "Gheorghe Asachi" Technical University of Iasi, Str. Prof. Dr. Doc. D. Mangeron, No. 59A, 700050 Iasi, Romania
e-mail: dnedelcu@tuiasi.ro

A.-D. Mazurchevici
e-mail: andrei0maz@yahoo.com

R.-I. Popa
e-mail: popa_ramona83@yahoo.com

C. Carausu
e-mail: c_carausu@yahoo.com

D. Nedelcu
Academy of Romanian Scientists, Str. Ilfov, Nr. 3, Sector 5, Bucharest, Romania

H. K. Dave and D. Nedelcu (eds.), *Advances in Manufacturing Processes*, Lecture Notes in Mechanical Engineering, https://doi.org/10.1007/978-981-15-9117-4_20

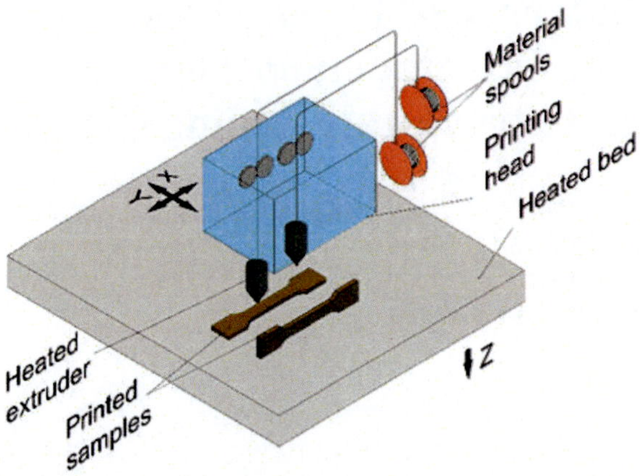

Fig. 1 Orientation of the part on the printing table

The FDM technology, used for prototyping the parts of this study, can be characterized by dimensional and shape accuracy, [3], technical indicators such as tensile, bending and impact strengths, respectively, economic indicators such as printing time and quantity of material used. This type of technology is significantly influenced by the sets of process parameters, which can be controlled by the equipment operator directly or indirectly. The process parameters that have the greatest impact on the characteristics of the part are the sample orientation on the printing table, which can be modified directly [4].

The sample orientation on the printing table can be considered as representing the angular difference between the basic plane of the prototyped object and the plane that determines the direction of the object's division into layers (Fig. 1). Deposition around the vertical direction (Z axis) does not show significant influence from the technical indicators point of view, as it does not favor in any way how the part is divided (sliced) into layers [1].

The dependence between the sample orientation during printing process and technical indicators is studied globally in order to identify and characterize the relationships between them. The purpose of the studies is to allow the control of the mechanical responses by setting the optimal values of the process parameters [1, 4].

One of the purposes of this paper is to determine the influence of the sample orientation on tensile strength for the biodegradable materials. The other purposes of the paper were to present the results of the simulations for the basic mechanical behavior (tensile and bending strength), results for heat transfer during the printing process and the deformations that occur during the printing process.

The experimental research was based on complete factorial plan. The sample orientation on the printing table was made on two different printing directions, "on the edge" and "flat". The other two parameters considered, deposited layer thickness and infill rate also varying on two levels.

2 Systemic Analysis of the 3D Printing Process

In order to identify the parameters involved in the AM process, a systemic analysis was performed, which tracked the influences between the input parameters, the hard controllable factors and the output parameters of the process, as shown in Fig. 2.

According to some studies in the technical literature [5–12], the input parameters that significantly influence the characteristics of the parts obtained by 3D printing are:

- *The number of outer layers (shell)* has a pretty big influence on the tensile strength of the samples that do not have full filling degree; whereas increasing the number of layers, the raster area becomes very narrow, the part being almost entirely made up of filaments continuously deposited from the edge towards the center of the model; it is behaving like a solid entity. This outer area has a tensile strength greater than the standard printed samples (outer layers—the model shell; inner layers—the infilling area), resulting in a tensile strength similar to samples with 100% infilling degree [5];

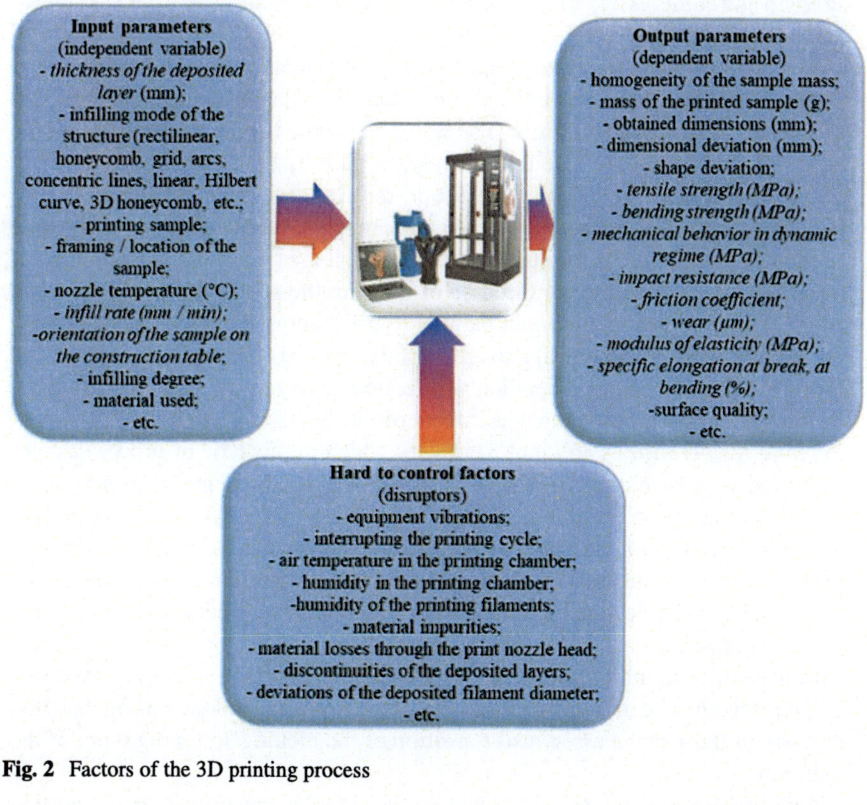

Fig. 2 Factors of the 3D printing process

- *Infilling degree* of the part directly influences the mechanical characteristics of the printed parts; thus, the lower the degree of infilling the lower the mechanical properties of the part will decrease [6];
- *Infilling type of the part*: Usually, the raster area cannot withstand tensions as well as the outer area/wall (shell) of the part, but depending on its complexity, it can save time and material. According to some simulations belonging to the present paper authors, it was found that regardless of the infilling type initially selected, if the infilling of the part is set to be complete, then the program will no longer take into account the selected infilling type, passing automatically to the rectilinear infilling type [6]. The complete infilling is used when it is desired to obtain a part that will be subjected to different mechanical test types (tensile, bending, impact) whereas the part will have a more compact, more homogeneous and implicitly more resistant structure. If the infilling is not complete, the honeycomb filling type can be considered the best model to maximize the tensile strength of 3D printed parts [5].
- *The thickness of the deposited layer* influences both the mechanical characteristics of the part and the surface; the thinner the layer is, the density of the part and the surface quality increases. Mechanical properties (tensile strength, bending strength and compressive strength) increase with decreasing layer thickness [7, 8].
- *Sample orientation*: According to our own studies and not only [9, 10], the most used directions of the part orientation on the printing table are "flat", "on edge" and "vertical". The weakest mechanical properties are obtained for the parts printed in a vertical direction, and the best ones are registered for the orientation "on the edge" of the part. In the case of the analyzed part, by the authors of the present paper, samples dumbbell type, specific to tensile tests, their thickness (4 mm) becomes the printing plane, the filling area is much reduced compared to the case of the flat orientation of the sample, and the shell area occupies a percentage of over 70% of its structure. In the case of printing the specimens placed "on the edge", the filaments are oriented parallel to the loading direction, thus carrying out tests with high mechanical strength. For the vertical orientation of the part, the layers are deposited in a perpendicular direction to the request direction (traction, bending), thus facilitating the detachment of the layers much faster.
- *Printing temperature* is an input parameter which in this type of process cannot be varied very much, maximum 10 °C in the case of biodegradable polymeric materials. Failure to observe the optimum printing temperature will either lead to a solidification of the material in the extrusion head, blocking the printing process or to the destruction/damage of the intermolecular bonds due to the overheating/carbonization of the material. For the correct selection/setting of the printing temperature, a study of the material thermal behavior can be performed, by differential scanning calorimetry (DSC) and differential thermal analysis (DTA), which reveals the transitions suffered by the material during heating, and one of these is the associated transition of the melting/softening point of the material.
- *Infilling rate*: a too high printing speed can lead to deformations of the deposited filament because when depositing a layer, the previously deposited thread fails to

fully solidify until the time of depositing a new filament. Thus, due to the weight and the holes specific to the selected type of filling, the filaments will be deformed. In the 3D printing process, the infilling rate has an impact on the height and width of the part formation, and both dimensions decrease with the infilling rate, but the influence on the width is higher than in the case of height [11, 12].

- *Material selection* plays a fundamental role in determining the part properties, but also on the "layer thickness" parameter. As an example, the minimum layer thickness for acrylonitrile butadiene styrene (ABS) thermoplastic is about 0.13 mm, while for polycarbonate (PC) it is 0.18 mm, which results in a direct influence on the quality of the part surface [13].

According to the mentioned, it is appreciated that the printing parameters that affect influence significantly the quality and functionality of a part used in applications involving various mechanical demands which are the part orientation, the layer thickness and the infilling rate. These printing parameters will also be taken into account when planning the experimental plan for this study.

3 Materials and Methods

Tensile tests were performed on specimens conforming to the ISO 527 standard. A complete factorial plan was used for these tests, which included three process parameters (layer thickness, infilling rate, part orientation on the printing table), each of these having two variation levels. The complete factorial plan, 2^3 (eight experiments), is presented in Table 1.

The materials used for samples printing are in the category of eco-friendly, biodegradable materials. For the study, a number of six such polymers were selected:

Table 1 Complete factorial experimental plan

Exp. No.	Input parameters		
	Layer thickness (mm)	Infilling rate (mm/min)	Part orientation on the printing table
1	0.1	40	Flat
2	0.1	40	On the edge
3	0.1	80	Flat
4	0.1	80	On the edge
5	0.2	40	Flat
6	0.2	40	On the edge
7	0.2	80	Flat
8	0.2	80	On the edge

- *Polylactic acid* (PLA) filament, produced by Raise3D, natural biodegradable polymer produced from starch by a direct condensation process. It is used in many industrial applications, having properties similar to those of synthetic polymers [14, 15].
- *HD PLA Green* filament, produced by Fiberlogy, product from renewable raw materials, being suitable for applications involving contact with food, in accordance with the European Union directive and Food and Drug Administration regulations, is also certified by the Restriction of Hazardous Substances Directive [16].
- *Impact PLA Gray* filament, produced by Fiberlogy, it offers properties comparable to those of the ABS filament while retaining the typical printing properties of the PLA. It has high impact resistance, thus ensuring an increased resistance to cracking, breaking and use at higher temperatures, and the obtained marks can be used in extreme conditions [16];
- *Extrudr BDP Pearl* filament, produced by Extrudr, a biodegradable polymer (according to DIN EN ISO 14,855), 100% from renewable resources, mechanically and thermally comparable with PLA [17, 18];
- *Extrudr BDP Flax* filament, produced by Extrudr, biodegradable polymeric material that allows a prototyping high speed, due to the mineral filler addition. It is suitable for rapid prototyping of complex models [17].
- *Fiber Wood* filament, produced by Fiberlogy, biodegradable that has natural wood compounds in its structure [16].

3D printing was done on the Raise3D Pro2 PLUS equipment (FDM process), within laboratory of Fine Mechanics and Nanotechnologies, "Gheorghe Asachi" Technical University of Iasi, Romania. The print nozzle diameter was 0.4 mm, and the samples were printed with three outer layers, 100% infilling degree and rectilinear infilling type. The extrusion and the printing bed temperatures for the used materials are shown below:

- *PLA*, print nozzle temperature 220 °C, printing table temperature 60 °C;
- *HD PLA Green*, print nozzle temperature 215 °C, printing table temperature 65 °C;
- *Impact PLA Gray*, print nozzle temperature 230 °C, printing table temperature 65 °C;
- *Extrudr BDP Pearl*, print nozzle temperature 180 °C, printing table temperature 55 °C;
- *Extrudr BDP Flax*, print nozzle temperature 190 °C, printing table temperature 55 °C;
- *Fiber Wood*, print nozzle temperature 200 °C, printing table temperature 55 °C.

The prototyped samples were subjected to uniaxial tensile test on Instron 3382 equipment, with a constant crosshead speed of 5 mm/min according to ISO 527–3: 2003 recommendation.

An universal test machine WTW 50 was used to determine the bending strength, using a load of 1kN and a load rate of 2 mm/min.

4 Results and Discussions

4.1 Results Regarding the Simulation of Mechanical and Thermal Tests

Modeling of 3D Printed Samples

In order to obtain 3D models of the samples used in various mechanical simulations, the SolidWorks modeling and simulation program was used. The models made are in accordance with the standards in force and shoulder as much as possible the physical structure of the model made by 3D printing, using the Fused Deposition Modeling (FDM) method. The realized models are represented in Fig. 3.

In the case of the specific sample to the tensile test, three variants of it were modeled, Figs. 3a–c, with the sample orientation in the horizontal plane. Also, the model with the orientation of the sample on "the edge" was considered, but taking into account the very small length of the deposited wire and the 4 mm thickness of the part, the 3D model could not be made due to the appearance of numerous model geometry errors, which made it impossible simulation.

The model from Fig. 3a best coincides with the sample physically printed by the FDM method, comprising three outer layers (rectilinear deposition) and twelve overlapping layers (grid-type filling, −45°/+45°). Taking into account the deformations that appear in the moment of filament deposition (its flattening), it was taken into account that in the section, the wire has an elliptical shape with a length of 0.52 mm and 0.37 mm width. The adhesion between the layers was achieved by the interpenetration of the filaments in all directions by 0.01 mm. This model could not be

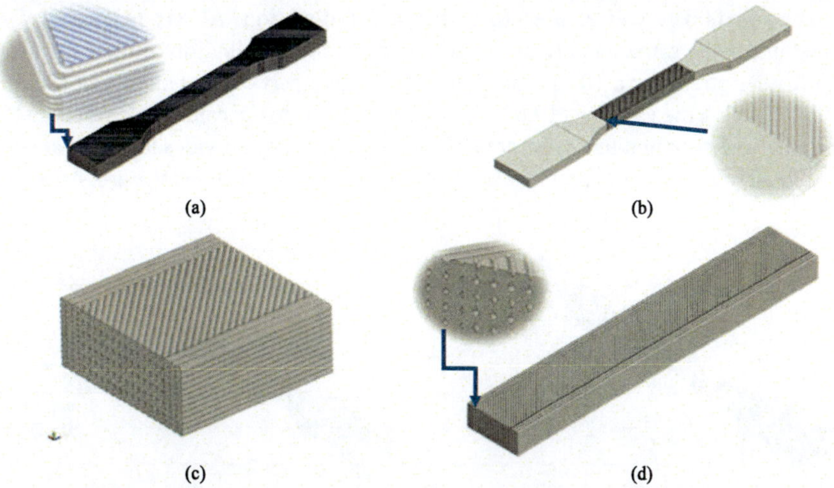

(a) (b) (c) (d)

Fig. 3 3D models: **a** complete tensile test sample; **b** sample with solid ends for tensile test; **c** simplified model of the tensile test sample; **d** bending test sample used in the simulation process due to its complexity but also due to the insufficient computing power of the workstation. In this situation, a new model was made with the test area (length 60 mm) that respects the printed physical model, but with solid ends, Fig. 3b. This model could be used to simulate the tensile test, but it showed many errors. The third model, Fig. 3c, is a simplified one and represents only a portion of the test area of the printed physical model having a length of 10 mm, 10 mm width and 4 mm thickness. This model could be used in the simulation of tensile testing of biodegradable polymer samples.

Figure 3d presents the three-dimensional model of the three-point bending test sample, with the following dimensions: length 60 mm, width 10 mm and 4 mm thickness. This model could be used in simulations

Tensile Test Simulations

The simulation of the tensile test on the sample model with solid ends and test area according to the printed model was performed with the SolidWorks software. The proposed model was discretized with a maximum element size of 0.4 mm resulting in a total number of 5,526,159 elements and 7,957,856 nodes. The discretization time was 8 min and 10 s. The sample was recessed at one end and at the other end was applied a force of 1000 N in the –Z direction.

The values obtained from the tensile test simulation, equivalent von Misses stress (773.35 MPa) and the maximum linear displacement (1.59 mm) can be seen in Fig. 4.

In the case of this simulation, it was observed that due to the model complexity, the elements maximum distortion is high (422.28). Due to this, the sample model has been simplified so that the simulation results do not be affected.

The simplified model was discretized with a maximum element size of 0.3 mm resulting in a total number of 546,867 elements and 851,789 nodes. The discretization time was 1 min and 27 s. The sample was recessed at one end, and at the other end, a force of 1000 N was applied to it in the −Z direction. All six materials were tested under the same conditions. Figure 5 shows the distribution of the equivalent von Misses stress for a 10 mm portion of the active part of the tensile test specific sample, having the value of 119.63 MPa and which due to the calculation relationship (ratio between applied force and area of force application) provides the same answer for all materials. The maximum value of the equivalent von Misses stress distribution

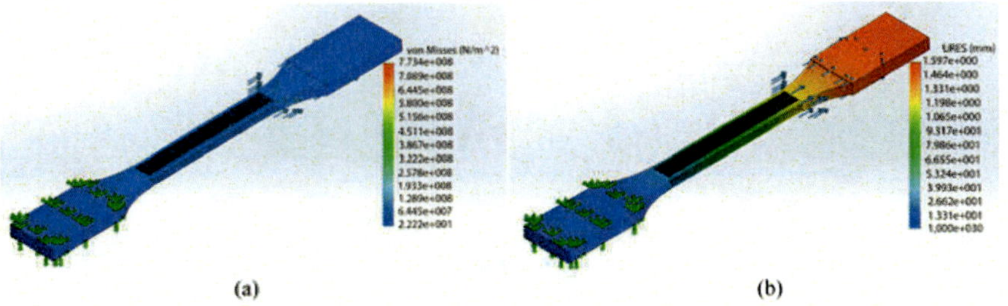

Fig. 4 Values resulting from the simulation: tensile test of the PLA sample: **a** equivalent von Misses stress; **b** maximum linear displacement

is in the middle of the simplified model, and the minimum is at the recessed end of the model.

The highest value of maximum linear displacement was recorded for Impact PLA Gray and Extrudr BDP Pearl materials, 1.86 mm (0.31 mm/simplified model × 6 lengths), Fig. 6 (for Extrudr BDP Pearl material), with the location of this value in the force application area, and the minimum of linear displacement in the sample recessed area.

For the equivalent strain, the highest value was recorded by the Extrudr BDP Flax material, Fig. 7; the equivalent strain is 0.120, for the entire active test area reaching

Fig. 5 Distribution of the equivalent von Misses stress for a portion of the active part of the tensile test specific sample

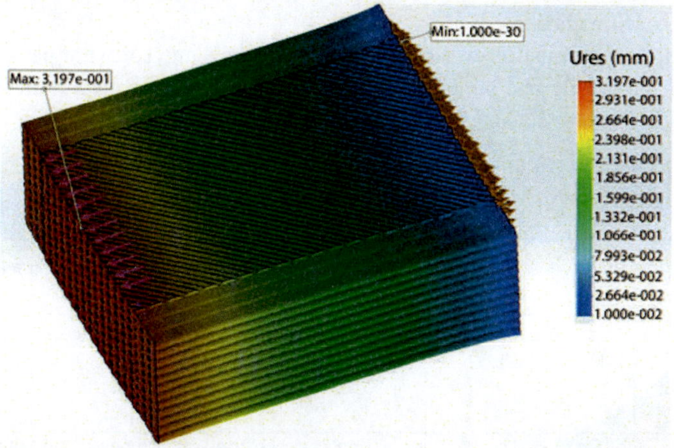

Fig. 6 Linear displacement resulted from the tensile test simulations for PLA material

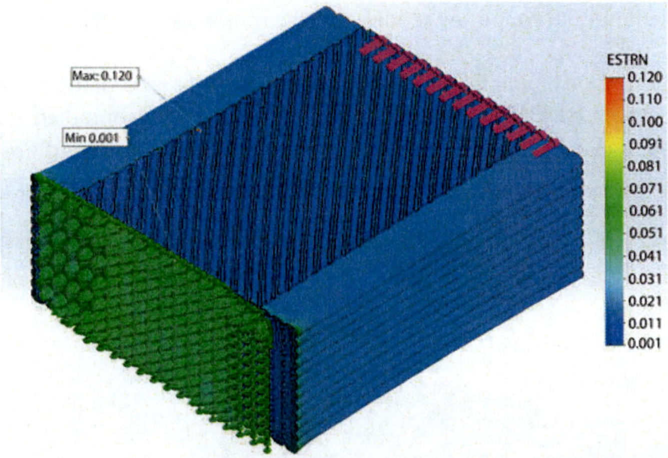

Fig. 7 Equivalent strain of Extrudr BDP Flax material

a value of 0.72. The maximum of this value is located in the tested model middle area, and the minimum is in the fixed surface vicinity.

In order to be able to find out the complete values of the maximum linear displacement and equivalent strain, it was necessary to multiply the result by six, thus reaching values for the entire test length of the standard sample, 60 mm. The results obtained from this calculation are highlighted in Table 2.

According to the obtained results, the materials that showed the most rigid behavior were Fiber Wood and Extrudr BDP Flax, and the most elastic–plastic behavior was highlighted by the Impact PLA Gray and Extrudr BDP Pearl materials.

Bending Test Simulations

The bending test simulation was performed using the same software package as in the case of the tensile test simulation. The materials used for the simulation were PLA, HD PLA Green, Impact PLA Gray, Extrudr BDP Flax, Extrudr BDP Pearl and Fiber Wood.

Table 2 Linear displacement and equivalent strain, results from tensile test simulations

Material characteristic	PLA	HD PLA Green	Impact PLA Gray	Extrudr BDP Flax	Extrudr BDP Pearl	Fiber Wood
Maximum linear displacement (mm)	0.24 × 6 = 1.44	0.27 × 6 = 1.62	0.31 × 6 = 1.86	0.22 × 6 = 1.32	0.31 × 6 = 1.86	0.21 × 6 = 1.26
Equivalent strain	0.065 × 6 = 0.39	0.073 × 6 = 0.43	0.082 × 6 = 0.49	0.120 × 6 = 0.72	0,084 × 6 = 0.50	0.056 × 6 = 0.36

The used model is the one presented in Fig. 3(d). It was discretized, with a maximum element size of 0.3 mm resulting in a total number of 3,373,752 elements, respectively, 5,212,939 nodes. The discretization time was 8 min and 27 s. The sample was recessed at both ends, and on the sample center, in a transverse direction, a linear force of 700 N was applied and distributed over the entire sample width. All six materials were tested under the same conditions.

As in the case of tensile test simulations, very close results were obtained for the distribution of the equivalent von Misses stress value. The maximum equivalent stress according to the von Misses criterion was recorded in the middle of the outer face [in the contact area of the shell (outer layers) with the raster (sample infill area)], where the tangential stress is maximum and has the value of 3668 MPa for the Impact PLA Gray material, Fig. 8. It is a value that exceeds the tensile stress of the material. The minimum value of the equivalent von Misses stress distribution is registered in the deposited filaments inflection zone, marked in Fig. 8. For the maximum displacement, different values were obtained with their maxims in the sample middle area, the area where the force was applied and minimums on the sample ends. The maximum displacement has the highest value for Extrudr BDP Pearl (18.74 mm), Fig. 9, and the lowest value for Fiber Wood (6.95 mm). The material that recorded the highest stiffness was Fiber Wood. According to the simulations, the areas located on the outer faces at the ends and the specimen middle are the most strained which corresponds to the theory, Fig. 10 (Extruder BDP Flax material). The maximum value of the displacement is a focus in node 253,668 where the equivalent strain is 0.35, and the minimum value is in the deposited filaments inflection area, in node 444,935, with the value of 0.00014. The areas of neutral fiber from the middle of the sample, along the axis, are the least strained. It is also observed that on the outer faces, the bending stress changes into the compression stress.

The obtained result values from the bending simulations for the studied biodegradable materials are presented in Table 3.

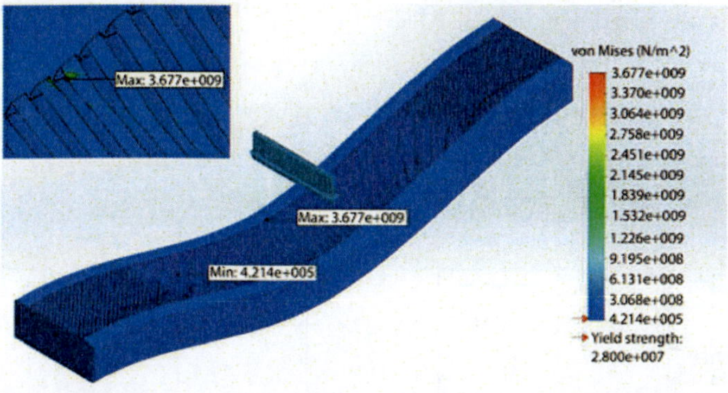

Fig. 8 Equivalent von Misses stress distribution for the Impact PLA Gray material

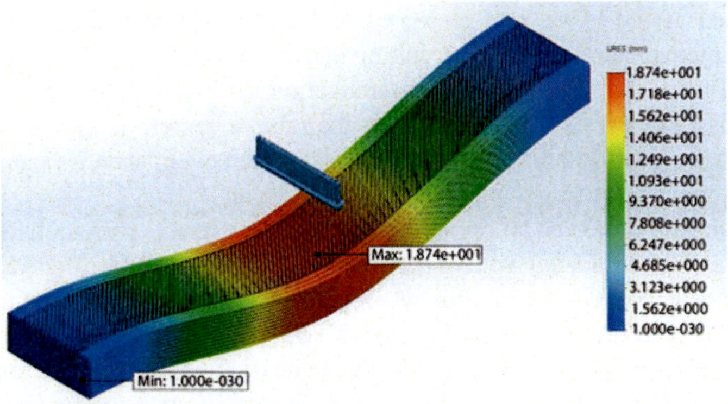

Fig. 9 Maximum displacement for Extruder BDP Pearl material

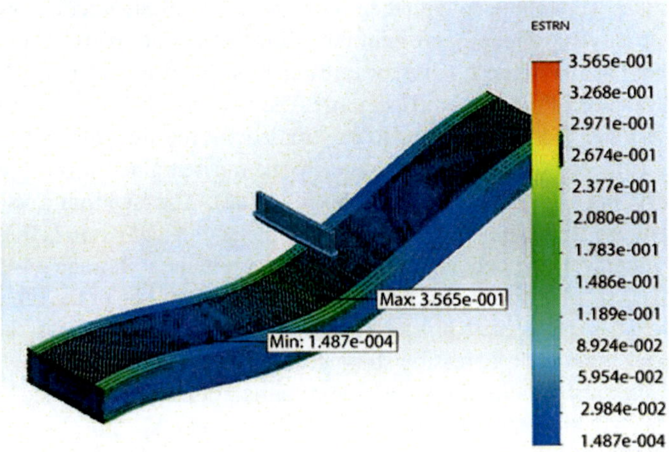

Fig. 10 Equivalent strain of Extrudr BDP Flax

Table 3 Results of bending simulations

Material characteristics	PLA	HD PLA Green	Impact PLA Gray	Extrudr BDP Flax	Extrudr BDP Pearl	Fiber Wood
Equivalent von Misses stress distribution (MPa)	3668.04	3668.05	3676.58	3668.04	3675.33	3675.84
Maximum displacement (mm)	14.48	13.71	18.37	12.60	18.74	6.95
Equivalent strain	0.40	0.38	0.51	0.35	1.08	0.41

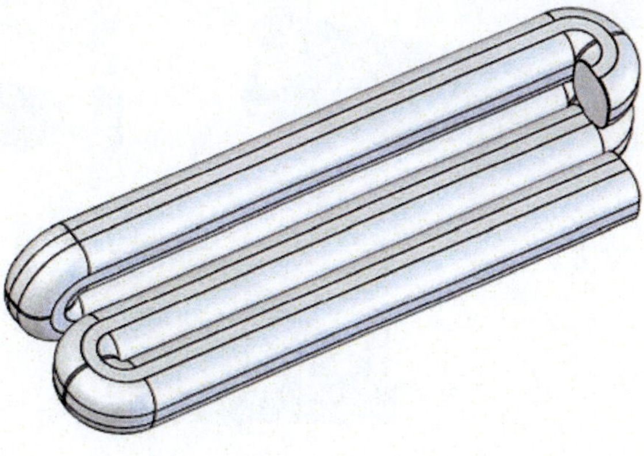

Fig. 11 3D model used to simulate heat transfer and deformation during 3D printing by using FDM method

Simulations of thermal Transfer and Deformations During the Printing Process

The heat transfers and deformation simulations that occur during the printing process were performed with the ANSYS program. The model used for this simulation was made in SolidWorks and consists of overlapping two material layers. The model is shown in Fig. 11. The deposited filament dimensions are identical to those used for tensile and bending models. The material used in these simulations was PLA.
Heat transfer during the printing process

Figure 12 shows the simulation of the mode in which the heat transfer takes place between two successively deposited layers. It was considered for the simulation the printing temperature of 230 °C (Fig. 12a) (maximum printing temperature according to the manufacturer's specifications), the printing table temperature of 70 °C (Fig. 12b). It is observed that the newly deposited layer transfers a very large heat amount (204 °C), Fig. 12c which leads to layers' adhesion and implicitly to the gaps infilling specific to 3D printing by the FDM method. The higher the heat transfer is, the higher the adhesion between the layers will be, and the smaller the voids between the deposited filaments, thus giving the sample better mechanical properties. Also, the printing table temperature helps to better layers' adhesion and to reduce the size of the voids.

Simulation of deformations that occur during the printing process

The deformation and the residual stresses of a printed material are of interest because the shape and the strength of the printed part can be influenced. The simulation results regarding the deformations that appear during the printing process are highlighted in Fig. 13. The instantaneous layers' deposition was considered for this simulation. Consequently, the resulting temperature gradient was taken into account only for the deposition direction of the extrusion head. For a better understanding of the deformations that occur during printing, some numerical approaches are needed. Figure 13a, b presents the value of maximum and minimum deformations resulting during material deposition, namely 0.043 mm.

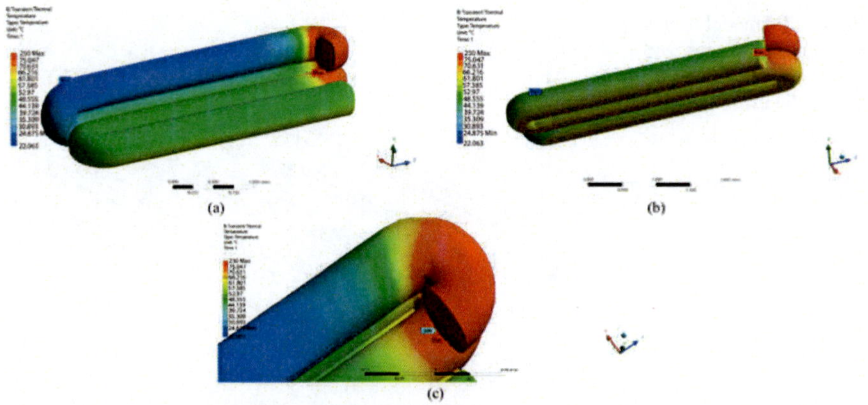

Fig. 12 Heat transfer simulation: **a** maximum printing temperature; **b** printing table temperature; **c** detail of heat transfer between layers

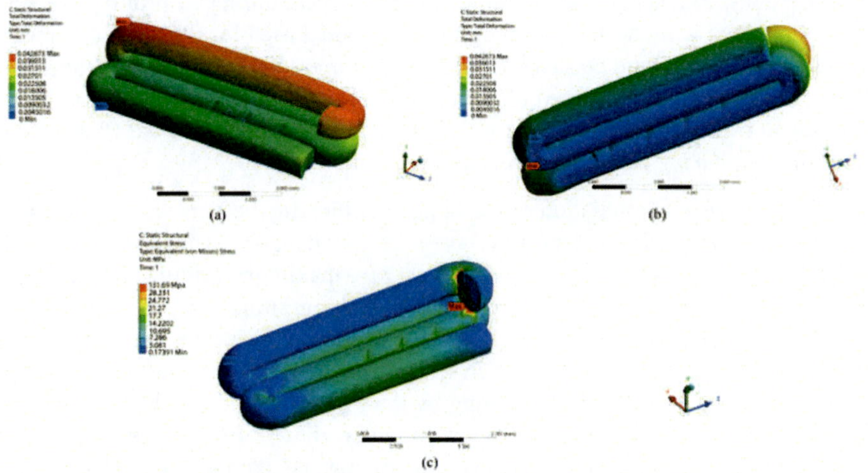

Fig. 13 Deformations and stresses simulation that occur during the 3D printing process: **a** maximum deformation; **b** the minimum deformation; **c** equivalent von Misses stress distribution

The *thermal expansion coefficient* influences the 3D printing process by inducing stresses between adjacent layers of deposited material, which already has a low temperature, depending on the printing medium (if closed, it can be relatively controlled). When a new layer is deposited, after cooling, the material undergoes a contraction that induces in the new layer tensile stresses, Fig. 13c, with a value of 131.69 MPa. If the *adhesion* between the layers is low, the internal stresses and forces that are generated by the printing process (even if they are relatively small) can lead to the exfoliation of some layers.

For certain geometries or arrangements of the samples on the printing table, thermal shrinkage can lead to deformations that prevent the deposition of the following layers. One way to avoid these defects would be to move the print head only in the opposite direction to the peak.

4.2 Experimental Results of Mechanical Tests of Printed Samples Made from Biodegradable Materials

Tensile Tests

Following the uniaxial tensile tests, were obtained results regarding tensile strength (σ [MPa]), elongation (ε [%]) and modulus of elasticity (E [MPa]) for each of the six materials analyzed. For the mechanical strength of the studied materials, Fig. 14, the PLA biodegradable material offers the best response, with a maximum average value of 51 MPa, for experiment number 4, "on the edge" part orientation. The next material that showed a response similar to PLA is HD PLA Green polymer (46 MPa, experiment number 6), then Impact PLA Gray (34 MPa, experiment number 6), Extrudr BDP Flex (33 MPa, experiment number 6), Extrudr BDP Pearl (28 MPa, experiment number 6) and Fiber Wood (18 MPa, experiment number 4). Also, it can be easily observed that, for all the studied materials, the highest values are obtained in the case of even number experiments, experiments in which the sample orientation was "on the edge". The differences between the results of the odd experiments, with the "flat" placement of the parts on the printing table, and the even experiments, with the placement "on the edge" of the parts, are very large, even up to 50% of the mechanical response value.

Regarding the elongation of the studied materials, Fig. 15, this is visibly influenced by the process factor, part orientation. The material with the highest response

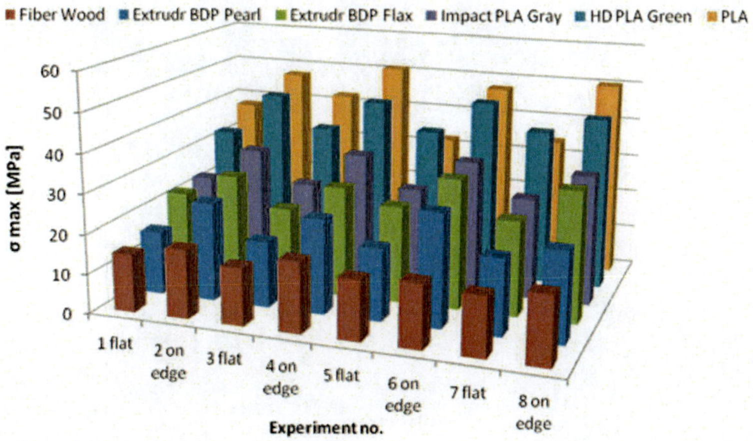

Fig. 14 Results regarding the tensile strength

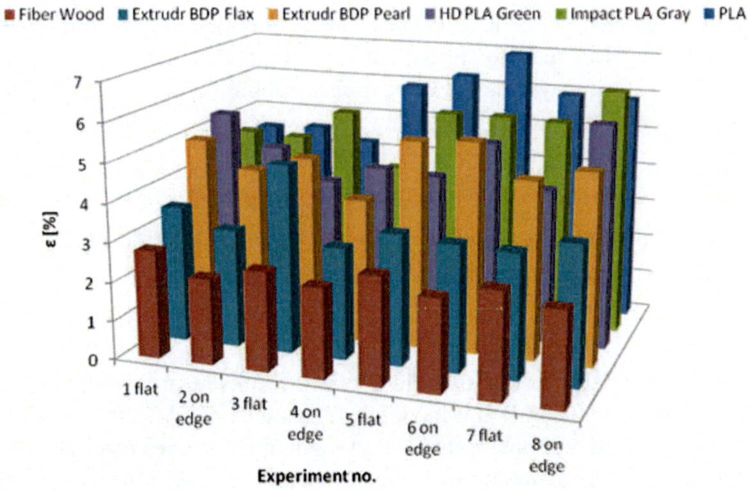

Fig. 15 Results regarding the elongation

regarding the elastic deformation was PLA (approximately 7%, experiment number 6), and the most rigid material is Fiber Wood (approximately 2.2%, experiment number 2).

The modulus of elasticity of the analyzed polymers, Fig. 16, is higher for "on the edge" oriented parts during the prototyping process. The highest values are presented by Fiber Wood and Extrudr BDP Flax, approximately 1580 MPa, experiment number 8, which denotes that these two materials are stiffer than the other four, which deform more because they have a lower modulus of elasticity.

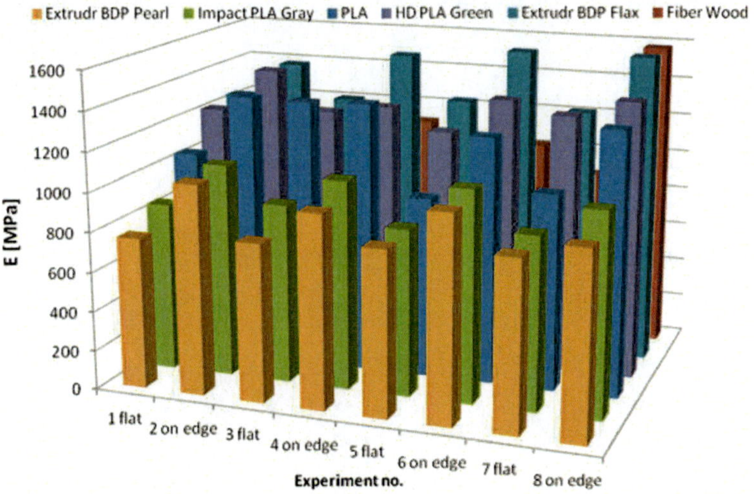

Fig. 16 Results regarding the modulus of elasticity

The mechanical properties obtained for the studied materials are comparable to those of the synthetic materials (such as acrylonitrile butadiene styrene (ABS)—40 MPa tensile strength, flexible—26–43 MPa tensile strength, high impact polystyrene (HIPS)—32 MPa tensile strength, polypropylene (PP)—32 MPa tensile strength and metal filled—tensile strength 20–30 MPa) [19], thus finding its applicability in various industries.

Bending Tests

Before starting the bending test, certain conditions have been established according to the technical literature. Thus, it was considered that for materials with the maximum arrow in the middle of the sample (linear displacement in the direction of application force) greater than 5 mm, the results obtained cannot be taken into account for this type of test.

In order to achieve three-point bending test, three specimens were prepared (to demonstrate that the printing process is stable and reproducible) from each biodegradable material. A single test was performed on materials with a deformation greater than 5 mm. The samples were placed "flat" on the printing table, the infill type was grid for the raster area and rectilinear type in the shell area, and the samples infill was 100% and were determined the distance between the sample placement supports, $L =$ 60 mm, the length of the tested sample 50 mm, the thickness $h = 4$ mm and its width b = 10 mm. Following the bending test of the samples from the studied biodegradable materials, a displacement of less than 5 mm was registered for only two of them, namely for PLA and Fiber Wood. Therefore, only the materials mentioned above are suitable for this type of mechanical test. The highest average value of bending strength was recorded by the PLA material (82.77 ± 1.97)MPa and the lowest by the Fiber Wood material which recorded an average value of (24.56 ± 8.35)MPa. Regarding the deformations of the two materials, the material with the most elastic behavior was the PLA with an average value of (3.67 ± 0.32)mm, followed by the Fiber Wood material (3.03 ± 0.24)mm. Figure 17 presents the values recorded by the six studied materials in terms of bending strength. The notations used in Fig. 17 were: δ—displacement, mm; σ—bending strength, MPa.

The other four biodegradable materials, HD PLA Green, Impact PLA Gray, Extrudr BDP Flax and Extrudr BDP Pearl according to the obtained results from the three-point bending test are not suitable for this type of test as they do not yield easily under the action of a progressive load, because they are elastic–plastic materials.

The plastics that can be successfully replaced by the three biodegradable materials for this type of mechanical test are polypropylene (PP) (approximately 26 MPa), polypropylene reinforced with short glass fibers (approximately 8 MPa) [20], ABS (between (36.8–59.6)MPa) [21], polymeric materials reinforced with metal and wood powders [22], and others.

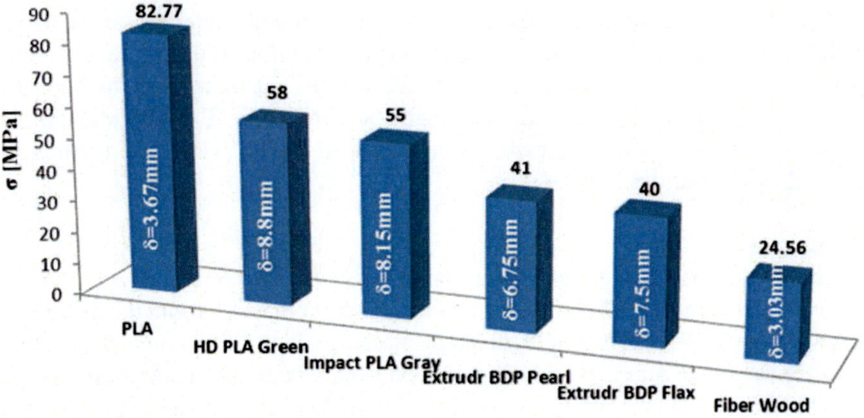

Fig. 17 Results on bending strength and displacement

5 Conclusions

The parts obtained by using the FDM 3D printing technology are significantly influenced by the process parameters that can be controlled directly or indirectly before starting the model printing. Selecting the optimum printing parameters is quite difficult, if it is to achieve a part with the high mechanical characteristics.

Comparing the simulation results with those actually obtained during mechanical tests, it was observed that approximately all materials showed the same behavior, the differences being most likely due to printing parameters, which during the simulations were not taken into account: sample orientation—simulations were made for planar landmarks; the deposition speed was not taken into account during the simulations. Also, the adhesion of the layers in the case of the 3D model was 100%, while in the case of the printed samples this was no longer valid, thus the values of the mechanical characteristics being lower.

Following tensile tests, mechanical property often used to evaluate the usefulness of the product manufactured by prototyping, the best values of tensile strength were obtained for "on the edge" oriented parts. The values of the tensile strength in the case of the "flat" sample orientation have decreased compared to the other type of orientation with up to 50%. Plan-oriented samples have a more fragile behavior, giving lower elongation results. The best mechanical behavior was highlighted by the biodegradable PLA material, and the weakest results were obtained for the Fiber Wood material. From tensile test point of view, the obtained results for the studied materials are comparable to those of the synthetic materials such as ABS—40 MPa tensile strength, flexible—(26–43)MPa tensile strength, HIPS—32 MPa tensile strength, PP—32 MPa tensile strength and metal filled—tensile strength 20–30 MPa).

For the bending test, only two of the six biodegradable materials studied proved to be suitable for this type of test (elongation displacement less than 5 mm), namely PLA and Fiber Wood. The thermoplastic materials that can be substituted by these two materials for this type of mechanical test are polypropylene (approximately 26 MPa), short glass fiber reinforced polypropylene (approximately 8 MPa), ABS (between 36.8 and 59.6 MPa), polymeric materials reinforced with metal and wood powder, and others.

Trebuie sa mentionam faptul ca comparand rezultatele obtinute in urma simularilor cu cele obtinute efectiv pe probele printate se poate observa faptul ca comportamentul probelor a fost in proportie destul de mari acelasi. Valorile obtinute insa difera intrucat in timpul simularilor nu au fost luate in calcul erorile de printare. Astfel, in cazul simularilor adeziuna dintre straturile depuse este perfecta si uniforma, pe cand in cazul epruvetelor printate efectiv apar defecte precum lipsa adeziunii intre straturile sau intre firele depuse, goluri de dimensiuni diferite (specifice FDM) in zona de raster si in zona de shall.

It must be mentioned the fact that comparing the obtained results from the simulations with those obtained actually on the printed samples, it can be concluded that the sample's behavior was in a fairly large proportion the same. However, the values obtained differ because printing errors were not taken into account during the simulations. Thus, in the case of the simulations, the adhesion between the deposited layers is perfect and uniform, while in the case of printed samples there are defects such as lack of adhesion between the layers or between the deposited filaments, filament interruptions, gaps of different sizes (specific to FDM) in the raster and shell areas. One of the future directions of research consists of searching reinforcement solution in order to obtain biodegradable composite materials with higher properties compared with base material.

Acknowledgements This work was supported by a grant of the Romanian Ministry of Research and Innovation, CCCDI–UEFISDI, project number PN-III-P1.2-PCCDI-0446/82PCCDI/2018, acronym TFI PMAIAA/FAMCRIA, within PNCDI III.

References

1. Gorski F, Kuczko W, Wichniarek R (2014) Impact strength of ABS parts manufactured using fused deposition modeling technology. Arch Mech Technol Autom 34(1):3–12
2. Gorski F, Kuczko W, Wichniarek R (2013) Influence of process parameters on dimensional accuracy of parts manufactured using fused deposition modelling technology. Adv Sci Technol Res J 7(19):27–35. https://doi.org/10.5604/20804075.1062340
3. Gorski F, Wichniarek R, Andrzejewski J (2012) Influence of part orientation on strength of ABS models manufactured using fused deposition modeling technology. Przetworstwo Tworzyw 9
4. Cumin J, Raos P, Grizelj B (2008) Rapid prototyping–3D printing. Manuf Eng 7(2):40–42
5. Ćwikła G, Grabowik C, Kalinowski K, Paprocka I, Ociepka P (2017) The influence of printing parameters on selected mechanical properties of FDM/FFF 3D-printed parts. In: IOP conference series: materials science and engineering 227:012033. https://doi.org/10.1088/1757-899X/227/1/012033
6. Mazurchevici A-D, Carausu C, Ciofu C, Popa R, Mazurchevici S-N, Nedelcu D (2019) Infill and type influence on tensile strength of PLA biodegradable material using FDM technology. Int J Mod Manuf Technol XI(2):44–49

7. Wu W, Geng P, Li G, Zhao Di, Zhang H, Zhao Ji (2015) Influence of layer thickness and raster angle on the mechanical properties of 3D-printed PEEK and a comparative mechanical study between PEEK and ABS. Materials 8:5834–5846. https://doi.org/10.3390/ma8095271
8. Vyavahare S, Teraiya S, Panghal D, Kumar S (2019) Fused deposition modelling: a review. Rapid Prototyping J. https://doi.org/10.1108/RPJ-04-2019-0106
9. Mazurchevici S-N, Pricop B, Istrate B, Mazurchevici A-D, Cârlescu V, Cărăușu C, Nedelcu D (2020) Technological parameters effects on mechanical properties of biodegradable materials using FDM. Materiale Plastice 52(2):215–227. https://doi.org/10.37358/MP.20.2.5368
10. Wu W, Ye W, Wu Z, Geng P, Wang Y, Zhao J (2017) Influence of layer thickness, raster angle, deformation temperature and recovery temperature on the shape-memory effect of 3D-printed polylactic acid samples. Materials 10:970. https://doi.org/10.3390/ma10080970
11. Chen C, Ran Y, Huang J, Hu Q, Wang X (2017) Research of the influence of 3D printing speed on printing dimension. Curr Trends Comput Sci Mech Autom 1:157–163
12. Carausu C, Mazurchevici A, Ciofu C, Mazurchevici S (2018) The 3D printing modelling of biodegradable material. In: IOP conference series: materials science and engineering 400:042008. https://doi.org/10.1088/1757-899X/400/4/042008
13. Introducere în imprimarea 3D (Introduction to 3D printing). https://3d-p.eu/wp-content/uploads/2018/08/IO3_3DP-courseware_RO.pdf. Accessed 09 Jan 2020
14. Wu D, Hakkarainen M (2015) Recycling PLA to multifunctional oligomeric compatibilizers for PLA/starch composites. Eur Polym J 64:126–137. https://doi.org/10.1016/j.eurpolymj.2015.01.004
15. Raise3D manufacturer. https://www.raise3d.com/filaments/. Accessed 19 Jan 2020
16. Fiberlogy manufacturer. https://fiberlogy.com/en/. Accessed 19 Jan 2020
17. Mazurchevici S-N, Mazurchevici A-D, Nedelcu D (2020) Dynamical mechanical and thermal analyses of biodegradable raw materials for additive manufacturing. Materials 13:1819. https://doi.org/10.3390/ma13081819
18. Extrudr manufacturer. https://www.extrudr.com/en/products/catalogue. Accessed 25 Feb 2020
19. Simplify3D: Filament Properties Table. https://www.simplify3d.com/support/materials-guide/properties-table/. Accessed 25 Feb 2020
20. Nedelcu D, Stefan A, Mîndru TD, Plavanescu S (2012) Flexural properties of samples obtained from "liquid wood." Selected Eng Probl 3:151–154
21. Sodeifian G, Ghaseminejad S, Yousefi AA (2019) Preparation of polypropylene/short glass fiber composite as fused deposition modeling (FDM) filament. Results Phys 12:205–222. https://doi.org/10.1016/j.rinp.2018.11.065
22. Comparison of typical 3D printing materials. https://2015.igem.org/wiki/images/2/24/CamJIC-Specs-Strength.pdf. Accessed 19 Jan 2020

Experimental Investigation on Influence of Process Parameters on Mechanical Properties of PETG Parts Made by Fused Deposition Modelling

Soham Teraiya, Swapnil Vyavahare, and Shailendra Kumar

1 Introduction

Additive manufacturing (AM), also known as 3D printing, is one of the most popular modern manufacturing techniques which provide the highest design flexibility. Nowadays, it is widely used in various applications such as architectural, medical, dental, aerospace, automotive, furniture, tooling, and jewellery [1, 2]. AM is a process of producing a part layer by layer. According to the use of power source and material, additive manufacturing processes are classified into seven basic types, viz. vat polymerization, binder jetting, material jetting, material extrusion, powder bed fusion, sheet lamination, and directed energy deposition [3]. Fused deposition modelling (FDM) is one of the most popular and economic additive manufacturing processes. In FDM, a polymer in a filament form is used as a feedstock material. It is extruded through a heated nozzle to form a raster. A combination of such adjacent rasters forms a layer. Further, the whole part is printed by stacking each layer on the previous layer. The products of FDM have started to penetrate the consumer market as a functional part along with prototypes.

Recently, various researchers have performed experimental investigation in the field of fused deposition modelling. Ahn et al. [4] studied influence of raster angle, air gap, bead width, and model temperature on tensile and compressive strengths of 3D printed specimens. Based on experimental results, several build rules were formulated for designing FDM parts. Bellini and Güçeri [5] observed the influence of raster angle and build orientation on tensile and flexural strength using analytical and experimental approach. Chin Ang et al. [6] experimentally investigated the effects of raster width, air gap, build layer, build orientation, and build profile, on the porosity and mechanical properties (compressive strength and compressive

S. Teraiya · S. Vyavahare · S. Kumar (✉)
Sardar Vallabhbhai National Institute of Technology, Surat, Gujarat, India
e-mail: skbudhwar@gmail.com

H. K. Dave and D. Nedelcu (eds.), *Advances in Manufacturing Processes*, Lecture Notes in Mechanical Engineering, https://doi.org/10.1007/978-981-15-9117-4_21

modulus) of ABS scaffold structures. They determined that the air gap and raster width are the most significant parameters. Sun et al. [7] studied the effect of deposition strategy (longitudinal or lateral), envelope temperature, and liquefier temperature on the mesostructure, cooling characteristics, flexural strength, and overall bond strength between layers.

Bakar et al. [8] experimentally investigated the effect of internal raster width, layer height, and contour width on surface finish and dimensional accuracy of the parts. They further used 3D printed part as master pattern for silicon rubber moulding. Chang and Huang [9] observed the effect of contour depth, contour width, raster angle, and raster width on extruding aperture and profile error of the specimens. Sood et al. [10] explained the influence of part build orientation, layer thickness, raster width, raster angle, and air gap on the compressive properties of the specimen. Also, they developed a statistically validated predictive equation through quantum-behaved particle swarm optimization (QPSO).

Croccolo et al. [11] observed the effects generated by the number of contours and build orientation on the tensile properties of ABS-M30 parts. They predicted the mechanical behaviour of the 3D printed parts using raster pattern. Magalhães et al. [12] suggested that appropriate selection of raster angles in a structure with sandwich-like configuration can considerably improve the final strength or stiffness of parts. Carniero et al. [13] investigated the effect of orientation, layer thickness, and infill degree, on mechanical properties of a glass fibre reinforced and non-reinforced polypropylene parts. They found that 3D printed parts can not only be used as prototypes but also as functional goods. Chockalingam et al. [14] studied the reliance of raster angle, orientation, air gap, and raster width on tensile strength and density by using non-dominated sorting genetic algorithm (NSGA-II). Cantrell et al. [15] performed experiments by varying the raster angle and build orientations to determine the directional properties of the materials.

Arif et al. [16] experimentally investigated the effect of the raster angle and build orientation on flexural strength, tensile strength, and fracture toughness of the parts. Due to the presence of interfacial voids, stick-slip fracture and lower Poisson's ratio were observed for vertically built specimens. Gebisa and Lemu [17] investigated the effect of raster width, air gap, contour number, raster angle, and contour width on the flexural properties of ULTEM 9085 material. Rajpurohit and Dave [18] investigated the influence of layer thickness, raster angle, and raster width on the tensile properties of PLA specimens. They found that layer thickness and raster angle are the most significant parameters. Srinivasan et al. [19] used response surface methods in order to understand the influence of infill density, infill pattern, and layer thickness on tensile strength and hardness 3D printed ABS specimens. They found that layer thickness and infill density are the most significant factors.

From the literature survey, it has been found that the majority of the research work has been done in analysing the influence of build orientation, raster angle, and layer thickness on the mechanical properties of most popular thermoplastics such as ABS and PLA. Very less amount of work has been reported, which investigates the influence of process parameters on tensile properties of co-polyester such as polyethylene terephthalate glycol PETG (PETG-modified) parts. Taguchi's orthogonal array L18

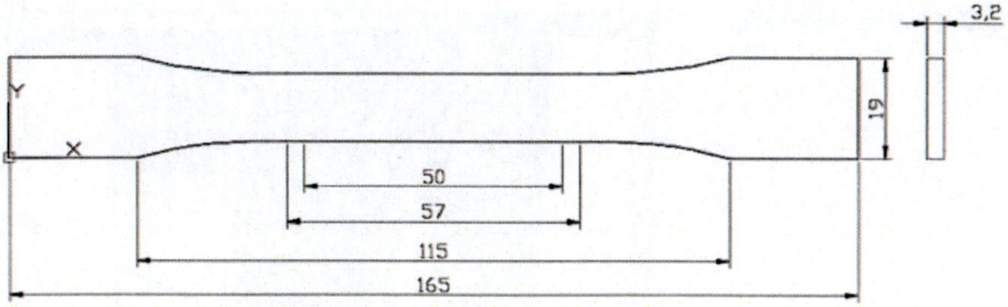

Fig. 1. 2D CAD model of specimen

has been used for experimentation. The measured value of mechanical properties is further studied by ANOVA. The optimum parameter settings have been suggested for obtaining the maximum tensile strength and modulus of PETG parts (Fig. 1).

2 Materials and Methods

In the present work, specimens are produced by a fused deposition modelling process. Delta WASP 2040 FDM machine has been used for printing the parts, as shown in Fig. 2. The machine has a build volume of Ø200 mm × 400 mm. It can print the parts of various materials such as ABS, PLA, HIPS, PETG, and Nylon. The maximum extrusion temperature and maximum bed temperature that can be achieved are 260 °C and 120 °C, respectively. The extruder of the machine is equipped with a nozzle of 0.4 mm diameter.

PETG feedstock material is used for printing the specimen. The CAD model is created in AutoCAD 2020 software. The dimensions of the specimen are decided according to ASTM D638. The 2D CAD model of the specimen is shown in Fig. 1. The 3D CAD model is converted into stl file. The ULTIMAKER CURA 4.0 software slices the stl file into number of layers. The software exports g-code file, which is fed to FDM machine using SD card. PETG is used as feedstock material. The value of density, bulk tensile strength, glass transition temperature, and print temperature range are 1.27 g/cm^3, 50 MPa, 80 °C, and 230–250 °C, respectively (Table 1).

In the present study, five process parameters are selected for experimentation, such as print speed, raster angle, raster width, layer thickness, and print temperature. Table 2 lists the process parameters along with their respective levels. The level and range of these parameters are selected based on the literature survey and machine setup range. Other process parameters such as build orientation, number of contours, infill percentage, and nozzle diameter are kept constant. Their values are stated in Table 1. In order to reduce the number of experiments, Taguchi's orthogonal array L18 is used for experimental design. Thus, 18 samples were fabricated by varying process parameters according to the layout shown in Table 3.

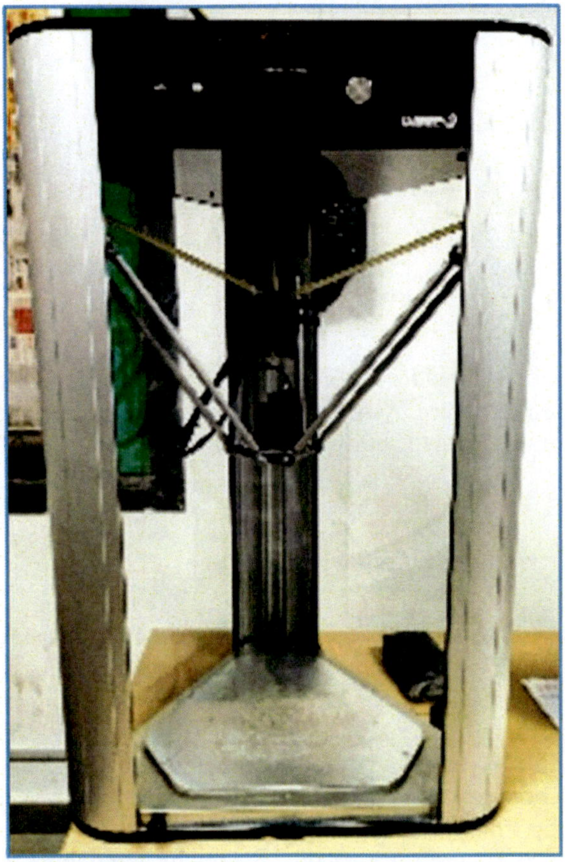

Fig. 2 FDM machine setup

Table 1 Constant process parameters of FDM

Parameter	Value
Build orientation	Flat (*XY* plane)
Bed temperature	100 °C
No. of contour	01
Infill	100%
Nozzle diameter	0.4 mm

Table 2 Process parameters and their range

Parameters	Level 1	Level 2	Level 3
Print speed (mm/s)	35	45	–
Raster angle (°)	0	45	90
Raster width (mm)	0.4	0.5	0.6
Layer thickness (mm)	0.15	0.25	0.35
Printing temperature (°C)	230	240	250

Table 3 Experimental design and results

Run	Print speed (mm/s)	Raster angle (°)	Raster width (mm)	Layer thickness (mm)	Print temp (°C)	Tensile strength (MPa)	Young's modulus (MPa)
1	35	0	0.4	0.15	230	42.5	258.12
2	35	0	0.5	0.25	240	38.7	279.68
3	35	0	0.6	0.35	250	38.2	203.47
4	35	45	0.4	0.15	240	41.1	240.05
5	35	45	0.5	0.25	250	38.2	230.12
6	35	45	0.6	0.35	230	31.8	218.02
7	35	90	0.4	0.25	230	39.6	227.11
8	35	90	0.5	0.35	240	36.5	218.04
9	35	90	0.6	0.15	250	40.2	240.37
10	45	0	0.4	0.35	250	36.5	185.11
11	45	0	0.5	0.15	230	44.4	250.42
12	45	0	0.6	0.25	240	40.2	263.78
13	45	45	0.4	0.25	250	35.6	192.05
14	45	45	0.5	0.35	230	33.9	230.21
15	45	45	0.6	0.15	240	40.1	276.39
16	45	90	0.4	0.35	240	34.1	147.05
17	45	90	0.5	0.15	250	42.3	217.22
18	45	90	0.6	0.25	230	32.1	215.4

The specimens are clamped in a tensometer and allowed to elongate until it reaches a breakage point. The rate of movement of the non-stationary jaw was 5 mm/min. Loading capacity of the equipment was 20 kN. The load–displacement curve is achieved after each tensile testing. It is further processed using Microsoft Excel in order to get the values of strength and modulus. For the linear portion of the load–displacement curve, the ratio of stress to strain is considered as value of young's modulus.

3 Results and Discussion

The measured value of tensile strength and modulus for each experiment is depicted in Table 3. Figure 3 shows all the tested specimens under tensile loading. It was observed that some samples are broken out of guage length with brittle behaviour. This may happen due to the presence of voids and stress concentration within the layers of printed specimen. Analysis of variance (ANOVA) is performed in order to understand the statistical significance of process parameters on the mechanical

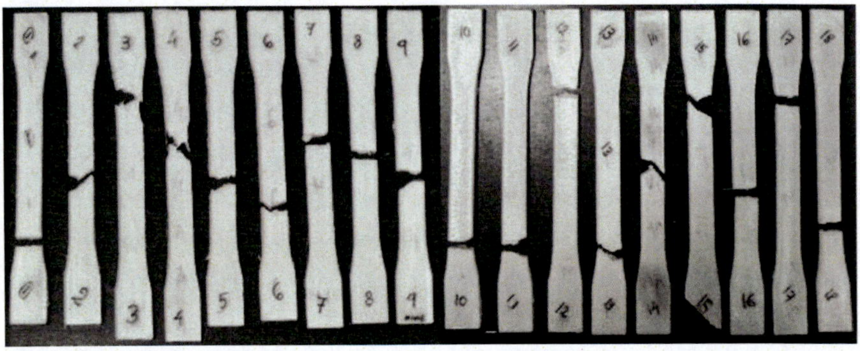

Fig. 3 Tested specimens

properties of the part. Tables 4 and 5 show ANOVA with *F*-ratio and *p*-values for the tensile properties of printed samples.

Figure 4 depicts the stress–strain curve for a run no. 05. The results of tested specimens showed that almost all PETG specimens exhibit identical brittle behaviour under tensile loading. Maximum tensile strength and modulus are 44.4 MPa and 279.68 MPa, respectively. Due to anisotropy and microscopic voids, the values of

Table 4 ANOVA for tensile strength

Source	DF	Adj SS	Adj MS	F-value	P-value	Comment
Print speed	1	3.209	3.209	0.82	0.391	
Raster angle	2	36.40	18.204	4.66	0.046	Significant
Raster width	2	10.96	5.482	1.40	0.300	
Layer thickness	2	135.23	67.616	17.31	0.001	Significant
Print temp	2	4.77	2.387	0.61	0.566	
Error	8	31.25	3.906			
Total	17	221.83				

Table 5 ANOVA for modulus of elasticity

Source	DF	Adj SS	Adj MS	F-value	P-value	Comment
Print speed	1	1048	1048.1	3.60	0.094	
Raster angle	2	2692	1345.8	4.62	0.042	Significant
Raster width	2	3295	1647.7	5.66	0.029	Significant
Layer thickness	2	7047	3523.6	12.10	0.004	Significant
Print temp	2	2353	1176.3	4.04	0.061	
Error	8	2329	291.1			
Total	17	18,764				

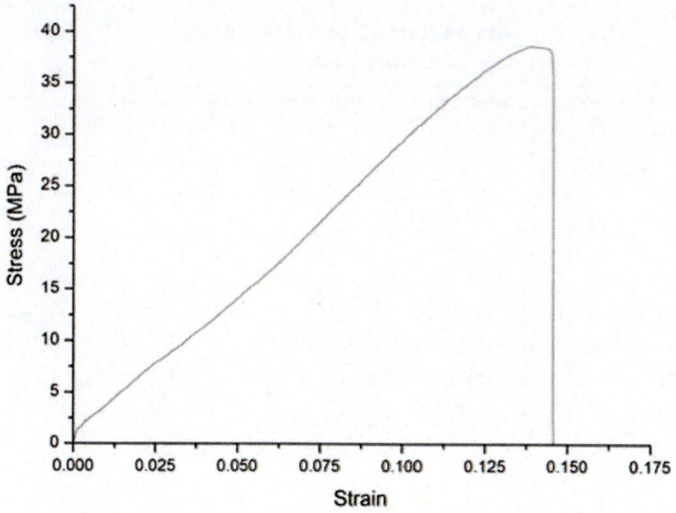

Fig. 4 Stress–strain curve for run 05

mechanical properties are less than that of injection moulded parts. Further, the influence of each parameter on response characteristics is briefly discussed below.

3.1 Effect of Process Parameters on Tensile Strength and Modulus of Elasticity of Parts

3.1.1 Effect of Raster Angle

As ANOVA table suggests, the influence of raster angle on tensile strength is significant, as the value of p is less than 0.05 (Table 4). From Fig. 5b, it is found that specimen printed at 0° raster angle has maximum tensile strength. As the raster angle increases, the tensile strength decreases. The reason is that, at 0° raster angle, the tensile load is applied along the raster for each layer. It results in better tensile strength as compared to the load applied in the direction of normal to the rasters. Further, it is observed that higher strength is achieved at 90° as compared to that of 45°. This result is contradicting the findings of previous researchers. Further, this behaviour may be explained better with the help of interaction effect of parameters. Limitation of Taguchi's experimental design is that it does not explain the interaction effect of process parameters on response characteristics. For the modulus of elasticity, the raster angle is significant because the *p*-value is less than 0.05 (Table 5). From the main effect plot of S/N ratio (Fig. 6b), the trend is similar to that of the tensile strength.

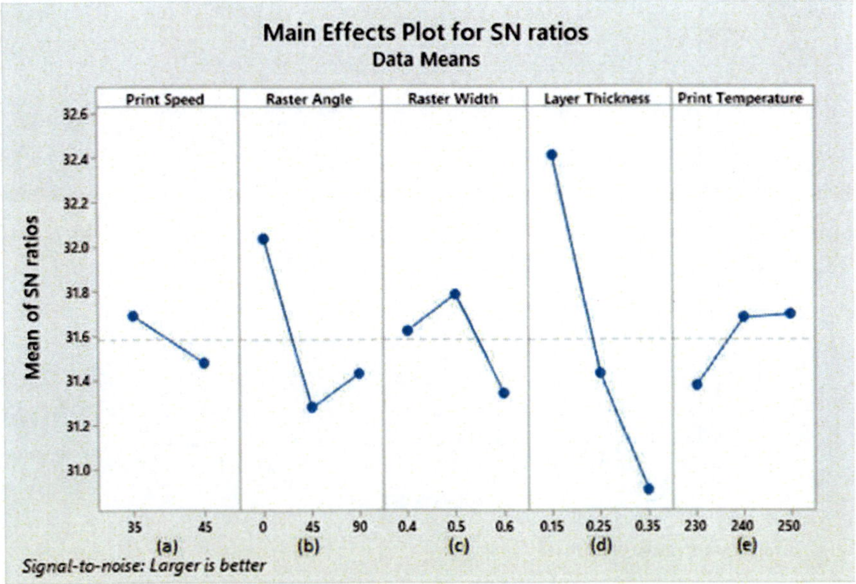

Fig. 5 Main effect plot of S/N ratio for tensile strength

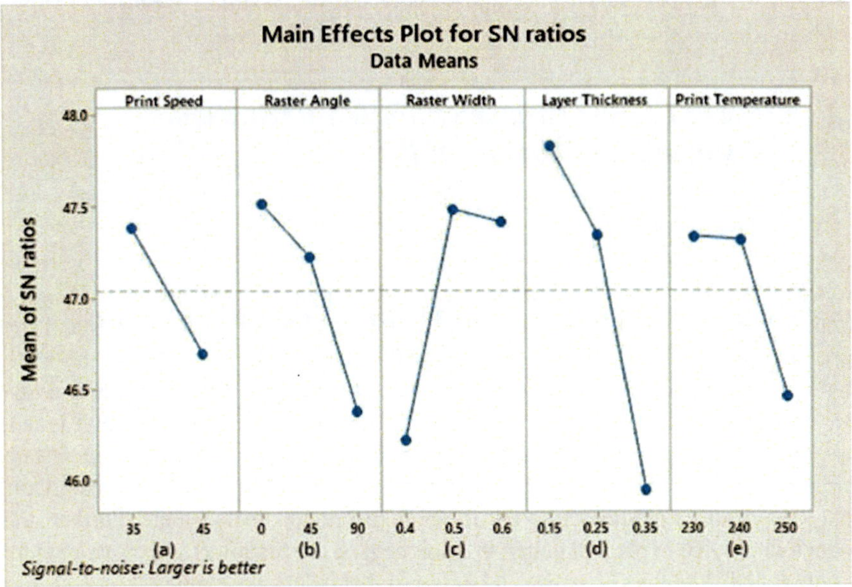

Fig. 6 Main effect plot of S/N ratio for modulus of elasticity

3.1.2 Effect of Raster Width

From ANOVA table, it is found that the influence of raster width of tensile strength is insignificant (Table 4). Raster width of 0.5 and 0.6 mm is achieved from the nozzle of 0.4 mm diameter. The heated filament is squeezed by controlling the extruder motor along each raster. According to the main effect plot of S/N ratio, it is observed that an increase in the raster width causes the first increase and then decrease in the tensile strength as shown in Fig. 5c. Maximum tensile strength is attained at 0.5 mm raster width. As raster width increases, the heat transfer rate of the raster reduces. It gives more time to the raster for diffusion with previous layers. With further increases in raster width, the strength decreases. It may happen due to the presence of voids and internal print defects. For the modulus of elasticity, the influence of this parameter is significant (Table 5). According to the main effect plot of S/N ratio (Fig. 6c), the trend for modulus of elasticity is similar to that of the tensile strength.

3.1.3 Effect of Layer Thickness

According to ANOVA table, it is observed that the influence of layer thickness on tensile strength and modulus of elasticity is significant (Tables 4 and 5). An increase in the layer thickness causes a decrease in the tensile strength, as shown in Fig. 5d. Maximum tensile strength is attained at minimum layer thickness. With a reduction in layer thickness, the total number of layers needed to print the whole part increases. It causes an increase in reheating of previous layers, which promotes better layer diffusion. Further, the modulus of elasticity is maximum at minimum at layer thickness (Fig. 6d). Similar results were observed by Durgashyam et al. [20].

3.1.4 Effect of Print Speed

From ANOVA table, it is seen that the p-value for the print speed is more than 0.05, which shows that its effect is insignificant for tensile strength and modulus of elasticity (Tables 4 and 5). From Fig. 5a, it is found that specimens printed at minimum print speed have maximum tensile strength. The reason is that lower print speed allows enough time for a material to get extruded and bonded to the previous raster and layer. From Fig. 6a, it is seen that modulus of elasticity reduces with an increase in print speed.

3.1.5 Effect of Print Temperature

From ANOVA table, it is seen that its effect of the print temperature is insignificant for tensile strength and modulus of elasticity (Tables 4 and 5). From Fig. 5e, it is found that with an increase in print temperature, tensile strength also increases. This happens because, at a lower temperature, extruded material adhere less to the previous layer.

From Fig. 6e, it is found that change in modulus of elasticity is inversely proportional to the variation in print temperature. However, maximum print temperature depends upon for the material to be used for FDM fabrication.

In the present study, efforts are applied to optimize the tensile strength and modulus of printed specimens. A set of optimum parameters is defined based on the average S/N ratio for each parameter. For tensile strength, the optimum combination of parameters is 35 mm/s print speed, 0° raster angle, 0.5 mm raster width, 0.15 mm layer thickness, and 250 °C print temperature. For modulus of elasticity, the optimum combination of parameters is 35 mm/s print speed, 0° raster angle, 0.5 mm raster width, 0.15 mm layer thickness, and 230 °C print temperature.

4 Conclusion

An experimental investigation has been carried out to study the influence of process parameters on the mechanical properties of PETG parts printed by FDM. From the analysis of experimental results, the following observations are made:

1. Raster angle and layer thickness are the most significant process parameters that influence the tensile strength of the specimen.
2. For modulus of elasticity, raster angle, raster width, and layer thickness are the most significant process parameters.
3. Tensile strength increases with a decrease in raster angle; increase raster width; decrease in layer thickness; decrease in print speed; and increase in print temperature.
4. Modulus of elasticity increases with a decrease in raster angle; increase raster width; decrease in layer thickness; decrease in print speed; and decrease in print temperature.

A set of optimum parameter has been obtained on the basis of experimental results. The findings of the present paper can be useful for further investigation on understanding the influence of the interaction of process parameters on mechanical properties of 3D printed PETG specimens.

References

1. Korpela J, Kokkari A, Korhonen H, Malin M, Närhi T, Seppälä J (2012) Biodegradable and bioactive porous scaffold structures prepared using fused deposition modeling. J Biomed Mater Res B Appl Biomater 101B:610–619
2. Duda T, Raghavan L (2018) 3D metal printing technology: the need to re-invent design practice. AI Soc 33:241–252
3. Li H, Wang T, Sun J (2018) Yu Z (2018) The effect of process parameters in fused deposition modelling on bonding degree and mechanical properties. Rapid Prototyping J 24:80–92

4. Ahn S, Montero M, Odell D, Roundy S, Wright P (2002) Anisotropic material properties of fused deposition modeling ABS. Rapid Prototyping J 8:248–257
5. Bellini A, Güçeri S (2003) Mechanical characterization of parts fabricated using fused deposition modeling. Rapid Prototyping J 9:252–264
6. Chin Ang K, Fai Leong K, Kai Chua C, Chandrasekaran M (2006) Investigation of the mechanical properties and porosity relationships in fused deposition modelling-fabricated porous structures. Rapid Prototyping J 12:100–105
7. Sun Q, Rizvi G, Bellehumeur C, Gu P (2008) Effect of processing conditions on the bonding quality of FDM polymer filaments. Rapid Prototyping J 14:72–80
8. Bakar N, Alkahari M, Boejang H (2010) Analysis on fused deposition modelling performance. J Zhejiang Univ-Sci A 11:972–977
9. Chang D, Huang B (2010) Studies on profile error and extruding aperture for the RP parts using the fused deposition modeling process. Int J Adv Manu Tech 53:1027–1037
10. Sood A, Ohdar R, Mahapatra S (2012) Experimental investigation and empirical modelling of FDM process for compressive strength improvement. J Adv Res 3:81–90
11. Croccolo D, De Agostinis M, Olmi G (2013) Experimental characterization and analytical modelling of the mechanical behaviour of fused deposition processed parts made of ABS-M30. Comput Mater Sci 79:506–518
12. Magalhães L, Volpato N, Luersen M (2013) Evaluation of stiffness and strength in fused deposition sandwich specimens. J Brazilian Soc Mech Sci Eng 36:449–459
13. Carneiro O, Silva A, Gomes R (2015) Fused deposition modeling with polypropylene. Mater Des 83:768–776
14. Chockalingam K, Jawahar N, Praveen J (2015) Enhancement of anisotropic strength of fused deposited ABS parts by genetic algorithm. Mater Manuf Processes 31:2001–2010
15. Cantrell J, Rohde S, Damiani D, Gurnani R, DiSandro L, Anton J, Young A, Jerez A, Steinbach D, Kroese C, Ifju P (2017) Experimental characterization of the mechanical properties of 3D-printed ABS and polycarbonate parts. Rapid Prototyping J 23:811–824
16. Arif M, Kumar S, Varadarajan K, Cantwell W (2018) Performance of biocompatible PEEK processed by fused deposition additive manufacturing. Mater Des 146:249–259
17. Gebisa A, Lemu H (2018) Investigating effects of fused-deposition modeling (FDM) processing parameters on flexural properties of ULTEM 9085 using designed experiment. Materials 11:500
18. Rajpurohit S, Dave H (2018) Analysis of tensile strength of a fused filament fabricated PLA part using an open-source 3D printer. Int J Adv Manu Tech 101:1525–1536
19. Srinivasan R, Pridhar T, Ramprasath L, Charan N, Ruban W (2020) Prediction of tensile strength in FDM printed ABS parts using response surface methodology (RSM). Mate Today Proc
20. Durgashyam K, Indra Reddy M, Balakrishna A, Satyanarayana K (2019) Experimental investigation on mechanical properties of PETG material processed by fused deposition modeling method. Mater Today Proc 18:2052–2059

Analysis of Compressive Strength of 3D Printed PLA Part

Ashish R. Prajapati, Shilpesh R. Rajpurohit, Naushil H. Patadiya, and Harshit K. Dave

1 Introduction

3D printing process forms the objects from three-dimensional virtual models, which can be obtained from computer-aided design software. It is suitable for the construction of particular objects with complex geometries. 3D printing is widely used in various fields of application, for example, in the aerospace and defense, in automotive and electronics industry, and in the medical sector, which includes dental applications, prosthetics and bone fixation implants. Consumer industries such as sports, furniture, jewelry and the food industry have also started the implementation of 3D printing technologies [1].

In the fused deposition modeling (FDM) process, a thermoplastic material is extruded from a nozzle and makes the part layer upon layer. The feedstock material is supplied as a solid polymer filament. The nozzle contains a resistive heater, which is used to heat the polymer material such that it runs smoothly from the nozzle and forms the layer. FDM process has significant advantages in terms of eliminating costly tools, flexibility and the ability to produce very complex parts and shapes. The limitations of the FDM are the quality of the parts fabricated. It makes it necessary to understand the performance of parts from the FDM process with the variation in processing parameters so that they are reliable for different applications. There are vast applications of FDM in medical industries, viz. dental and bone implants. Compressive strength is playing a fundamental role in these applications. Numerous researchers have studied the effect of different process parameters of FDM printed parts on the compression strength of the parts.

A. R. Prajapati (✉) · S. R. Rajpurohit · N. H. Patadiya · H. K. Dave
Department of Mechanical Engineering, S. V. National Institute of Technology, Surat 395007, India
e-mail: iamarp5155@gmail.com

H. K. Dave and D. Nedelcu (eds.), *Advances in Manufacturing Processes*, Lecture Notes in Mechanical Engineering, https://doi.org/10.1007/978-981-15-9117-4_22

Chin et al. [2] investigated the compression properties of the FDM parts using different materials. They observed that parts printed using PCL resulted in higher strength as compared to other materials. Ahn et al. [3] found that parts printed horizontally on the print bed are resulted in higher compressive strength. Lee et al. [4] observed that the compressive strength of the part printed in axial direction has more strength than that of the part printed in the transverse direction. Sood et al. [5] observed that as the layer height decreases and air gap increases, the compressive strength decreases for the FDM printed ABS specimens. Hernandez et al. [6] studied the compression strength of FDM made components using different build orientations. They have observed that samples have higher strength in *XY*-plane as compared to *Z*-plane. Motapatri et al. [7] studied different FDM processing parameters, viz. air gap, angle of raster and part orientation for the compression testing of ABS made parts. They have observed that as compared to vertically made parts, horizontally build parts resulted in higher strength. Upadhyay et al. [8] reported that higher compressive strength can be achieved by depositing layers in perpendicular to the testing. Mishra et al. [9] investigated the impact of built orientation, a number of perimeters and air gap on compression strength of the FDM printed parts. Jami et al. [10] stated that parts having solid build have higher strength as compared to porous specimens. Dave et al. [11], in their experiments, studied the effect of infill density, layer thickness and print speed on the compressive strength of FDM printed polylactic acid (PLA) parts.

From the literature study, it is observed that effect of the infill parameters, viz. density and pattern, is not thoroughly examined for the mechanical properties mostly for compression strength. In the present work, the effect of three process parameters, viz. build orientation, infill density and infill pattern on the compressive strength, has been investigated.

2 Experimental Details

2.1 Material

Polylactic acid (PLA) is used in the present work as a raw material. It is a thermoplastic polymer made from various renewable resources. Solid PLA filament having 1.75 mm diameter supplied by eSUN is used to 3D print the compression specimens. Open source omega dual extruder 3D printer is used in the present work.

2.2 Specimen Preparation

Higher values of compressive strength are very important for parts made of FDM. In this study, the compressive testing method ASTM D695 was applied for the testing.

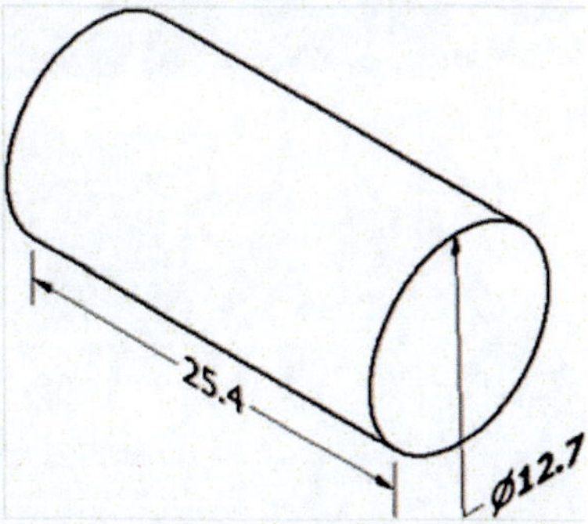

Fig. 1 Compressive test specimen as per ASTM D695 standard (all dimensions are in mm)

Table.1 Parameter levels and values

Parameters	Levels
Orientation	Flat, On-edge
Infill pattern	Rectilinear, Concentric, Hilbert curve
Infill density	60, 80, 100

The dimensions of the compressive part, according to standard, is shown in Fig. 1. The 3D models of specimens are prepared using Autodesk Inventor software. After the test specimen is modeled, it is then exported into slicer software for slicing. After that, the g-code file is imported into the machine.

2.3 *Process Parameters*

In this investigation, three different processing parameters, viz. part build orientation, infill density and infill pattern, were selected. Two types of part build orientation flat and on-edge were selected. Rectilinear, concentric and Hilbert curve infill patterns were used for the fabrication of specimen. Three ranges of infill density 60, 80 and 100% were considered in the present work. Process parameters and their range are tabulated in Table 1.

Specific parameters, viz. nozzle and bed temperature, layer thickness, raster width, raster angle and speed, were fixed for all the samples.

2.4 *Compression Testing*

Compressive strength at break is determined in accordance with ASTM D695 (determination of plastic compression properties) using the PC2000 electronic tensometer as shown in Fig. 2. It is an automatic material testing system with a crossbar speed of 1.3 mm/min and a full-scale load range of 20 KN.

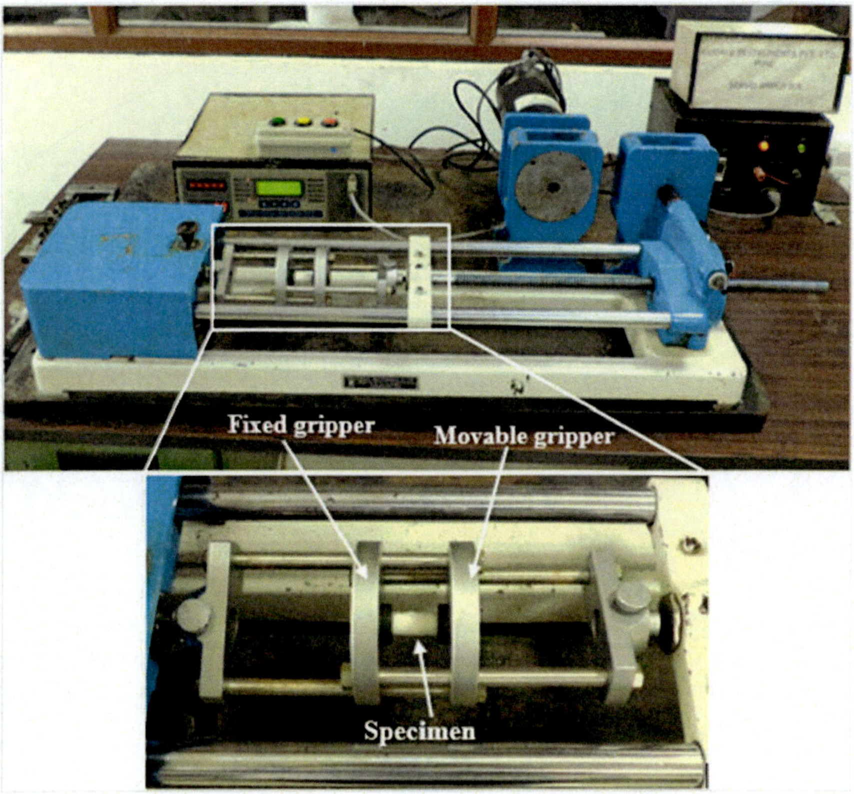

Fig. 2 Compression testing machine

In the present investigation, the full factorial design was selected, and as per the design, there were a total of 18 combinations of different processing conditions. To ensure the repeatability of the testing of components, all 18 specimens having different parameter combinations were printed twice and mean values of the compressive strength were considered for further investigation.

3 Results and Discussion

The full factorial design of 18 samples and compression strength values derived from both tests with mean and standard deviation values are reported in Table 2.

Analysis of variance (ANOVA) is used to identify FDM processing parameters that have a significant impact on the compression strength of FDM made parts. Table 3 shows the analysis of variance for compressive strength. ANOVA analysis represents the percentage contribution of process parameters on the compressive strength, where infill density had a maximum contribution of 85.22%. And the contribution

Table 2 Results of the compression test

Sr. No.	Orientation	Pattern	Density	Compressive strength (MPa)		Mean compressive strength (MPa)	SD
				Test 1	Test 2		
1	Flat	R	60	19.6	18.8	19.2	0.57
2	Flat	R	80	31.7	32.8	32.25	0.78
3	Flat	R	100	57.4	56.6	57	0.57
4	Flat	C	60	20.3	19.4	19.85	0.64
5	Flat	C	80	34.3	34.5	34.4	0.14
6	Flat	C	100	62.4	62.1	62.25	0.21
7	Flat	H	60	13.6	15	14.3	0.99
8	Flat	H	80	28.4	29.8	29.1	0.99
9	Flat	H	100	53.5	50.8	52.15	1.91
10	On-edge	R	60	28.1	24.7	26.4	2.40
11	On-edge	R	80	45.7	44.1	44.9	1.13
12	On-edge	R	100	67	67.4	67.2	0.28
13	On-edge	C	60	25.6	27.8	26.7	1.56
14	On-edge	C	80	40.4	39.9	40.15	0.35
15	On-edge	C	100	57.9	57.8	57.85	0.07
16	On-edge	H	60	28.6	25.5	27.05	2.19
17	On-edge	H	80	43.5	44.9	44.2	0.99
18	On-edge	H	100	58.8	55.2	57	2.55

*R-Rectilinear pattern, C-Concentric pattern, H-Hilbert curve pattern

Table 3 Analysis of variance for compressive strength

Factor	DOF	SS	MS	F	P	Contribution (%)
Orientation	1	22.703	22.70	20.42	0.001	8.55
Infill pattern	2	3.212	1.60	1.44	0.274	1.21
Infill density	2	226.31	113.15	101.8	0	85.22
Error	12	13.33	1.11			5.02
Total	17	265.56				100.00

of orientation and infill pattern was 8.55% and 1.21%, respectively. The *p*-values from the analysis indicate the significance of the individual process parameter. For the build orientation and infill density, the *p*-value for compressive strength is less than 0.05, which suggests that orientation and infill density are significant process parameters for compressive strength.

3.1 *Effect of Process Parameters on Compressive Strength*

Figure 3 shows the graphical representation of the correlation between three process parameters and the strength of the compressive test specimen.

The graphs from Fig. 3 reveal that parts printed with 100% infill density resulted in higher compressive strength as compared to other infill density. From the graph, it is clear that Hilbert curve pattern shows the lowest strength values for compressive strength. The results showed that the specimens printed in on-edge orientation have higher compressive strength in comparison with the flat orientation specimens. Figure 4 shows the stress–strain behavior of flat and on-edge orientation at 100% infill density and rectilinear pattern. It has been observed that, after the elastic limit, the amount of plastic deformation increases when the layers are pressed together and slows down when the layers are fully squeezed.

The reason behind the higher compressive strength in the parts printed in on-edge orientation is the arrangement of the layers in the sample. The layers in the on-edge orientation stacked one above the other, as shown in Fig. 5, and all layers were arranged perpendicular to the applied load during the compression testing, while, in the flat orientation, layers were arranged parallel to the applied compression load. Compression specimens built-in flat orientation fail under compressive load due to the buckling of layers (Fig. 6), whereas the parts built-in on-edge direction fail under a higher compressive load due to inter-layer sliding as shown in Fig. 5. Hence, the on-edge orientation specimen has better mechanical properties in comparison with flat orientation specimen.

It is observed that the parts printed with higher infill density resulted in higher compressive strength. The compression specimens having 100% infill density are fully dense, and rasters are connected to each other as shown in Fig. 7a. While part having 80% infill density, there are minor gaps between the rasters (Fig. 7b). However, in the case of 60% infill density, these gaps are wider as shown in Fig. 7c. As the infill density reduces, the gap between the rasters increases and bonding between rasters became weaker. This was the reason behind lower compression strength at lower value of infill density as these gaps reduce the ability of the part to resist more compressive load.

The compression strength of parts printed with rectilinear pattern and concentric patterns was the same as the rasters are continuous in both of the patterns. However, in the case of parts printed with the Hilbert curve pattern, the compressive strength was slightly low as compared to the other two patterns. This may be due to the fact that in Hilbert curve pattern, there is no longer raster in particular as only short beads were used, and after few millimeters, bead changes its direction at 90 as shown in Fig. 8c. Due to this sudden change in direction, extruder experienced jerk and vibration which leads to uneven deposition and variation in material flow rate and ultimately resulted in lower values of compression strength.

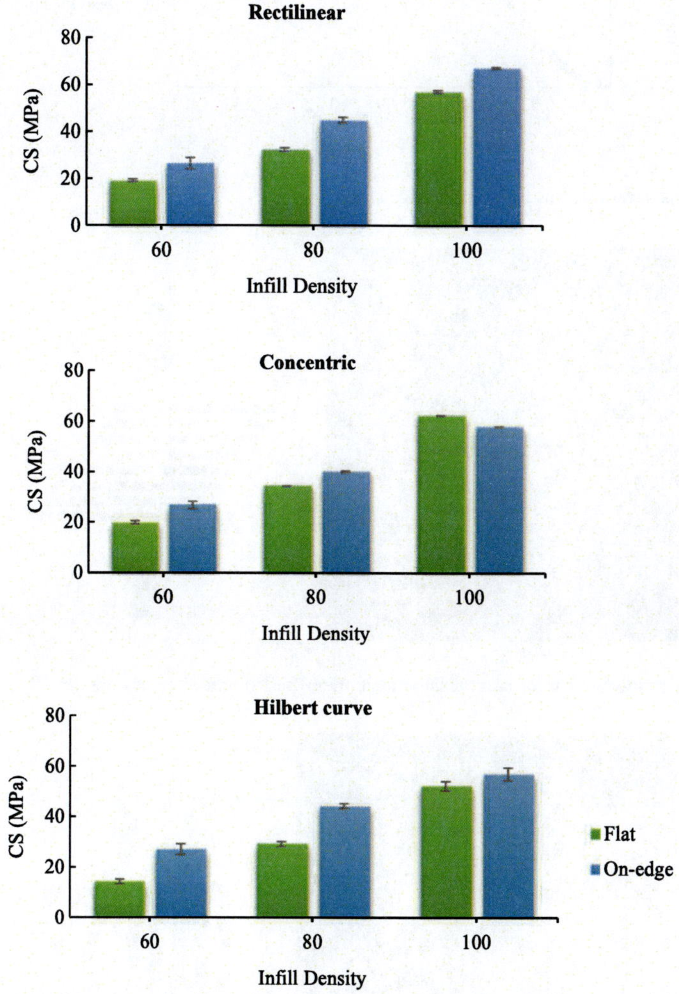

Fig. 3 Effect of process parameters on compressive strength

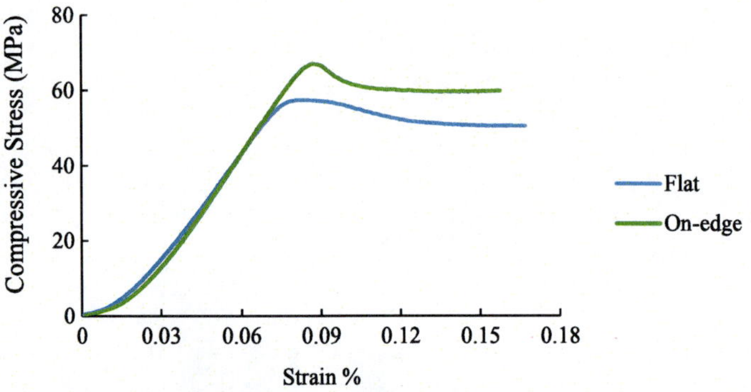

Fig. 4 Stress–strain curves for compressive stress

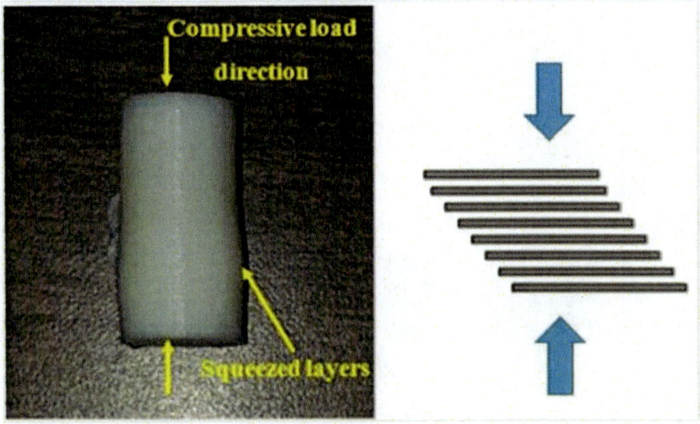

Fig. 5 Failure mechanism of compression specimens built-in on-edge orientation

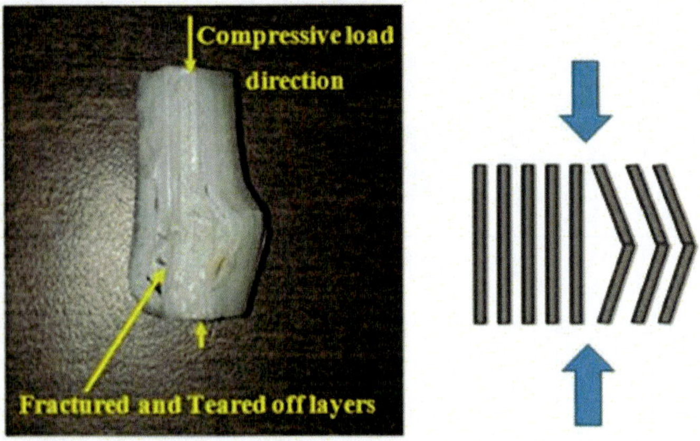

Fig. 6 Failure mechanism of compression specimens built-in flat orientation

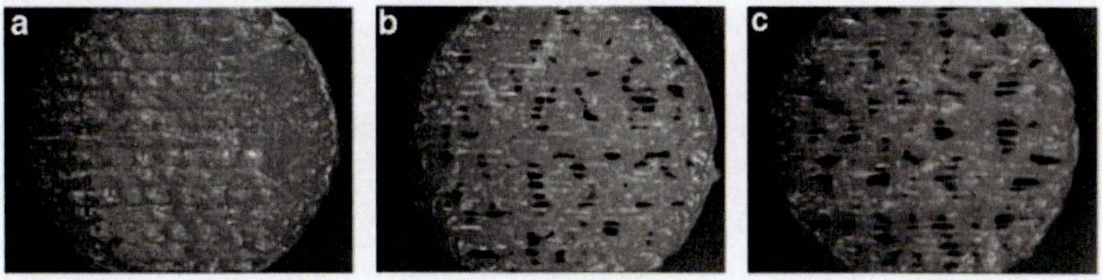

Fig. 7 Parts printed with **a** 100% **b** 80% **c** 60% infill density

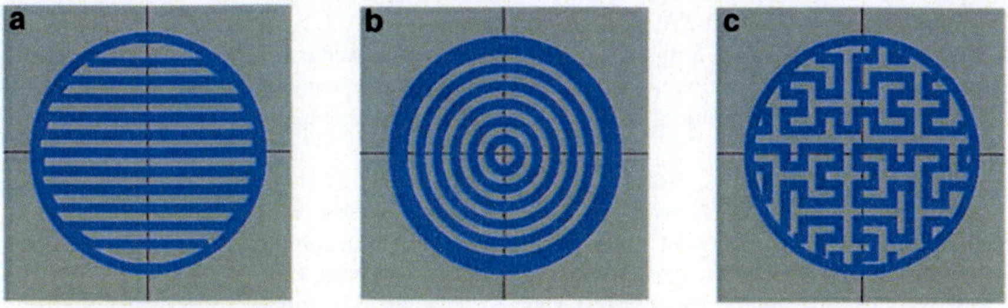

Fig. 8 Infill pattern **a** rectilinear **b** concentric **c** Hilbert curve

4 Conclusion

In the present study, a full factorial experiment was carried out to investigate the effects of part build orientation, infill pattern and infill density on compression properties of the FDM printed specimen. PLA material was used to fabricate specimen with variations in build parameters on omega dual extruder FDM machine. It was observed that on-edge orientation shows the higher values of compressive strength as the compressive load is taken by layer, not by the individual rasters. Concerning testing results, a decrease in strength with the decrement of infill density was observed. The parts printed with Hilbert curve pattern resulted in lower values of compression strengths compared to rectilinear and concentric patterns.

References

1. Bikas H, Stavropoulos P, Chryssolouris G (2016) Additive manufacturing methods and modelling approaches: a critical review. Int J Adv Manu Technol 83(1–4):389–405
2. Chin Ang K, Fai Leong K, Kai Chua C, Chandrasekaran M (2006) Investigation of the mechanical properties and porosity relationships in fused deposition modelling-fabricated porous structures. Rapid Prototyping J 12(2):100–105
3. Ahn SH, Montero M, Odell D, Roundy S, Wright PK (2002) Anisotropic material properties of fused deposition modeling ABS. Rapid Prototyping J 8(4):248–257

4. Lee CS, Kim SG, Kim HJ, Ahn SH (2007) Measurement of anisotropic compressive strength of rapid prototyping parts. J Mater Process Technol 187:627–630. https://doi.org/10.1016/j.jmatprotec.2006.11.095
5. Sood AK, Ohdar RK, Mahapatra SS (2012) Experimental investigation and empirical modelling of FDM process for compressive strength improvement. J Adv Res 3(1):81–90
6. Hernandez R, Slaughter D, Whaley D, Tate J, Asiabanpour B (2016) Analyzing the tensile, compressive, and flexural properties of 3D printed ABS P430 plastic based on printing orientation using fused deposition modeling. 27th Annual International Solid Freeform Fabrication Symposium. Austin, TX, pp 939–950
7. Motaparti KP, Taylor G, Leu MC, Chandrashekhara K, Castle J, Matlack M (2016) Effects of build parameters on compression properties for ULTEM 9085 parts by fused deposition modeling. In: Proceedings of the 27th annual international solid freeform fabrication symposium. Austin, TX, USA, pp 8–10
8. Upadhyay K, Dwivedi R, Singh AK (2017) Determination and comparison of the anisotropic strengths of fused deposition modeling P400 ABS. In: Wimpenny DI, Pandey PM, Kumar LJ (eds) Advances in 3D printing & additive manufacturing technologies. Springer, Singapore, pp 9–28
9. Mishra SB, Abhishek K, Satapathy MP, Mahapatra SS (2017) Parametric appraisal of compressive strength of fdm build parts. Mater Today Proc 4(9):9456–9460
10. Jami H, Masood SH, Song WQ (2018) Dynamic stress-strain compressive behaviour of FDM made ABS and PC parts under high strain rates. IOP Conf Series Mater Sci Eng 377(1):012153
11. Dave HK, Rajpurohit SR, Patadiya NH, Dave SJ, Thambad SS, Srinivasn VP, Sheth KV (2019) Compressive strength of PLA based scaffolds: effect of layer height, infill density and print speed. Int J Mod Manu Technol 11(1):21–27

An Experimental Study of Influence of Gradient Parameters on Compressive Strength, Stiffness, and Specific Energy Absorption (SEA) of Auxetic Structures Fabricated by FDM

Swapnil Vyavahare, Soham Teraiya, and Shailendra Kumar

1 Introduction

Structures whose properties are determined by their geometric parameters are widely used in recent times [1]. Negative Poisson's ratio structures or auxetic structure are one of such meta-materials or structures. They exhibit desirable mechanical properties like high strength-to-weight ratio, indentation resistance, impact resistance, bending stiffness, fracture toughness, vibration damping, hardness, excellent shock absorption capacity than the conventional positive Poisson's ratio structures [2, 3]. They have applications in field of aerospace, automotive, biomedical, composites, defense, sensors, and textile [4–6]. With recent advances in fused deposition modeling (FDM) which is one technique of additive manufacturing (AM), fabrication of varying density parts has become possible [7–11]. A remarkable advantage of this technique is that it is economical than other AM techniques. Also, the strength of the parts made of FDM is sufficiently high so that they can replace parts made of conventional plastic manufacturing techniques like injection molding. AM is capable to fabricate structures according to functional requirement. Therefore, it is important to realize relation between gradient parameters of a structure and responses like strength, stiffness, specific energy absorption (SEA) to fully utilize the potential of this technique. Many researchers have made efforts to investigate in this domain [12–17]. Also, various researchers have studied gradient structures made by different AM techniques for different loading conditions. Lira et al. [12] studied gradient auxetic structures for use as cores in aero-engine fan blade. Their natural frequencies and mode shapes for first three fundamental modes are studied. The optimized configuration resulted in decrease in mass of fan blade, reduction in dynamic modal displacement, and decrease in first three natural frequencies. Boldrin

S. Vyavahare · S. Teraiya · S. Kumar (✉)
Sardar Vallabhbhai National Institute of Technology, Surat 395007, Gujarat, India
e-mail: skbudhwar@gmail.com

H. K. Dave and D. Nedelcu (eds.), *Advances in Manufacturing Processes*, Lecture Notes in Mechanical Engineering, https://doi.org/10.1007/978-981-15-9117-4_23

et al. [13] studied visco-acoustic behavior of auxetic gradient honeycomb structures. Two types of gradient cellular configurations are studied by varying internal cell wall angle and by varying cell wall aspect ratio across surface of honeycomb. It was found that only specific gradient topology shows significant sensitivity of dynamic properties versus change in gradient geometry. Hou et al. [14] proposed novel design for energy dissipating structures using gradient auxetic configuration. Optimization is also performed to minimize mass of crash box with constraints of crash-resistant force and energy absorption. Chen et al. [15] studied cylindrical structures with triangular and hexagonal configurations for impact loading. They observed that normalized plastic energy absorption is affected by relative density. Ratio of cell wall to skin thickness is found a vital factor for determining SEA and deformation mode. By keeping positive density gradient along crushing direction enhances energy absorption at early stage. Ituarte et al. [16] proposed a workflow for design and manufacturing that simultaneously integrates material design, structural design, and product fabrication of functionally graded materials (FGMs). The proposed approach is also validated on FGM tensile structures with different material gradient. Jefferson et al. [17] studied effect of cell wall thickness gradient, impact energy and cell topology on energy absorption capacity, and damage mechanism of honeycombs. Geometrical tailoring resulted in change in failure mode from brittle fracture to progressive damage of cell walls.

The above literature review reveals that although efforts have been made to study influence of FDM gradient parameters on mechanical properties of standard components, limited investigations have been done to study the influence of gradient parameters of auxetic structures on combined mechanical and physical properties with manufacturing constraints. Therefore, the present study is focused on investigating the influence of gradient parameters on compressive strength, stiffness, and SEA of auxetic structures of ABS and PLA fabricated by FDM. Experiments are designed on the basis of response surface methodology, and analysis of experimental results is performed using ANOVA.

2 Methodology

Methodology of present study involves experimental design, CAD modeling, fabrication of parts, and measurement of responses of auxetic structures. These steps are briefly described as under.

2.1 *Experimental Design*

Three gradient parameters of auxetic structure, namely thickness coefficient (α), length coefficient (β), and height ratio (h_r), are considered in the present study. Levels of these selected parameters as given in Table 1 are decided on the basis

Table 1 Gradient parameters and their range

Gradient parameters	Range	
Thickness coefficient (α)	0.125	0.25
Length coefficient (β)	0.5	1
Height ratio (h_r)	0.8	1.2

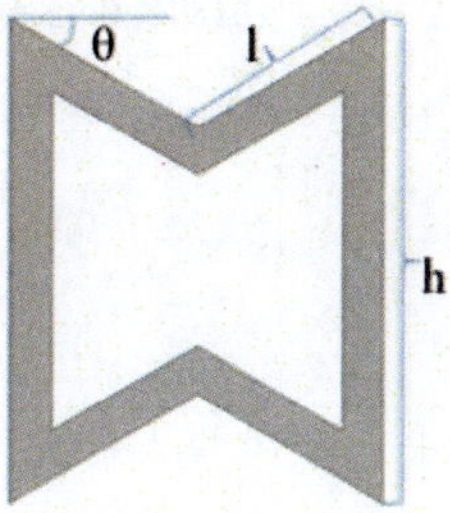

Fig. 1 Re-entrant auxetic unit cell

of literature review. Re-entrant auxetic unit cell is shown in Fig. 1. On the basis of gradient parameters, geometrical parameters (angle, height, and arm-length) are determined (as given in Table 2) using Eq. 1.

$$\begin{aligned} &t = \alpha \times h, \\ &l = \beta \times h, \\ &\text{Height of next row} = h_r \times (\text{height of current row}) \end{aligned} \tag{1}$$

where t = thickness of cell wall, l = arm-length of unit cell, h = height of unit cell, and θ = re-entrant angle of unit cell as shown in Fig. 1.

Table 2 lists gradient parameters and corresponding geometrical parameter of each row of unit cell. Using central composite design (CCD) of response surface methodology (RSM), experiments are designed to investigate the influence of factors on strength, stiffness, and SEA. Table 2 gives design of experiments. According to respective gradient parameter of a structure, relative density is varied within the auxetic structure in range of 0.206–0.69. Relative density is important factor for cellular solid and is calculated using following Eq. 2 [18].

$$\left(\frac{\rho *}{\rho s}\right) = \frac{\left(\frac{t}{l}\right) \times \left(\left(\frac{h}{l}\right) + 2\right)}{2 \times \cos\theta \times \left(\left(\frac{h}{l}\right) + \sin\theta\right)} \tag{2}$$

where $\rho*$ = density of auxetic structure,

ρ_s = density of solid material from which struts re made,

$\rho*/\rho_s$ = relative density of auxetic structure.

Table 2 Design of experiments

Run no.	Gradient parameter (α, β, h)	Row no.	Angle (°)	*h* (mm)	*l* (mm)	*t* (mm)	Relative density
1	0.25, 0.5, 08	1	−5	8.000	4	2.000	0.525
		2	−5	6.400	4	2.000	0.597
		3	−5	5.120	4	2.000	0.690
2	0.1875, 0.75, 0.8	1	−5	8.000	6	1.500	0.336
		2	−5	6.400	6	1.500	0.393
		3	−5	5.120	6	1.500	0.467
3	0.125, 1, 0.8	1	−5	8.000	8	1.000	0.206
		2	−5	6.400	8	1.000	0.246
		3	−5	5.120	8	1.000	0.300
4	0.25, 1, 0.8	1	−5	8.000	8	2.000	0.412
		2	−5	6.400	8	2.000	0.493
		3	−5	5.120	8	2.000	0.599
5	0.25, 0.5, 1.2	1	−5	8.000	4	2.000	0.525
		2	−5	9.600	4	2.000	0.477
		3	−5	11.52	4	2.000	0.438
6	0.1875, 0.75, 1	1	−5	8.000	6	1.500	0.336
		2	−5	8.000	6	1.500	0.336
		3	−5	8.000	6	1.500	0.336
7	0.25, 1, 1.2	1	−5	8.000	8	2.000	0.412
		2	−5	9.600	8	2.000	0.361
		3	−5	11.52	8	2.000	0.319
8	0.25, 0.75, 1	1	−5	8.000	6	2.000	0.448
		2	−5	8.000	6	2.000	0.448
		3	−5	8.000	6	2.000	0.448
9	0.1875, 0.75, 1.2	1	−5	8.000	6	1.500	0.336
		2	−5	9.600	6	1.500	0.299
		3	−5	11.52	6	1.500	0.268
10	0.1875, 1, 1	1	−5	8.000	8	1.500	0.309
		2	−5	8.000	8	1.500	0.309
		3	−5	8.000	8	1.500	0.309
11	0.1875, 0.5, 1	1	−5	8.000	4	1.500	0.394
		2	−5	8.000	4	1.500	0.394
		3	−5	8.000	4	1.500	0.394
12	0.125, 0.75, 1	1	−5	8.000	6	1.000	0.224
		2	−5	8.000	6	1.000	0.224

(continued)

Table 2 (continued)

Run no.	Gradient parameter (α, β, h)	Row no.	Angle (°)	*h* (mm)	*l* (mm)	*t* (mm)	Relative density
		3	−5	8.000	6	1.000	0.224
13	0.125, 0.5, 1.2	1	−5	8.000	4	1.000	0.262
		2	−5	9.600	4	1.000	0.239
		3	−5	11.52	4	1.000	0.219
14	0.125, 1, 1.2	1	−5	8.000	8	1.000	0.206
		2	−5	9.600	8	1.000	0.180
		3	−5	11.52	8	1.000	0.160
15	0.125, 0.5, 0.8	1	−5	8.000	8	1.000	0.206
		2	−5	6.400	8	1.000	0.246
		3	−5	5.120	8	1.000	0.300

Fig. 2 Auxetic structure

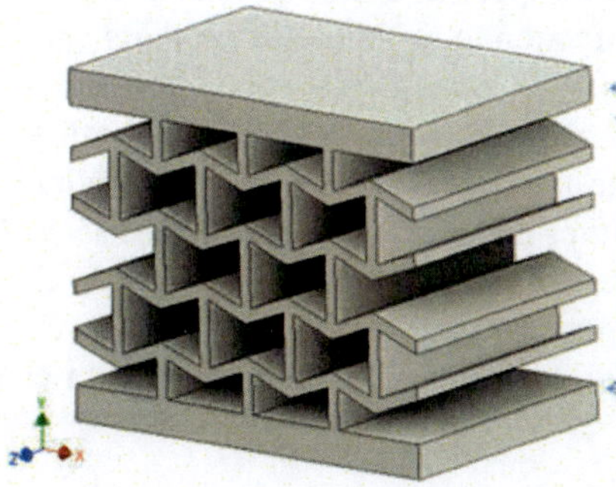

2.2 *CAD Modeling*

As per experimental design, fifteen different configurations of auxetic structure are modeled. Unit cells of each configuration are modeled using Autodesk Inventor 2020 software, arrayed in XY plane, and then extruded along Z-direction. To avoid out-of-plane buckling and to focus on 2D in-plane compressive loading, the thickness of sandwich structure is taken as 40 mm [19]. The thickness of upper and lower face sheet is 5 mm for all configurations. Figure 2 shows auxetic structure. All geometrical files are saved in STL format.

2.3 *Fabrication of Parts*

STL files of modeled structures are imported in CURA 4.2.1 software. GCODE files are prepared and then fed to FDM machine (Model: Delta 2040, M/s. WASP, Italy)

Fig. 3 Fabricated auxetic structures

one-by-one for fabrication of specimens. It is provided with a nozzle of 0.4 mm diameter. For fabrication of structures, layer thickness is kept constant as 0.2 mm. The print temperatures are taken as 240 and 210 °C, and bed temperatures are 100 and 60 °C for ABS and PLA material, respectively. Infill percentage is kept as 100% with inner and outer wall speed as 60 and 30 mm/s, respectively. Specimens are manufactured along Z-direction to avoid use of supports and impact of build orientation on mechanical properties. Also, *Z*-axis as build orientation and *Y*-axis as loading direction produce components with maximum strength as load is carried axially along fibers [20]. Time for fabrication for each configuration is noted using stopwatch. Weight of structures is measured using a digital scale. Quasi-two-dimensional specimens are fabricated with same number of unit cells (3 $\times$ 3) which gives cellular structure with approximate size of 50 $\times$ 40 $\times$ 40 mm^3. Thus, total 30 structures (15 structures each of ABS and PLA material) are fabricated as shown in Fig. 3.

2.4 Measurement of Responses of Auxetic Structures

Quasi-static uniaxial compressive tests are performed on universal testing machine (UTM) at a crosshead speed of 5 mm/min. Compressive loading is applied in downward-direction along *Y*-axis. Load-displacement curves, initial stiffness, and energy absorbed are assessed. Failure mechanisms of specimens are evaluated. Applied force is measured by machine load cell. Tests are conducted according to ASTM C365 which is a standard test method for flatwise compression properties of sandwich cores [21, 22]. Load-displacement curve measured from uniaxial compression test is converted in stress-strain curve by measuring dimensions of specimens. Stiffness is calculated from initial linear region of stress-strain curve. Structures are tested till 60% strain [19]. Nominal stress and strain are computed according to dimensions of the structure. The energy absorbed during compression is calculated for each structure by integrating the stress-strain curve (i.e., area under the curve) as Eq. (3). SEA of cellular material is given by Eq. (4). Table 3 lists the values of responses.

Table 3 Experimental results

	ABS structures			PLA structures		
Run no.	Strength MPa	Stiffness MPa	SEA J/gm	Strength MPa	Stiffness MPa	SEA J/gm
1	6.21	41.46	128.6	12.17	147.24	160.2
2	6	60	149.91	9.8	117.38	401.3
3	5.59	84.35	226.22	12.89	159.59	267.38
4	3.88	27.66	87.21	5.11	88.37	384.13
5	5.45	60.7	212.06	7.89	166.19	616.81
6	4.7	56.65	146.16	8.28	126.67	283.5
7	1.2	32.5	199.94	0.967	18.11	143.54
8	4.5	45	164.74	5.24	111.47	330.14
9	3.61	80.222	154.43	5.01	55.67	426.35
10	3.12	45.7	157.812	5.82	64.67	126.35
11	6.12	61.2	185.23	10.2	113.33	136.81
12	5.59	85.85	169.786	10.42	115.78	164.51
13	5.42	130.6	177.59	9.75	60.16	107.24
14	3.05	60.1	60.45	5.06	37.69	137.26
15	7.22	56.39	157.15	11.13	127.2	68.9

$$W = \int_{0}^{\varepsilon} \sigma(\varepsilon)\mathrm{d}\varepsilon \tag{3}$$

$$\mathrm{SEA} = \frac{W}{\Delta\rho \times \rho_s} \tag{4}$$

where $\Delta\rho =$ relative density of structure, $\rho_s =$ density of bulk material

3 Results and Discussion

Summary of means and standard deviation of responses to factors is given in Table 4. Analysis of variance (ANOVA) is performed to determine significance of gradient parameters on responses. Table 5 lists main and interaction effect of factors on the responses from ANOVA.

The main effect plots for the compressive strength are shown in Fig. 4. It is observed that length coefficient and height ratio are most significant parameters for compressive strength of auxetic structures. Compressive strength increases with decrease in length coefficient and height ratio. As length coefficient decreases, arm-length decreases which results in increase in relative density of unit cell, and thus, it increases compressive strength of structure. With decrease in height ratio, relative

Table 4 Summary of means and standard deviation of responses to factors

		ABS structures						PLA structures					
Summary of responses	Levels	Strength		Stiffness		SEA		Strength		Stiffness		SEA	
Factor		Mean	SD	Mean	SD	Mean	SD	Mean	SD	Mean	SD	Mean	SD
Thickness gradient	0.125	5.374	1.491	83.458[a]	29.611	158.239	60.608	9.85	2.614	100.084	44.743	149.058	67.159
	0.1875	4.71	1.359	60.754	12.482	158.708	15.471	7.822	2.083	95.544	29.343	274.862	126.589
	0.25	4.248	1.922	41.464	12.781	158.51	51.481	6.275	3.687	106.276	51.747	326.964[a]	172.478
Length gradient	0.5	6.084[a]	0.733	70.07	34.771	172.126[a]	31.285	10.228[a]	1.434	122.824	36.091	217.992	201.733
	0.75	4.88	0.942	65.544	17.027	157.005	9.972	7.75	2.254	105.394	25.352	321.16	93.301
	1	3.368	1.586	50.062	22.944	146.326	71.167	5.969	3.860	73.686	49.114	211.732	100.304
Height ratio	0.8	5.78	1.219	53.972	21.295	149.818	50.648	10.22	2.759	127.96[a]	24.707	256.382	127.939
	1	4.806	1.149	58.88	16.614	164.745	14.479	7.992	2.151	106.384	21.513	208.262	82.758
	1.2	3.746	1.781	72.824	36.495	160.894	60.316	5.735	2.983	67.564	51.501	286.24	201.746
	Total	4.78	0.219	61.89	7.02	158.49	10.79	7.98	0.614	100.63	15.83	250.29	5.86

[a]Maximun mean value for each response for all levels of all factors, SD—standard deviation

Table 5 Main and interaction effect of factors for responses from ANOVA

Factors	ABS structures						PLA structures					
	Strength		Stiffness		SEA		Strength		Stiffness		SEA	
	F (df = 1)	p	F (df = 1)	p	F (df = 1)	p	F (df = 1)	p	F (df = 1)	p	F (df = 1)	p
(α)	66.1	<0.05	89.45	<0.05	0.0016	0.9695	84.86	<0.05	0.3824	0.556	2305.72	<0.05
(β)	384.57	<0.05	20.31	<0.05	14.29	<0.05	120.44	<0.05	24.08	<0.05	2.85	0.166
(h_r)	215.69	<0.05	18.03	<0.05	2.63	0.1487	133.57	<0.05	36.38	<0.05	64.95	<0.05
$(\alpha \times \beta)$	17.35	<0.05	0.0007	0.9791	0.0318	0.8636	40.57	<0.05	23.46	<0.05	831.69	<0.05
$(\alpha \times h_r)$	2.11	0.1895	1.70	0.2337	125.17	<0.05	0.2057	0.6639	9.45	<0.05	345.09	<0.05
$(\beta \times h_r)$	18.44	<0.05	32.30	<0.05	26.43	<0.05	13.23	<0.05	10.35	<0.05	2729.54	<0.05
$(\alpha \times \beta \times h_r)$	3.62	0.0984	17.92	<0.05	49.83	<0.05	14.41	<0.05	0.5885	0.4681	1018.31	<0.05

α = thickness gradient, β = length gradient, h_r = height ratio

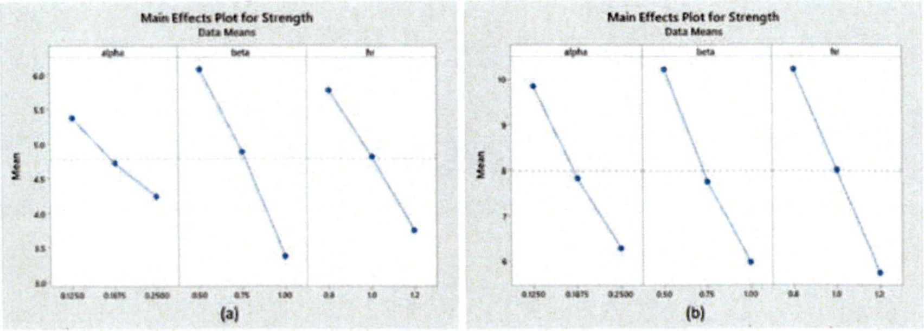

Fig. 4 Main effect plots for compressive strength of structures, **a** ABS, **b** PLA

density of bottom rows of unit cell increases which results in increase in compressive strength. Figure 5 shows the interaction effect plots for compressive strength of structures. It is observed that interactions are present between all the three factors as the lines are not parallel [23]. From the interaction plots, it is observed that compressive strength decreases with increase in length coefficient and height ratio for all values of thickness coefficient (0.125, 0.1875, and 0.25 shown by blue, brown, green lines, respectively). Also for all values of length coefficient (0.5, 0.75, and 1.00), compressive strength decreases with increase in height ratio. Compressive strength decreases with increase in height ratio for all values of thickness coefficient and length coefficient. This is due to increase in relative density of structure.

The main effect plots for the compressive stiffness of structures are shown in Fig. 6. It is observed that thickness coefficient and length coefficient are the most significant parameters influencing compressive stiffness of ABS structures, while length coefficient and height ratio are the most significant for PLA structures. Compressive stiffness increases with decrease in values of above significant parameters. With decrease in values of these parameters, relative density of the auxetic structure increases which results in increase in compressive stiffness. Synergistic interaction is also observed between thickness coefficient and length coefficient and thickness coefficient and height ratio as shown in Fig. 7. Interaction between thickness coefficient and length

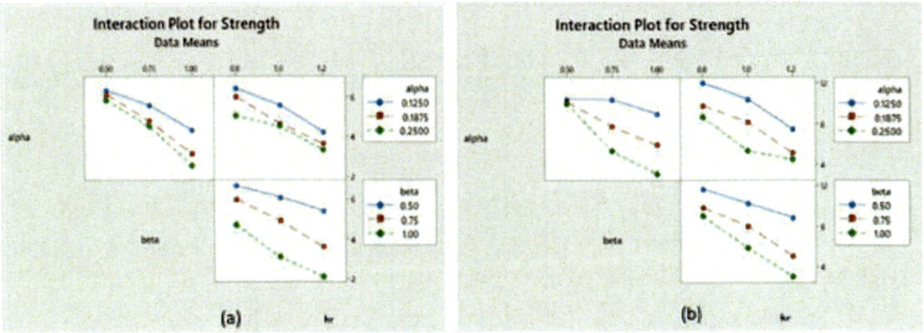

Fig. 5 Interaction effect plots for compressive strength of structures, **a** ABS, **b** PLA

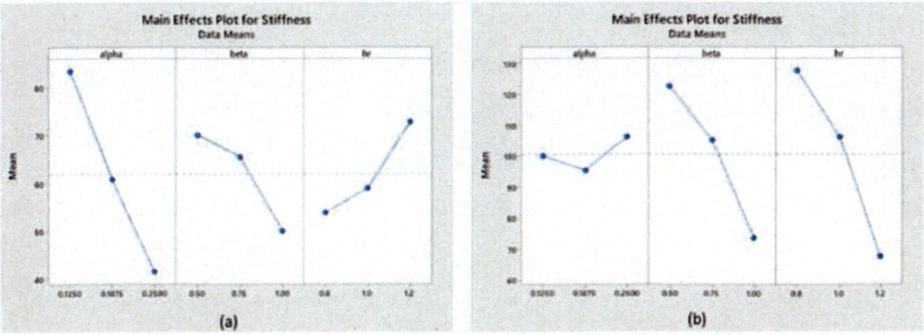

Fig. 6 Main effect plots for compressive stiffness of structures, **a** ABS, **b** PLA

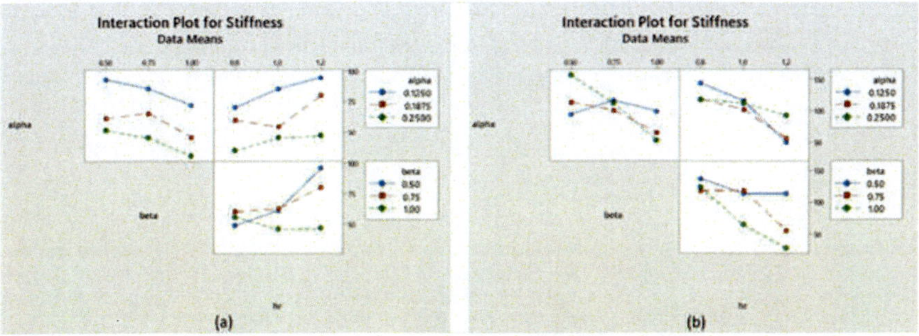

Fig. 7 Interaction effect plots for compressive stiffness of structures, **a** ABS, **b** PLA

coefficient indicates that compressive stiffness decreases with increase in length coefficient for all values of thickness coefficient (0.125, 0.1875, and 0.25) except 0.125 thickness coefficient in case of PLA structures. From interaction between thickness coefficient and height ratio, it is observed that for all values of thickness coefficient compressive stiffness increases in case of ABS structures, while it decreases in case of PLA structures with increase in height ratio.

Also, for all values of length coefficient, compressive stiffness increases in case of ABS structures, while it decreases in case of PLA structures with increase in height ratio. Such difference between the behavior of stiffness of structures of ABS and PLA materials is because of inherent ductile and brittle nature, respectively.

The main effect plots for the SEA of auxetic structures are shown in Fig. 8. It is observed that length coefficient and thickness coefficient are the most influential parameters for SEA of ABS and PLA structures, respectively. SEA increases with decrease in length coefficient for ABS structures and increase in thickness coefficient for PLA structure. In case of ABS structures, with decrease in length coefficient, energy applied by compressive loading gets transferred to adjacent cell walls which results in increase in SEA. In case of PLA structure, with increase in thickness coefficient, more material is available inside unit cell of auxetic structure for energy absorption, thus resulting in increase in SEA. The difference in energy absorption is

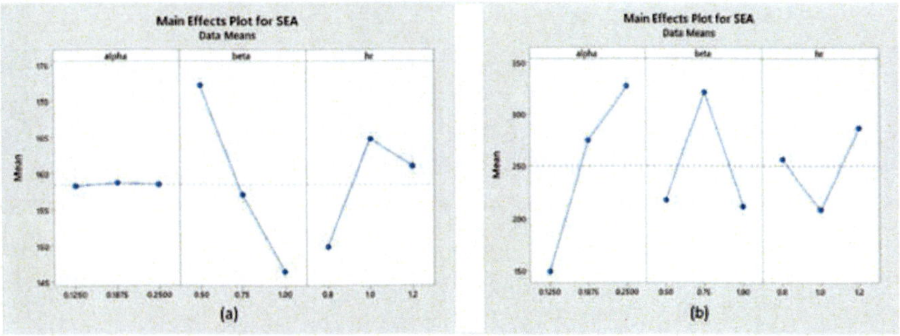

Fig. 8 Main effect plots for SEA of structures, **a** ABS, **b** PLA

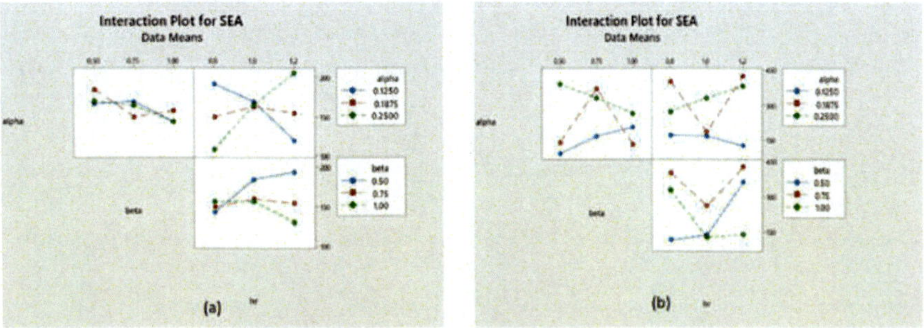

Fig. 9 Interaction effect plots for SEA of structures, **a** ABS, **b** PLA

observed in auxetix structures because of inherent deformation mechanism of ABS structures (i.e., elastic-plastic) and PLA structures (i.e., brittle) [15]. Synergistic interaction is also observed between all three gradient parameters as shown in Fig. 9. As shown in interaction plots, difference in SEA behavior is observed for structures made of both materials due to variation in deformation mechanism.

Failure mechanisms of auxetic structures made of ABS and PLA materials depict distinct behavior due to inherent ductile and brittle nature of the materials, respectively. For ABS structures, initially with downward displacement of top platen of UTM, vertical cell walls of unit cells gradually buckle. This is followed by bending of inclined cell wall till start of densification. While in PLA structures, no such buckling of vertical cell walls is observed. PLA structures resist load with minimum strain giving higher value of stiffness for most of the structures and fail by abrupt rupture of vertical cell walls.

4 Conclusion

The present paper describes an experimental study of influence of gradient parameters on compressive strength, stiffness, and SEA of auxetic structures fabricated by FDM. Followings are the major findings of present study:

1. Length coefficient and height ratio are most significant parameters for compressive strength of auxetic structures. Compressive strength increases with decrease in length coefficient and height ratio.
2. In ABS structures, stiffness increases with decrease in thickness coefficient and length coefficient, while in PLA structures, stiffness increases with an decrease in length coefficient and height ratio.
3. SEA of ABS structures increases with decrease in length coefficient, while it increases with increase in thickness coefficient in PLA structure.

The findings of the present study are useful for selection of gradient parameters to improve compressive strength, stiffness, and SEA of auxetic structures mainly used in aerospace and automotive sectors. Future work will be focused on investigating effect of gradient on behavior of auxetic structure under shear and flexural loading.

References

1. Yu X, Zhou J, Liang H, Jiang Z, Wu L (2018) Mechanical metamaterials associated with stiffness, rigidity and compressibility: a brief review. Prog Mater Sci 94:114–173. https://doi.org/10.1016/j.pmatsci.2017.12.003
2. Álvarez Elipe J, Díaz Lantada A (2012) Comparative study of auxetic geometries by means of computer-aided design and engineering. Smart Mater Struct 21:105004. https://doi.org/10.1088/0964-1726/21/10/105004
3. Zhang X, Yang D (2016) Mechanical properties of auxetic cellular material consisting of re-entrant hexagonal honeycombs. Materials 9:900. https://doi.org/10.3390/ma9110900
4. Durgun I (2015) Sheet metal forming using FDM rapid prototype tool. Rapid Prototyping J 21:412–422. https://doi.org/10.1108/RPJ-01-2014-0003
5. Meier M, Tan K, Lim M, Chung L (2019) Unlocking innovation in the sport industry through additive manufacturing. Bus Process Manage J 25:456–475. https://doi.org/10.1108/BPMJ-10-2017-0285
6. García-García R, González-Palacios M (2018) Method for the geometric modeling and rapid prototyping of involute bevel gears. Int J Adv Manuf Technol 98:645–656. https://doi.org/10.1007/s00170-018-2246-9
7. Lam C, Mo X, Teoh S, Hutmacher D (2002) Scaffold development using 3D printing with a starch-based polymer. Mater Sci Eng C 20:49–56. https://doi.org/10.1016/S0928-4931(02)00012-7
8. Negis E (2009) A short history and applications of 3D printing technologies in Turkey. US-TURKEY Workshop Rapid Technol 24:23–30
9. Korpela J, Kokkari A, Korhonen H, Malin M, Närhi T, Seppälä J (2012) Biodegradable and bioactive porous scaffold structures prepared using fused deposition modeling. J Biomed Mater Res B Appl Biomater 101B:610–619. https://doi.org/10.1002/jbm.b.32863
10. Okwuosa T, Stefaniak D, Arafat B, Isreb A, Wan K, Alhnan M (2016) A lower temperature FDM 3D printing for the manufacture of patient-specific immediate release tablets. Pharm Res 33:2704–2712. https://doi.org/10.1007/s11095-016-1995-0

11. Bayar M, Aziz Z (2018) Rapid prototyping and its role in supporting architectural design process. J Archit Eng 24:05018003. https://doi.org/10.1061/(ASCE)AE.1943-5568.0000307
12. Lira C, Scarpa F, Rajasekaran R (2011) A gradient cellular core for aeroengine fan blades based on auxetic configurations. J Intell Mater Syst Struct 22:907–917. https://doi.org/10.1177/1045389X11414226
13. Boldrin L, Hummel S, Scarpa F, Di Maio D, Lira C, Ruzzene M, Remillat C, Lim T, Rajasekaran R, Patsias S (2016) Dynamic behaviour of auxetic gradient composite hexagonal honeycombs. Compos Struct 149:114–124. https://doi.org/10.1016/j.compstruct.2016.03.044
14. Hou W, Yang X, Zhang W, Xia Y (2017) Design of energy-dissipating structure with functionally graded auxetic cellular material. Int J Crashworthiness 23:366–376. https://doi.org/10.1080/13588265.2017.1328764
15. Chen L, Zhang J, Du B, Zhou H, Liu H, Guo Y, Li W, Fang D (2018) Dynamic crushing behavior and energy absorption of graded lattice cylindrical structure under axial impact load. Thin-Walled Struct 127:333–343. https://doi.org/10.1016/j.tws.2017.10.048
16. Ituarte I, Boddeti N, Hassani V, Dunn M, Rosen D (2019) Design and additive manufacture of functionally graded structures based on digital materials. Addit Manuf 30:100839. https://doi.org/10.1016/j.addma.2019.100839
17. Andrew J, Ubaid J, Hafeez F, Schiffer A, Kumar S (2019) Impact performance enhancement of honeycombs through additive manufacturing-enabled geometrical tailoring. Int J Impact Eng 134:103360. https://doi.org/10.1016/j.ijimpeng.2019.103360
18. Raeisi S, Tapkir P, Ansari F, Tovar A (2019) Design of a hybrid honeycomb unit cell with enhanced in-plane mechanical properties. SAE Tech Pap Ser. https://doi.org/10.4271/2019-01-0710
19. Ingrole A, Hao A, Liang R (2017) Design and modeling of auxetic and hybrid honeycomb structures for in-plane property enhancement. Mater Des 117:72–83. https://doi.org/10.1016/j.matdes.2016.12.067
20. Ahn S, Baek C, Lee S, Ahn I (2003) Anisotropic tensile failure model of rapid prototyping parts—fused deposition modeling (FDM). Int J Mod Phys B 17:1510–1516. https://doi.org/10.1142/S0217979203019241
21. Scarpa F, Blain S, Lew T, Perrott D, Ruzzene M, Yates J (2007) Elastic buckling of hexagonal chiral cell honeycombs. Compos A Appl Sci Manuf 38:280–289. https://doi.org/10.1016/j.compositesa.2006.04.007
22. Xu J, Wu Y, Wang L, Li J, Yang Y, Tian Y, Gong Z, Zhang P, Nutt S, Yin S (2018) Compressive properties of hollow lattice truss reinforced honeycombs (Honeytubes) by additive manufacturing: Patterning and tube alignment effects. Mater Des 156:446–457. https://doi.org/10.1016/j.matdes.2018.07.019
23. Antony J (2014) Design of experiments for engineers and scientists. Elsevier

Study on Compression Properties of ABS and FPU Parts Printed Using SLA

Ravi Teja Karumuri, Harshit K. Dave, Shilpesh R. Rajpurohit, and Ashish R. Prajapati

1 Introduction

Additive manufacturing (AM) is a technique in which material is deposited in a layer by layer approach to obtain the required part from the 3D CAD data. Stereolithography (SLA) process is the oldest among the AM processes. SLA is a vat photopolymerization technique [1], which uses light to cure the liquid monomer. Light initiates the polymerization reaction which cures the liquid monomer into a solid polymer. This light is scanned over the vat of liquid resin in the required shape, layer by layer. The parts that are fabricated using the SLA process have a better surface finish and higher dimensional accuracy as compared to parts produced using Fused Deposition Modelling (FDM) process. SLA process is widely used in the field of medicine [2], jewellery [3] and engineering [4] for producing parts with good dimensional tolerances. Because of slow and costly machines, SLA process is less widely utilized than the FDM process. Also, the post-processing operations such as cleaning of uncured resin and post-curing make the SLA process less encouraging to use the process.

Liquid crystal display-Stereolithography (LCD-SLA) process, a variant of Stereolithography (SLA), shares the same principle to cure the liquid resin through photopolymerization technique [1]. While the SLA process uses a single laser source, LCD-SLA cures entire layer by using a 2D array of light-emitting diodes (LEDs) placed below the LCD [5], which is used to display the image as per the slicing of the CAD file. As the entire layer is cured at a time, LCD-SLA process produces the part faster than the SLA process. The mechanical properties of the parts produced using the AM process vary with the process parameters used for the printing. In

R. T. Karumuri · H. K. Dave · S. R. Rajpurohit · A. R. Prajapati (✉)
Department of Mechanical Engineering, Sardar Vallabhbhai National Institute of Technology, Surat 395007, Gujarat, India
e-mail: iamarp5155@gmail.com

H. K. Dave and D. Nedelcu (eds.), *Advances in Manufacturing Processes*, Lecture Notes in Mechanical Engineering, https://doi.org/10.1007/978-981-15-9117-4_24

recent years work has been initiated to study the effect of process parameters on mechanical properties of additively manufactured parts.

In case of samples fabricated using FDM process, it was observed that compressive strength reduces with the decrease in the layer thickness [6], the compressive strength of the part printed in axial direction has more strength than that of the part printed in the transverse direction [7]. Dave et al. [8] in their experiments studied the effect of infill density, layer thickness and print speed on the compressive strength of FDM printed polylactic acid (PLA) parts and concluded that the infill density has the most significant effect on the compressive strength when compared to other parameters. Compressive strength of the samples fabricated using binder jetting process with a layer thickness of 0.1125 mm and printed in X direction showed good compressive strength and toughness [9]. Neuhäuserová et al. [10] in their experiments, studied the compressive behaviour of stainless steel 316L-0407, printed using the selective laser sintering (SLS) process. They conducted experiments on samples oriented in three directions viz., vertical, horizontal and tilted and found that samples printed in the horizontal direction exhibited the highest strength. The decrease of the values of powder layer thickness in SLS processing of Poly(3-hydroxybutyrate), increased compressive mechanical properties of scaffolds [12]. Chockalingam et al. [13] in their experiments studied the effect of layer thickness on tensile and impact strength of Epoxy resin in the SLA process. It was observed that tensile strength and impact strength decreases as the layer thickness increases. Watters and Bernhardt [13], in their work studies the effect of layer height, print orientation, ageing and post-curing height of the standard grey resin fabricated using Form 2 SLA system. Tests were conducted by following ASTM D695-15 standard. Their findings reveal that compressive strength decreases with increasing layer thickness. And the vertical orientation of printing showed more strength than the part printed in the diagonal orientation.

The literature shows that the compressive properties of the parts manufactured using the AM process were affected by the process parameters used to print the part. The present work aims to study the effect of layer thickness and build orientation on the compressive properties of the parts fabricated using LCD-SLA process. For this purpose, five layer thickness and two orientations were selected to study their influence on the energy absorbed by the specimen during compression testing.

2 Experimental Details

In the present work, parts were printed using Acrylonitrile butadiene styrene (ABS) and Flexible Polyurethane (FPU) material, and the printed parts were mechanically tested for the compressive behaviour of the material.

2.1 Fabrication of Samples

In the present work, samples were fabricated using Uniz Slash Plus, an LCD based SLA 3D printer. In this machine, an entire layer of liquid resin is cured by a blue light on to a print bed of 192 × 120 mm. This printer works on the bottom-up approach, so the part printed is hanging from the bed upside down. Uniz software was used for the slicing and to transfer the data to the 3D Printer. Test specimens were printed with five different layer thickness, i.e. 50, 75, 100, 150 and 200 μm and with two orientations, i.e. on long face and on short face as shown in Fig. 1. Two samples of each combination were printed to check the repeatability of the experiments. Total of 40 samples was printed by keeping following constant parameters, as shown in Table 1.

Exposure time used to obtain different layer thickness for ABS and FPU material was shown in Table 2. The printed parts were scrapped from the bed and cleaned in

Fig. 1 Orientation of sample

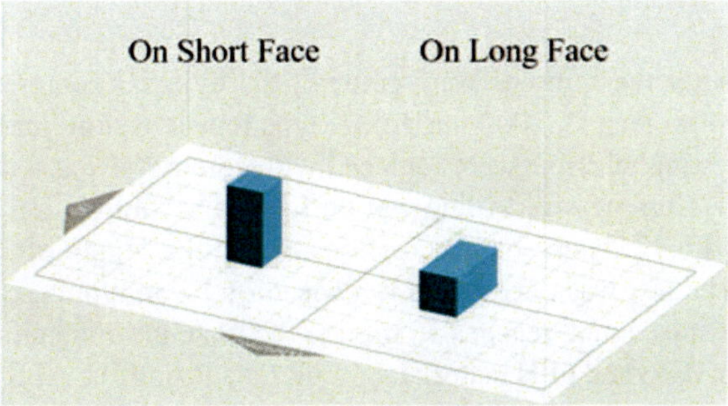

Table 1 Constant parameters during Printing

Parameter	Value
Infill density	100%
LED power	250
Motor Speed	Medium
Rise height	8 mm
Shrink	0 mm

Table 2 Exposure time

Layer thickness (μm)	ABS (s)	FPU (s)
50	2.2	6
75	2.5	6
100	3	6
150	5	9.6
200	6	11

Fig. 2 Compressive test attachment

isopropyl alcohol. Samples were dried using compressed air and stored in the dark place until testing was performed.

2.2 *Compressive Testing*

After the samples were printed, the uniaxial compressive test was performed on Mikrotech PC2000 horizontal type tensile testing machine. Grippers of the tensile testing machine were replaced with compressive test attachment as shown in Fig. 2

Compressive testing was performed according to ASTM D695 using a specimen with 12.7 mm square cross-section with a length of 25.4 mm, as shown in Fig. 3. The test was performed at a speed of 1.3 mm/min. The test was aborted when the displacement reaches 10 mm, to facilitate the comparison of compressive properties of ABS and FPU material. Samples made of FPU, ABS were tested with a load cell of capacity 2 and 20 kN, respectively. Within the displacement of 10 mm, samples made of ABS material reached the maximum capacity of the 20kN load cell, whereas the load in the FPU samples did not exceed 2kN. The load cell of 2kN capacity has lower least count, which results in smoother load vs displacement graph when compared to the 20kN load cell.

3 Results and Discussion

After the compression testing was carried out, energy absorbed by the specimen till 10 mm displacement was calculated by finding the area under the load vs displacement curve using MATLAB. Mean, and standard deviation (SD) of the energy absorbed for the two samples was calculated for ABS and FPU materials, as shown in Tables 3 and 4, respectively.

From Tables 3 and 4, it was clear that energy absorbed by the ABS specimens were more than the FPU samples. The higher energy absorbed may be due to the presence of polar bonding of the nitrile group in the ABS material. These bonds make ABS stronger, thus absorbing more energy.

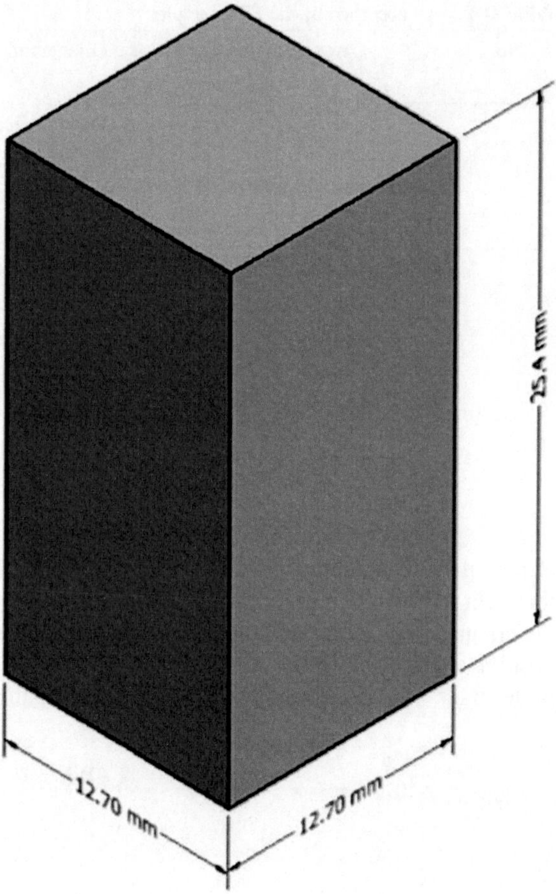

Fig. 3 Geometry of ASTM D695 specimen

Table 3 Energy absorbed by the ABS samples

S. No	Layer thickness (μm)	Orientation	Energy (J)	
			Mean	SD
1	50	On long face	73.21	1.76
2	50	On short face	72.52	1.93
3	75	On long face	70.48	15.30
4	75	On short face	83.15	16.09
5	100	On long face	78.15	4.16
6	100	On short face	89.04	14.54
7	150	On long face	90.67	5.99
8	150	On short face	100.56	10.58
9	200	On long face	93.24	3.40
10	200	On short face	104.15	1.57

Table 4 Energy absorbed by the FPU samples

S. No	Layer thickness (μm)	Orientation	Energy (J)	
			Mean	SD
1	50	On long face	1.83	0.12
2	50	On short face	2.34	0.03
3	75	On long face	1.91	0.02
4	75	On short face	2.37	0.17
5	100	On long face	1.63	0.14
6	100	On short face	2.12	0.11
7	150	On long face	1.79	0.22
8	150	On short face	2.06	0.06
9	200	On long face	2.33	0.30
10	200	On short face	2.43	0.15

Figure 4 shows the effect of layer thickness and build orientation on the energy absorbed by the samples fabricated using ABS material. From Fig. 4, it was observed that with an increase in layer thickness from 50 to 200 μm, energy increases in both the orientations. To obtain higher layer thickness, higher exposure time was used for printing, as shown in Table 2. Higher exposure time results in the larger cure depth in the resin. The larger cure depth may have led to the better bonding of the layers,

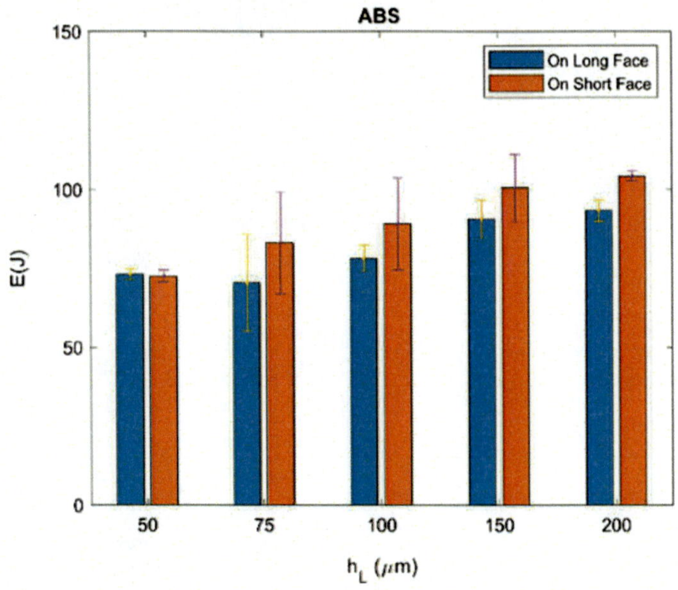

Fig. 4 Effect of layer thickness and build orientation of ABS samples

at higher layer thickness. Thus, requiring more force to compress through the same displacement.

Also, in the case of 'on long face' orientation, fracture of layer parallel to the applied load is observed because of the tension created in the direction perpendicular to the applied load. It may be noted that failure occurred between the two rows of light source out of the matrix of the light sources. So, it may be concluded that in the case of ABS bond between two layers is stronger than that of the bond between arrays of light sources of LCD, which may result in the fracture.

However, in the case of a sample printed on the short face, similar failure was observed, at higher loads, this may be due to more number of layers than the sample printed in long orientation, thus absorbing more energy. Figure 5 shows the fracture in long and short orientations of ABS samples printed with 200 μm layer thickness. From Fig. 5, it was clear that the ABS samples do not exhibit delamination.

Figure 6 shows the effect of layer thickness and build orientation on the energy absorbed by the samples fabricated using FPU material. From Fig. 6, it was observed that above 100 μm layer thickness, the energy absorbed increases with the increase in layer thickness. The higher energy absorbed for 50 and 75 μm layer thickness samples

(a)

(b)

Fig. 5 Fracture in the ABS samples printed in **a** on long face, **b** on short face orientation

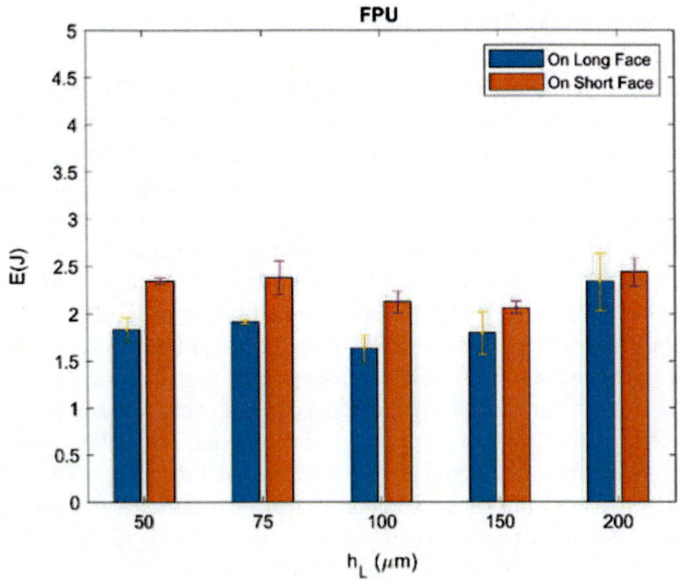

Fig. 6 Effect of layer thickness and build orientation of FPU samples

can be explained by the exposure time of 6 s used for printing of the material. In the case of 50 μm, the resin may be over cured and become brittle when compared to 75 μm. But in the case of 100 μm sample, as the exposure time is same, cure depth is also same. With the same cure depth and higher layer thickness may result in a weaker bond between the layers. This may result in the lower energy absorbed at 100 μm layer thickness among the 50, 75 and 100 μm layer thickness. As the exposure time increases after 100 μm layer thickness, the energy absorbed also increased.

Figure 7 shows the long and short orientations of FPU samples printed with 200 μm layer thickness. In the case of samples printed with FPU material, no visible fracture is observed even after compressing by 10 mm. But in the samples fabricated using 200 μm layer thickness, delamination was observed in samples printed on long face as shown in Fig. 7a.

4 Conclusion

In the present work, compression of LCD-SLA printed ABS and FPU parts having different layer thickness and orientations was carried out to investigate the energy absorbed by the test specimen up to a displacement of 10 mm. From the study, it can be concluded that

- ABS material absorbs more energy than that of the FPU material under similar conditions.

Fig. 7 FPU samples printed in **a** on long face, **b** on short face orientation

(a)

(b)

- Energy absorbed by the specimen is dependent on the build orientation. Specimen built on the short face absorbs more energy than the part built on the long face.
- Samples with 200 μm layer thickness absorb more energy than that of 50, 75, 100 and 150 μm layer thickness.

References

1. Gibson I, Rosen DW (2010) Stucker B (2010) Additive manufacturing technologies. Springer, New York
2. Liu Q, Leu MC, Schmitt SM (2006) Rapid prototyping in dentistry: technology and application. Int J Adv Manuf Technol 29(3–4):317–335. https://doi.org/10.1007/s00170-005-2523-2
3. Ferreira T, Almeida HA, Bártolo PJ, Campbell I (2012) Additive manufacturing in jewellery design. In: Advanced manufacturing processes; biomedical engineering; multiscale mechanics

of biological tissues; sciences, engineering and education; multiphysics; emerging technologies for inspection, vol 4. Presented at the ASME 2012 11th Biennial Conference on Engineering Systems Design and Analysis, ASME, Nantes, France, p 187. https://doi.org/10.1115/esda2012-82789

4. Chen X, Li D, Wu H, Tang Y, Zhao L (2011) Analysis of ceramic shell cracking in stereolithography-based rapid casting of turbine blade. Int J Adv Manuf Technol 55(5–8):447–455. https://doi.org/10.1007/s00170-010-3064-x
5. Wang Z, Martin N, Hini D, Mills B, Kim K (2017) Rapid fabrication of multilayer microfluidic devices using the liquid crystal display-based stereolithography 3D printing system. 3D Print Addit Manuf 4:156–164. https://doi.org/10.1089/3dp.2017.0028
6. Sood AK, Ohdar RK, Mahapatra SS (2012) Experimental investigation and empirical modelling of FDM process for compressive strength improvement. J Adv Res 3(1):81–90. https://doi.org/10.1016/j.jare.2011.05.001
7. Lee CS, Kim SG, Kim HJ, Ahn SH (2007) Measurement of anisotropic compressive strength of rapid prototyping parts. J Mater Process Technol 187:627–630. https://doi.org/10.1016/j.jmatprotec.2006.11.095
8. Dave HK, Rajpurohit SR, Patadiya NH, Dave SJ, Thambad SS, Srinivasn VP, Sheth KV (2019) Compressive strength of PLA based scaffolds: effect of layer height, infill density and print speed. Int J Modern Manuf Technol 11(1):21–27
9. Farzadi A, Solati-Hashjin M, Asadi-Eydivand M, Abu Osman NA (2014) Effect of layer thickness and printing orientation on mechanical properties and dimensional accuracy of 3D printed porous samples for bone tissue engineering. PLoS ONE 9(9):e108252. https://doi.org/10.1371/journal.pone.0108252
10. Neuhäuserová M, Koudelka P, Falta J, Adorna M, Fíla T, Zlámal P (2019) Strain-rate and printing direction dependency of compressive behaviour of 3d printed stainless steel 316l. Acta Polytech CTU Proc 25:68–72. https://doi.org/10.14311/APP.2019.25.0068
11. Pereira TF, Silva MAC, Oliveira MF, Maia IA, Silva JVL, Costa MF, Thiré RMSM (2012) Effect of process parameters on the properties of selective laser sintered Poly(3-hydroxybutyrate) scaffolds for bone tissue engineering: This paper analyzes how laser scan spacing and powder layer thickness affect the morphology and mechanical properties of SLS-made scaffolds by using a volume energy density function. Virtual Phys Prototyping 7(4):275–285. https://doi.org/10.1080/17452759.2012.738551
12. Chockalingam K, Jawahar N, Chandrasekhar U (2006) Influence of layer thickness on mechanical properties in stereolithography. Rapid Prototyping J 12(2):106–113. https://doi.org/10.1108/13552540610652456
13. Watters MP, Bernhardt ML (2018) Curing parameters to improve the mechanical properties of stereolithographic printed specimens. Rapid Prototyping J 24(1):46–51. https://doi.org/10.1108/RPJ-11-2016-0180

Experimental Investigation on Tensile Properties of Nylon Glass Fibre Material Made Using Fused Deposition Modelling Process

Mohit R. Chandravadia and M. K. Chudasama

1 Introduction

Additive manufacturing (AM) is highly advanced manufacturing technology. Several techniques have been introduced in the area of AM based on the method of manufacturing process. Fused deposition modelling (FDM) is one of the AM manufacturing techniques developed by Mr S. Scott crump in late 1980s and designed by Stratasys in 1990. FDM is filament extrusion-based process which fabricates 3D parts from its CAD model. In FDM process, filament made using thermoplastic material is fed in liquefier head with the help of stepper motor. In liquefier head, the filament is melted to semi-liquid state. After that, this melted filament material is extruded from nozzle to deposit roads/beads for filling each layer of the part on the platform which is placed in a chamber at predetermined temperature. The machine control head transfers in X-Y plane, whereas the platform (base) transfers in Z-direction according to selected layer thickness. Several process parameters are involved with FDM, and they have significant influence on part characteristics and efficiency of production. An operator must select process parameters while pre-processing the STL file in the FDM. In this work, literature survey is performed by referring papers on some important topics like optimization and analysis of process parameters, various materials used in FDM and mechanical behaviour of FDM parts.

Effect of FDM process variables (infill percentage, infill pattern, layer thickness and extrusion temperature) on mechanical properties and dimensional accuracy of PLA was studied. The Taguchi approach was used to analyse the influence of various

M. R. Chandravadia (✉)
M.E. (CAD/CAM), Government Engineering College, Dahod, Dahod, Gujarat, India
e-mail: chandravadiamohit@gmail.com

M. K. Chudasama
Mechanical Department, Government Engineering College, Dahod, Dahod, Gujarat, India
e-mail: mkcgecs@gmail.com

H. K. Dave and D. Nedelcu (eds.), *Advances in Manufacturing Processes*, Lecture Notes in Mechanical Engineering, https://doi.org/10.1007/978-981-15-9117-4_25

parameters [1]. Impact of FDM process variables (build orientation, layer thickness and feed rate) on mechanical behaviour of PLA was studied. ANOVA, regression model and response surface were used to perform parametric characterization [2]. Impact of process variable on the mechanical performance of 3D CFRTPCs (continuous fibre reinforced thermoplastic composites) was analysed. The parameters of the process considered for studies are build orientation, layer thickness and fibre volume content. Continuous carbon, glass and Kevlar reinforced nylon composite specimens were fabricated by FDM. SEM images of fractured surfaces were observed to determine failure mode and reason [3]. FDM process parameters (layer thickness, build orientation, infill density and no of contour) were optimized for better dimensional preciseness of ABS parts. Hybrid statistical tools RSM-GA, ANN, ANN-GA were used for optimization of process parameters. Validation test was also performed to verify the various derived models and optimum process parameters [4]. Influence of FDM process variables (layer thickness, feed rate and infill density) on the mechanical properties (tensile and flexural strength) of PETG material was analysed. The significance of the process variables was observed using ANOVA. Regression equation was generated for predicting the value of the response at considered process parameter [5]. Ultimate tensile strength and nominal strain at break of the FDM PLA parts were estimated. The parameters of the process considered for studies are layer thickness, infill orientation and number of shell perimeters. Experiments were carried out on the basis of the central composite design (CCD), and the second-order response surface model has been developed [6]. Impact of FDM process variables (raster angle, infill speed, nozzle temperature and layer thickness) on tensile properties of carbon fibre ABS was investigated. SEM images after tensile test were observed to determine failure mode and reason [7]. Systematic literature review was performed based on classification according to type of 3D printing polymer. Reviewed studies are mostly based on experimental results, but some of the papers in which analytical models have been developed [8]. Effect of three process parameters (raster angle, layer height and raster width) on the tensile properties of FDM PLA parts was quantified. The mean effect plot of each process parameters, effect of interaction and ANOVA were used for obtaining significant influence of process parameters. Fractography was performed on fractured specimen after tensile test for determining the mode of failure [9]. Tensile strength of FDM printed PLA part was studied by varying process parameters. The process parameters of the process considered for studies are raster angle, layer height and raster width. Images of fractured surfaces were used to know failure behaviour and its reasons [10]. Impact of FDM process variables (layer thickness, print temperature and infill pattern) on tensile strength was investigated. Material used in the study is carbon fibre PLA. Main effects and interaction between process parameters were estimated. ANOVA was performed to determine the most crucial process parameter impact on tensile strength. Second-order regression equation was generated for prediction purpose [11]. The relationship in between process parameters and the tensile strength using group method for data handling (GMDH) was obtained. Differential evaluation (DE) was used to determine the optimal set of the process parameter. The parameters of the process considered for studies are part orientation, raster angle, raster width and air gap [12]. Impact of FDM process variables (layer height, raster angle and infill density) on tensile properties of ABS was studied. Full factorial DOE was used for specimen preparation. RSM was used to derive the mathematical model and validation of experimental data [13]. Impact of process parameters (model orientation, layer thickness and shell thickness) on ultimate tensile strength and dimensional accuracy of FDM ABS and nylon materials was identified. Taguchi approach was used to investigate the impact of process parameters [14]. Influence of FDM process variables (extrusion temperature, printing speed and layer height) on the mechanical properties of filled polypropylene was studied. Specimens were prepared based on full factorial DOE, and ANOVA is used to analyse mechanical properties. The melting behaviour of the processed polypropylene and the morphological analysis was performed using DSC (differential scanning calorimetry) and SEM (scanning electron microscopy), respectively [15].

It is observed from the literature study that thermoplastics (ABS, PLA, PEEK, PC, nylon, PEI and PP) and reinforced thermoplastic composites (carbon fibre PLA, carbon fibre ABS, continuous carbon/glass/Kevlar fibre nylon) are mostly investigated polymers in the studies. In addition, process parameters such as raster and contour dimensions, build orientation, layer thickness and air gap are mostly considered in the studies and only a few studies focus on temperature and feed rate. Furthermore, less work has been carried out on the impact of FDM parameters on mechanical behaviour of nylon glass fibre material. For this reason, the influence on tensile properties of nylon glass fibre material using three process parameters, namely layer thickness, extrusion temperature and infill speed, is studied in this work.

2 Experimental Work

The experimental work carried out for investigation is described in the sections below and includes filament material, tensile test specimen, design of experiments, printing process and tensile testing.

2.1 Filament Material

In this work, nylon glass fibre (approximately 25% glass fibre reinforced nylon) is selected for the parts fabrication. It has high stiffness, high strength, low abrasive wear, low warping and excellent layer adhesion. The properties of nylon glass fibre material are density 1.18 kg/cm^3, melting point 180 °C, tensile strength 95 MPa, shore hardness 77 MPa and impact strength 80 kJ/m^2. 2.85-mm-diameter nylon glass fibre filament is used for this work.

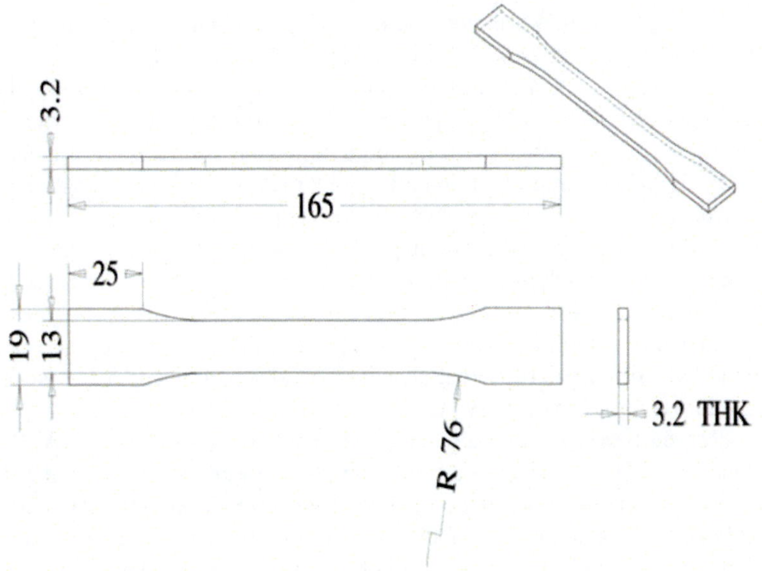

Fig. 1 Tensile test specimen

2.2 *Tensile Test Specimen*

Tensile test specimen is created using solid edge in accordance with ASTM D638 Type-1 standard (Fig. 1). It is most widely used in estimation of tensile properties for reinforced and unreinforced plastic and ensures that the break occurs at the centre rather than the gripping area of the specimen.

2.3 *Design of Experiments*

Design of experiments (DOE) is a field of applied statics used for scientific investigations of a system process or product in which process variables have been altered to study their influence on a measured response. In other words, it is a systematic approach used to evaluate the relationship in between process variables that influence the process and the process response. The properties of FDM parts usually depend on various process variables. In this study, layer thickness, extrusion temperature and infill speed are the three process parameters considered to determine its effect on tensile properties of nylon glass fibre FDM specimens. Table 1 represents the three levels of each considered process parameter and its notation. Taguchi L9 design of experiment (Table 2) is selected because it is unique and effective approach that allows optimization and investigation through minimum number of experiments.

Table 1 Levels of process parameters

Factors	Notation	Level 1	Level 2	Level 3
Layer thickness (mm)	L	0.1	0.2	0.3
Extrusion temperature (°C)	T	235	240	245
Infill speed (mm/s)	S	25	40	55

Table 2 Taguchi L_9 orthogonal array

Sample no.	Code	Layer thickness (mm)	Extrusion temperature (°C)	Infill speed (mm/s)
1	$L_1T_1S_1$	0.1	235	25
2	$L_1T_2S_2$	0.1	240	40
3	$L_1T_3S_3$	0.1	245	55
4	$L_2T_1S_2$	0.2	235	40
5	$L_2T_2S_3$	0.2	240	55
6	$L_2T_3S_1$	0.2	245	25
7	$L_3T_1S_3$	0.3	235	55
8	$L_3T_2S_1$	0.3	240	25
9	$L_3T_3S_2$	0.3	245	40

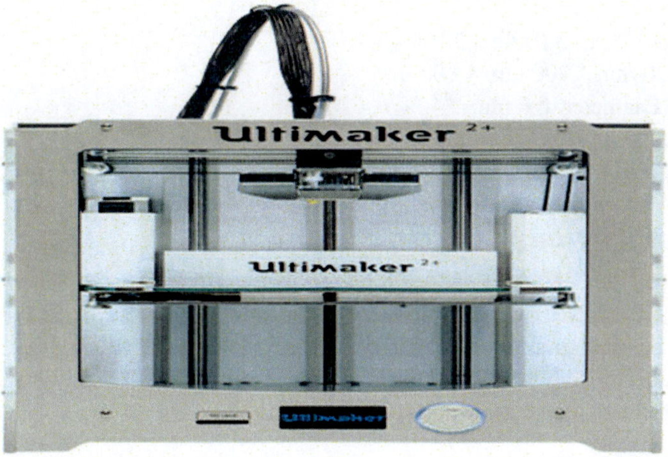

Fig. 2 Custom build 3D printer

Fig. 3 FDM printing process

2.4 *Printing Process*

Figure 2 shows the 3D printer used for specimen preparation, and Fig. 3 shows the printing head of a machine. The specification of FDM machine used for specimen preparation is

- Machine: Custom Build 3D Printer
- Size: 200 mm * 200 mm * 300 mm
- Nozzle Diameter: 0.6 mm
- Filament Diameter: 2.85 mm.

2.5 *Tensile Testing*

UTM DTRX 10kN is used to conduct the tensile test in accordance with ASTM D638. It has a maximum load capacity of 10kN. Load accuracy is ±0.05% of the applied load, from 2% to 100% of the load cell capacity. The UTM is shown in Fig. 4, and the specimen after tensile test is shown in Fig. 5.

3 Results and Discussion

The analysis of the experimental results is performed using the MIBITAB-19 trial version software. In this, the result obtained from experimental data is analysed based on main effect plot, S/N ratio plot, regression analysis and ANOVA. Also, the optimal process parameters selection is also carried out for superior tensile properties.

Fig. 4 UTM with loaded specimen

Fig. 5 Specimens after tensile test

3.1 Analysis of Tensile Strength

Tensile strength of all specimens is given in Table 3. It is larger the better type of response characteristics. Therefore, S/N ratio is evaluated using Eq. (1) and given in Table 3.

$$\eta \text{LB} = -10 \log 10\left(\frac{1}{n}\sum_{i=1}^{n}\frac{1}{Y\text{ij}^2}\right) \quad (1)$$

Main effect plot for tensile strength and S/N ratio Figure 6 represents the main effect plot for tensile strength of nylon glass fibre. From the graph, infill speed is

Table 3 Tensile strength result

Specimen	Tensile strength (MPa)	S/N ratio	Predicted strength (MPa)	Error (in %)
1	41.35	32.3295	41.66	0.74
2	43.31	32.7318	43.05	0.61
3	42.90	32.6491	44.43	3.58
4	38.97	31.8146	39.15	0.45
5	41.51	32.3631	40.53	2.35
6	44.29	32.9261	42.47	4.11
7	37.06	31.3781	36.63	1.15
8	36.92	31.3452	38.57	4.46
9	39.60	31.9539	39.96	0.90
			Average error	2.04

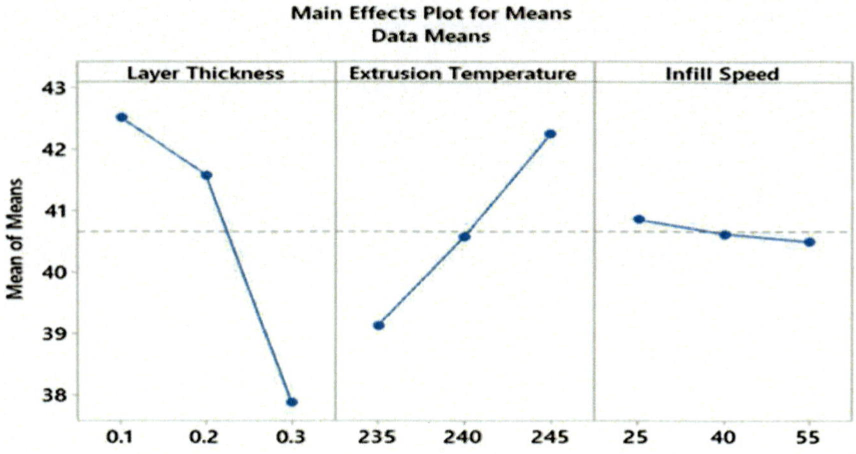

Fig. 6 Main effects plot for tensile strength

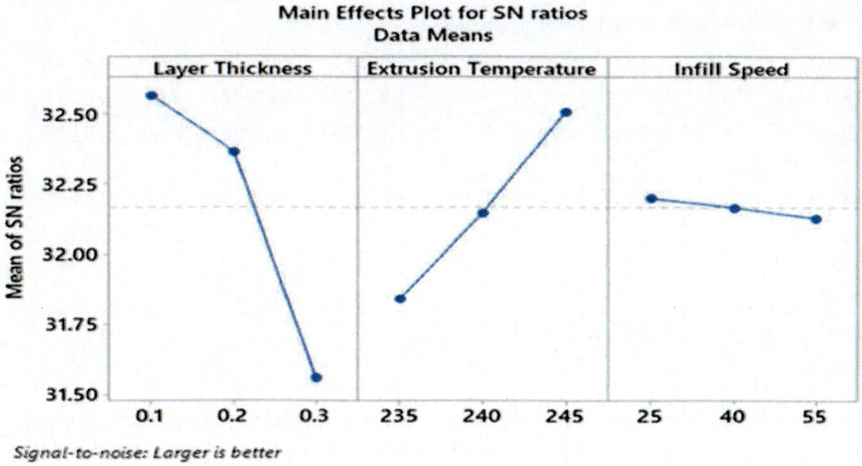

Fig. 7 Main effects plot for S/N ratios of tensile strength

noted to have slight impact on the tensile strength. Increasing infill speed may result in slight decrease in tensile strength. The extrusion temperature shows higher impact on tensile strength. As temperature increases, tensile strength is shown to increase. Additionally, tensile strength is majorly affected by layer thickness. Tensile strength decreases as layer thickness increases. However, the decreasing trend is not same that means the tensile strength slightly decreases from 0.1 to 0.2 mm layer thickness but then after from 0.2 to 0.3 mm sudden decrease takes place. Figure 7 represents the S/N ratio plot. It is observed that S/N ratio is maximizing by same process parameters as in main effect plot for tensile strength. From the S/N ratio graph, optimal condition for maximizing tensile strength is L1-T3-S1.

Regression analysis Equation (2) represents the regression equation for tensile strength. It is used for prediction of tensile strength for specimen at different combination of process parameters. Table 4 shows the summary of model for tensile strength which includes standard deviation (S), coefficient of determination (R^2), adjusted R^2 and predicted R^2.

$$\text{TS} = -29.5 - 23.30 * L + 0.314 * T - 0.0121 * S \quad (2)$$

Table 5 represents analysis of variance (ANOVA) results from regression analysis. It includes p-value and percentage contribution of each process parameters. If the

Table 4 Summary of tensile strength model

S	R^2	R^2 (adj)	PRESS	R^2 (pred)
1.40095	82.89%	72.62%	38.5402	32.79%

Table 5 Analysis of variance for tensile strength

Source	DF	Seq SS	Contribution (%)	Adj SS	Adj MS	F-value	P-value
Regression	3	47.5294	82.89	47.5294	15.8431	8.07	0.023
Layer thickness	1	32.5734	56.80	32.5734	32.5734	16.60	0.010
Extrusion temperature	1	14.7580	25.74	14.7580	14.7580	7.52	0.041
Infill speed	1	0.1980	0.35	0.1980	0.1980	0.10	0.764
Error	5	9.8134	17.11	9.8134	1.9627		
Total	8	57.3428	100.00				

p-value is lesser than α value (0.05), the process parameters are considered as significant. P-value for layer thickness and extrusion temperature is lesser than α value which means they are significant process parameters. Also, the model is regarded as statistically significant as its p-value is 0.023($<$0.05). Layer thickness contribution is greater among other parameters and it is 56.80% of the total.

Table 3 shows the predicted value calculated from regression equation for tensile strength. From comparison between experimental value and predicted value (specimen error is calculated using Eq. 3), it can be remarked that the average error between them is 2.04% which is not so considerable and suggest that the regression equation is significant.

$$\text{Error} = \left(\frac{|\text{Experimental value} - \text{Predicted value}|}{\text{Experimental value}}\right) * 100 \tag{3}$$

3.2 Analysis of Percentage Elongation

The results of the percentage elongation are given in Table 6. Percentage elongation is larger the better type of response characteristics so that the S/N ratio is evaluated using Eq. (1) and given in Table 6.

Main effects plot for percentage elongation and S/N ratio Figure 8 represents the main effect plot for percentage elongation of nylon glass fibre. From the graph, infill speed is noted to have slight impact on the percentage elongation. Increasing infill speed may result first slight increase and then decrease in percentage elongation with inflection point at 40 mm/s. The extrusion temperature shows higher impact on percentage elongation. As temperature increases, percentage elongation is shown to increase. Additionally, percentage elongation is majorly affected by layer thickness. Percentage elongation decreases as layer thickness increases. However, the decreasing trend is not same that means the percentage elongation is slightly decreased from 0.1 to 0.2 mm layer thickness but then after from 0.2 to 0.3 mm

Table 6 Percentage elongation result

Specimen	Percentage elongation	S/N ratio	Predicted percentage elongation	Error (in %)
1	19.36	25.7381	19.83	2.43
2	21.26	26.5513	21.28	0.09
3	21.48	26.6407	22.73	5.82
4	18.72	25.4461	18.16	3.01
5	20.94	26.4195	19.61	6.37
6	22.22	26.9349	21.16	4.79
7	15.98	24.0715	16.48	3.15
8	17.02	24.6192	18.03	5.95
9	19.26	25.6931	19.48	1.16
			Average Error	3.64

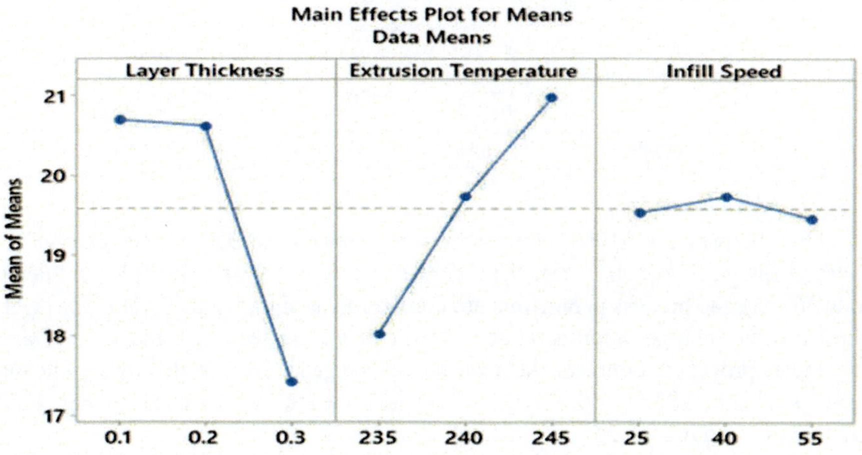

Fig. 8 Main effects plot for percentage elongation

sudden decrease takes place. Figure 9 represents the S/N ratio plot. It is observed that S/N ratio is maximizing by same process parameters as in main effect plot for percentage elongation. From the S/N graph, optimal condition for maximizing percentage elongation is L1-T3-S2.

Regression analysis Equation (4) represents the regression equation for percentage elongation. It is used for prediction of percentage elongation for specimen at different combination of process parameters. Table 7 shows the summary of model for percentage elongation which include standard deviation (S), coefficient of determination (R^2), adjusted R^2 and predicted R^2.

$$\mathrm{PE} = -48.2 - 16.40 * L + 0.2967 * T - 0.0022 * S \quad (4)$$

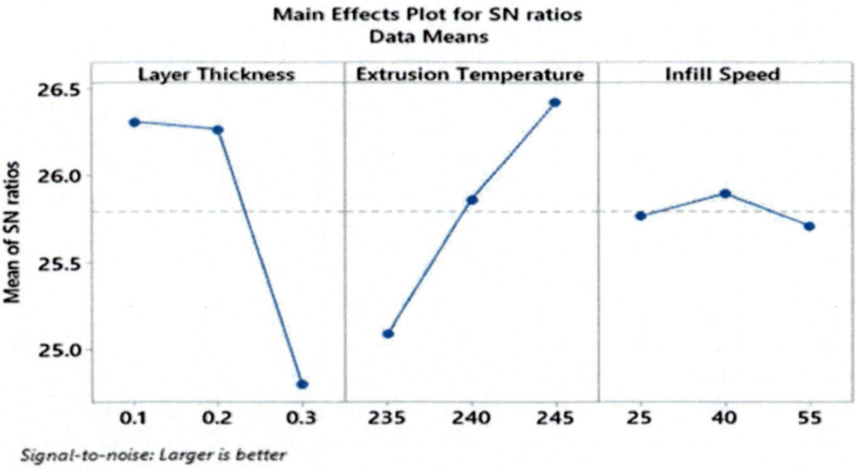

Fig. 9 Main effects plot for S/N ratios of percentage elongation

Table 7 Summary of percentage elongation model

S	R^2	R^2 (adj)	PRESS	R^2 (pred)
1.12335	82.30%	71.69%	23.4068	34.35%

Table 8 represents ANOVA results from regression analysis. It includes *p*-value and percentage contribution of each process parameters. If the *p*-value is lesser than α value (0.05), the process parameters are considered as significant. *P*-value for layer thickness and extrusion temperature is lesser than α value which means they are significant process parameters. Also, the model is regarded as statistically significant as its *p*-value is 0.025 (<0.05). Layer thickness contribution is greater among other parameters, and it is 45.26% of the total.

Table 6 shows the predicted value calculated from regression equation for percentage elongation. From comparison between experimental value and predicted

Table 8 Analysis of variance for percentage elongation

Source	DF	Seq SS	Contribution (%)	Adj SS	Adj MS	F-Value	*P*-Value
Regression	3	29.3459	82.30	29.3459	9.7820	7.75	0.025
Layer thickness	1	16.1376	45.26	16.1376	16.1376	12.79	0.016
Extrusion temperature	1	13.2017	37.03	13.2017	13.2017	10.46	0.023
Infill speed	1	0.0067	0.02	0.0067	0.0067	0.01	0.945
Error	5	6.3096	17.70	6.3096	1.2619		
Total	8	35.6556	100.00				

value (specimen error is calculated using Eq. 3), it can be remarked that the average error between them is 3.64% which is not so considerable and suggests that the regression equation is significant.

4 Conclusion

The impact of FDM process parameters, namely layer thickness, extrusion temperature and infill speed, on tensile properties of nylon glass fibre material has been investigated well by conducting tensile test. From investigation following conclusions are made,

- At minimum layer thickness and infill speed with higher extrusion temperature (in this study, these values are 0.1 mm, 25 mm/s and 245 °C, respectively), the material shows good tensile strength. From ANOVA results, layer thickness and extrusion temperature have significant impact on tensile strength. The *p*-value for tensile strength model is 0.023 (<0.05), and average error between predicted value by regression equation and experimental value for tensile strength is 2.04% which is not so considerable. Therefore, regression equation for tensile strength is significant and feasible for prediction of tensile strength at different combination of layer thickness, extrusion temperature and infill speed.
- Secondly, at smaller layer thickness, moderate infill speed and higher extrusion temperature (in this study, these values are 0.1 mm, 40 mm/s and 245 °C, respectively), the material shows good percentage elongation. From ANOVA results, layer thickness and extrusion temperature have significant impact on percentage elongation. The *p*-value for percentage elongation model is 0.025 (<0.05), and average error between predicted value by regression equation and experimental value for percentage elongation is 3.64% which is not so considerable. Therefore, regression equation for percentage elongation is significant and feasible for prediction of percentage elongation at different combination of layer thickness, extrusion temperature and infill speed.
- Layer thickness contribution is greater among other parameters, and it is 56.80 and 45.26% for tensile strength and percentage elongation, respectively.

References

1. Alafaghani A, Qattawi A (2018) Investigating the effect of fused deposition modeling processing parameters using Taguchi design of experiment method. J Manuf Proces 36:164–174
2. Chacon JM, Caminero MA, Garcia-Plaza E, Nunez PJ (2017) Additive manufacturing of PLA structures using fused deposition modelling: effect of process parameters on mechanical properties and their optimal selection. Mater Des 124:143–157
3. Chacon JM, Caminero MA, Nunez PJ, Garcia-Plaza E, Garcia-Moreno I, Reverte JM (2019) Additive manufacturing of continuous fibre reinforced thermoplastic composites using fused

deposition modelling: effect of process parameters on mechanical properties. Compos Sci Technol 181:107688
4. Deswal S, Narang R, Chhabra D (2019) Modeling and parametric optimization of FDM 3D printing process using hybrid techniques for enhancing dimensional preciseness. Int J Interact Des Manuf 13:1197–1214
5. Durgashyam K, Reddy MI, Balakrishna A, Satyanarayana K (2019) Experimental investigation on mechanical properties of PETG material processed by fused deposition modeling method. In: Reddy V (ed) 2nd International Conference on Applied Sciences and Technology (ICAST-2019): Material Science. Materials Today Proceeding, vol 18, pp 2052–2059
6. Lanzotti A, Grasso M, Staiano G, Martorelli M (2015) The impact of process parameters on mechanical properties of parts fabricated in PLA with an open-source 3-D printer. Rapid Prototyping J 21(5):604–617
7. Ning F, Cong W, Hu Y, Wang H (2017) Additive manufacturing of carbon fiber-reinforced plastic composites using fused deposition modeling: effects of process parameters on tensile properties. J Compos Mater 51(4):451–462
8. Popescu D, Zapciu A, Amza C, Baciu F, Marinescu R (2018) FDM process parameters influence over the mechanical properties of polymer specimens: a review. Polym Testing 69:157–166
9. Rajpurohit SR, Dave HK (2019) Analysis of tensile strength of a fused filament fabricated PLA part using an open-source 3D printer. Int J Adv Manuf Technol 101:1525–1536
10. Rajpurohit SR, Dave HK (2018) Effect of process parameters on tensile strength of FDM printed PLA part. Rapid Prototyping J 24(8):1317–1324
11. Rao VDP, Rajiv P, Geethika VN (2019) Effect of fused deposition modelling (FDM) process parameters on tensile strength of carbon fibre PLA. In: Reddy V (ed) 2nd International Conference on Applied Sciences and Technology (ICAST-2019): Material Science. Materials Today Proceeding, vol 18, pp 2012–2018
12. Rayegani F, Onwubolu GC (2014) Fused deposition modelling (FDM) process parameter prediction and optimization using group method for data handling (GMDH) and differential evolution (DE). Int J Adv Manuf Technol 73:509–519
13. Samykano M, Selvamani SK, Kadirgama K, Ngui WK, Kanagaraj G, Sudhakar K (2019) Mechanical property of FDM printed ABS: influence of printing parameters. Int J Adv Manuf Technol 102:2779–2796
14. Vishwas M, Basavaraj CK, Vinyas M (2018) Experimental investigation using taguchi method to optimize process parameters of Fused deposition modeling for ABS and Nylon materials. In: Sivaprasad K, Muthupandi V (ed) International Conference on Emerging Trends in Materials and Manufacturing Engineering (IMME17), March 10–12, 2017. Materials Today: Proceedings vol 5, no 2, pp 7106–7114
15. Wang L, Sanders JE, Gardner DJ, Han Y (2018) Effect of fused deposition modelling process parameters on the mechanical properties of a filled polypropylene. Progr Addit Manuf 3:205–214

Influence of Inter-Lamellar Spacing of Pearlite Phase on Spheroidized Annealed Structure

Partha Sarathi Sarkar, Bhavesh Rameshchandra Rana, and Pinkal D. Mistry

1 Introduction

Pearlite is one of the unique micro constituents in the iron-carbon phase diagram, which is composed of alternative layers of ferrite and cementite. The percentages of ferrite and cementite in pearlite are 88 and 12% respectively. The percentage of carbon majorly influences the final amount of pearlite. Eutectoid steel having 100% pearlite in the microstructure at room temperature. Figure 1 depicts the morphological image of the pearlite phase.

Beyond 0.80% C, Cementite phase remains in free form. Pearlitic steel has wide applications across various sectors, infrastructure, automobile, agricultural, electrical, etc. The application is dependent on the final properties of the steel, and it is designed through the microstructure. The below-mentioned parameters inter-lamellar spacing, size of pearlite patches, length of the pearlite, and prior austenite grain size, mentioned parameters are highly dependent on the cooling rate during the hot working process. The generation of the cooling rate is derived from Time—Temperature—Transformation (TTT) diagram and Continuous Cooling Transformation (CCT) diagram [2]. Pearlite can be achieved with isothermal cooling between 650 and 450 °C, below this temperature the formation of martensite and bainite will take

P. S. Sarkar
B.E Metallurgy, Technology Group, Theis Precision Steel India Pvt. Ltd., Navsari 396424, India
e-mail: parthass1974@gmail.com

B. R. Rana (✉)
B.E Metallurgy and Material Science, Technology Group, Theis Precision Steel India Pvt. Ltd., Navsari 396424, India
e-mail: ranabhaveshr@gmail.com

P. D. Mistry
Technology Group, Theis Precision Steel India Pvt. Ltd., Navsari 396424, India
e-mail: pinkalmistry22222@gmail.com

H. K. Dave and D. Nedelcu (eds.), *Advances in Manufacturing Processes*, Lecture Notes in Mechanical Engineering, https://doi.org/10.1007/978-981-15-9117-4_26

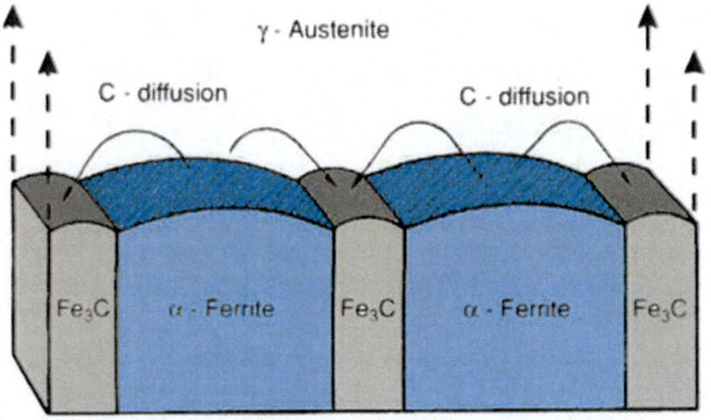

Fig. 1 Overview of Ferrite and Cementite lamellar morphology [1]

place. With increasing cooling rates in the pearlite continuous cooling transformation range, or with isothermal transformation temperatures approaching the pearlite nose of isothermal transformation diagrams, the interlamellar spacing of pearlitic ferrite and cementite becomes very fine [3, 4]. The pearlite lamellar spacing is one of the most important parameters that characterize pearlite because mechanical properties (Hardness, Tensile strength, Yield strength, and toughness) of pearlitic steel are strongly influenced by inter-lamellar spacing [4, 5]. For proper metallographic observations, the sectioning plane should be perpendicular to the lamellae so that the true spacing can be determined. Belaiew [6, 7] first considered the problems associated with interlamellar spacing measurement in pearlite structure. He concluded that the true interlamellar spacing of the colony remains constant within a sample and observed variation was solely due to the sectioning plane effect [7].

Belaiew assumed that if the true interlamellar spacing plane is smaller than the resolution limit of the optical microscope, about 0.2 μm, then the resolvable portion limit of the lamellae could be related to the true spacing through the distribution of apparent spacing created by the effect of the sectioning plane orientation. Certain samples consist of the true interlamellar spacing was less than the determination limit of the optical microscope, a few areas can be found where the pearlitic lamellae are resolvable [6, 7]. The apparent distance of these colonies is much larger than the true spacing. Belaiew determined that this frayed condition was obtained when the sectioning plane was 83° from the plane perpendicular to the lamellae.

To investigate the significant effect of interlamellar spacing on spheroidization annealing process, eutectoid steel is analyzed. Two different hot rolled samples of same chemical composition and dimensions but having different inter-lamellar spacing in pearlite, were taken. Further same spheroidization annealing treatment was given on these two samples. A detailed study of dimensional, chemical, mechanical, and metallurgical properties was carried out before and after spheroidization annealing. All the experimental results are depicted in the result and discussion chapter.

To increase the ductility of materials before cold forming, a spheroidizing annealing treatment is done on the medium and high carbon steel. During heat treatment, the cementite phase in the steel acquires a spherical morphology. There are two steps to the spheroidization process. In the first step, any carbides with high aspect ratios (such as cementite lamellae in pearlite) are broken into many small, spherical carbides. The spherical particles have lower surface area to volume ratios than the elongated structures. These small spherical particles are then coarsened by Ostwald ripening into large particles thus further decreasing the total surface area to volume ratio [8–10]. The kinetics of both stages of spheroidization is controlled by the diffusion of carbon and other alloying elements through the ferrite or austenite matrix.

FCR (final cooling rate) during the hot rolling process remains at 650–550 °C (depending upon final mechanical—metallurgical properties). So, the final hot rolling structure covers pearlite in ferrite matrix with an insignificant amount of lower bainite and some alloy carbides. Theoretically, the final hot rolling structure is to be 100% pearlite. Due to an insignificant amount of bainite and alloy carbide we are abandoning it in further discussion. All metallographic examinations are carried out with the metallurgical microscope at 1000X magnification except grain size measurement accomplished on 100X.

2 Materials and Methods

Eutectoid steel was considered for this study. Two hot rolled steel strips samples of same size, 4.30 mm thickness (tolerance ± 0.50 mm), 100.00 mm length (tolerance ± 2.00 mm), and 100 mm width (tolerance ± 2.00 mm) were taken for analysis. The chemical compositions of the samples are mentioned in Table 1.

Chemical compositions of both the samples confirm IS2507-80C6 grade. Optical emission spectrometer (model no.—Innovate T5, manufactured by JIEBO technology, China) was used to analyze the chemical composition (Accuracy—±0.005%, Resolution—1 nm). Mechanical properties were also inspected and mentioned in Table 2.

Tensile testing was carried out as per ASTM E8/E8M. Samples were taken in parallel to the rolling direction. The tensile tests were performed on the Universal Testing Machines (manufactured by Blue Star, Germany). Figures 3, 4 shows the stress v/s strain curve for the raw material i.e. in hot rolled condition and Figs. 5, 6 shows the stress v/s strain curve for the spheroidized annealed material. Hardness tests were done on Vickers hardness tester with the application of 20 kg load and 10 s of dwell time. Ten number indentations were carried out during hardness testing on each sample.

The determination of interlamellar spacing, ferritic grain size, and volume of ferrite and pearlite phase were carried out through metallographic analysis on raw material (before spheroidization annealing). Each sample was mounted in polycast type resin mould and the analyzed face was perpendicular to the rolling direction.

Table 1 Chemistry of the Investigated samples

Sample ID	C %	Mn %	S %	P %	Si %	Al %	Cr %	Ni %	Ti %	Fe %
Specification as per IS2507 80C6	0.75–0.85	0.50–0.80	0.05 max	0.05 max	0.10–0.35	–	–	–	–	Balance
Sample T1	0.81	0.746	0.005	0.015	0.229	0.028	0.11	0.02	0.002	Balance
Sample P1	0.80	0.694	0.005	0.013	0.216	0.016	0.12	0.01	0.001	Balance

Table 2 Mechanical properties of Raw material sample

Mechanical properties	Sample T1	Sample P1
Hardness (HV20)	306	276
Tensile strength (MPa)	977	864
Yield strength (MPa)	790	675
% Elongation	25	28
Yield ratio, Y_R	0.80	0.78

Grinding of resin mounted samples were done on successively decreasing grit sizes (used grit papers were 60, 240, 400, 600, 800, 1200, 1500, and 1900). Size of grit particles decreases with increasing grit numbers. Polishing of the samples was done with the Alumina paste (Al_2O_3). The polished samples were further etched with 2% Nital (2% nitric acid and 98% methanol) for 5–7 s to reveal the microstructure. Metallographic analysis was done by the metallurgical microscope at various magnifications, 100X, 200X, 500X, and 1000X.

Various shapes of the cementite particles like straight, curved, cranked, oval and elliptical, etc. were taken into account. The pearlite lamella spacing measurement method is based on specific geometrical assumptions. For every cementite particle, the length and width were measured according to Eqs. 1 and 2, respectively. The estimated length (Lr) and width (Wr) are mentioned in the below equation. A and P, are respectively the area and perimeter of the particle.

$$L_r = \frac{1}{2}\left(\frac{P}{2} - \sqrt{\left(\frac{P}{2}\right)^2 - 4A}\right) \tag{1}$$

$$W_r = \frac{1}{2}\left(\frac{P}{2} - \sqrt{\left(\frac{P}{2}\right)^2 - 4A}\right), \tag{2}$$

Following Chattopadhyay and Sellars [11, 12, 15, 16], the cementite particles were then classified as lamellar if their length to width aspect ratio was greater than 8 and as globular otherwise [14].

The results of the lamellar spacing, phase volume and grain size are mentioned in the result and discussion section. The same spheroidization annealing process was carried out on the samples T1 and P1. Both samples have different pearlitic microstructure configurations. The spheroidization annealing cycle (time-temperature-cooling rate) is mentioned in Fig. 2. The temperature of the heat treatment cycle is decided based on the chemical composition. The cycle time was decided based on standard thumb rule 25.4 mm/h. Spheroidization annealing process was carried out in bell annealing furnace. The heating was carried out by convection phenomena using hydrogen atmosphere, which also acted as a protective atmosphere. The purity of Hydrogen was 99.99%.

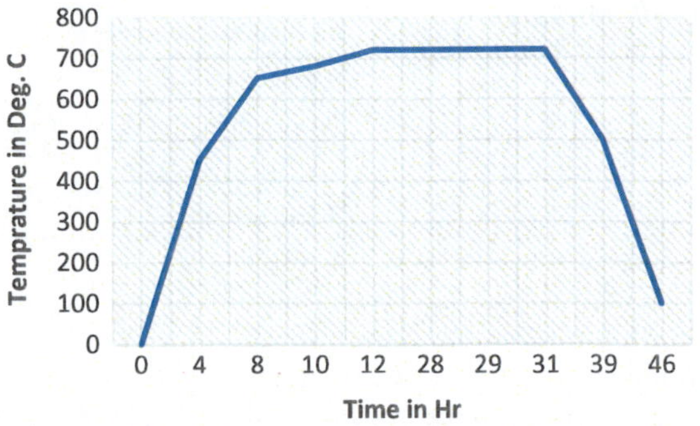

Fig. 2 Spheroidizing Heat treatment cycle for sample T1–P1

After the spheroidization annealing, mechanical and metallurgical testing were carried out on heat-treated samples. The effect of different prior interlamellar spacing on spheroidization annealing process has been compared and discussed in the result and discussion section.

3 Results and Discussion

3.1 Mechanical Properties

Mechanical testing of samples T1 and P1 was carried out at the raw material stage and after spheroidization annealing. The results are shown in Table 3,

Stress—strain curves of raw material samples T1 and P1 are shown in Figs. 3 and 4, respectively.

Stress—strain curves of Spheroidized annealed samples T1 and P1 are shown in Figs. 5 and 6, respectively.

Table 3 Mechanical properties of Raw material and Spheroidized annealed sample

Mechanical properties	Sample T1	Sample P1	Sample T1	Sample P1
	Properties at RM state		Properties after Spheroidization annealing	
Hardness (HV20)	306	276	206	180
Tensile strength (MPa)	977	864	664	583
Yield strength (MPa)	790	675	467	509
% Elongation	25	28	36	31
Yield ratio, Y_R (Y_s/T_s)	0.80	0.78	0.70	0.83

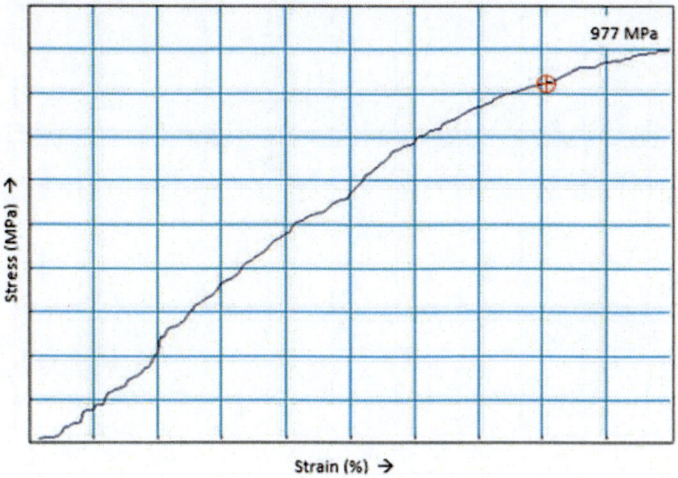

Fig. 3 Stress versus Strain diagram for T1—Raw material sample which is having fine inter-lamellar spacing.

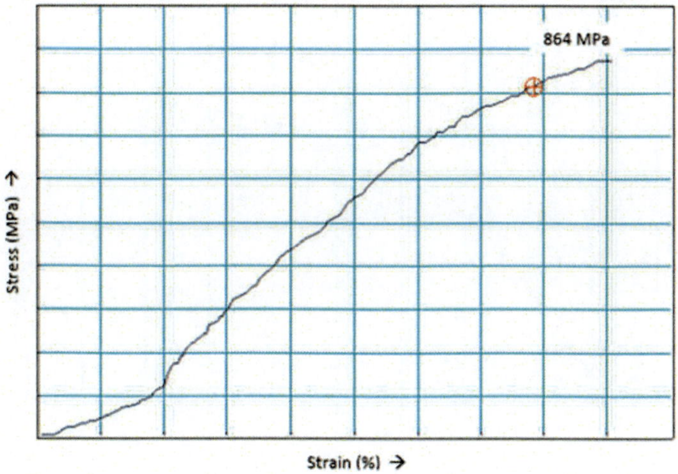

Fig. 4 Stress versus Strain diagram for P1—Raw material sample which is having fine interlamellar spacing

3.2 Metallurgical Properties

Interlamellar spacing of sample T1 and P1 at the raw material stage are shown in Figs. 7 and 8, respectively. The interlamellar spacing of samples T1 is 0.23 μm and the same for sample P1 is 0.51 μm in raw material stage.

After the lamellar spacing measurement, spheroidizing annealing treatment was accomplished on both samples T1 and P1 with the same cycle.

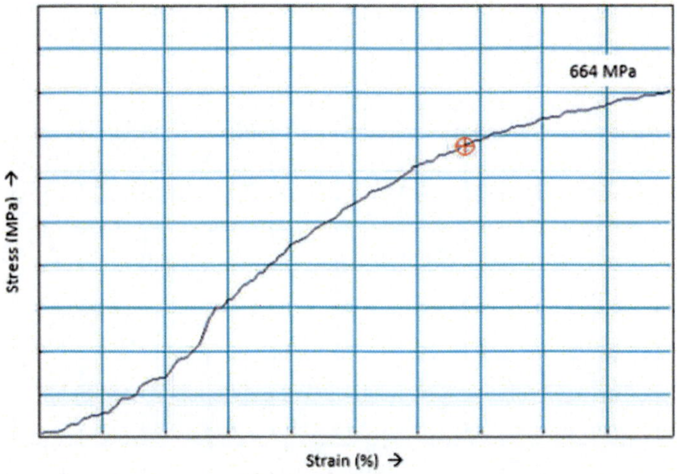

Fig. 5 Stress versus Strain diagram for T1—Annealed sample which is having fine interlamellar spacing

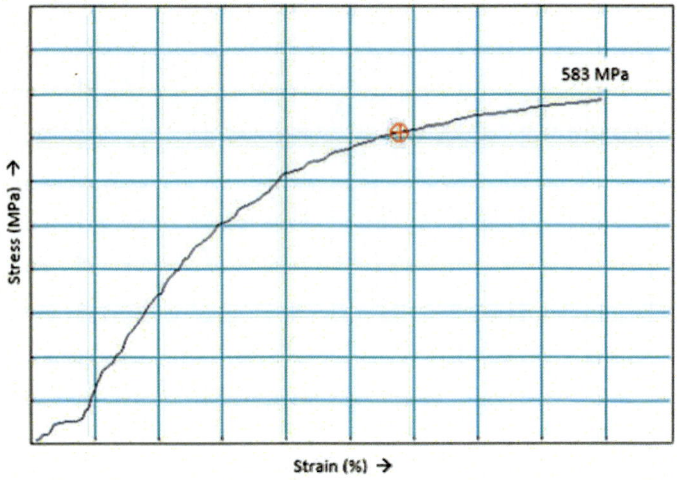

Fig. 6 Stress versus Strain diagram for P1—Annealed sample which is having coarse interlamellar spacing

Metallographic analysis was carried out by an optical metallurgical microscope, through Quantimet metallurgical software. Level of the spheroidization, volume of spheroids and the grain size (ferritic grain size) measurements were done on spheroidized annealed samples. The measurement of spheroidizing level is carried out as per standard SEP 1520 chart. Spheroidized micro-structure of samples T1 and P1 are shown Figs. 9 and 10.

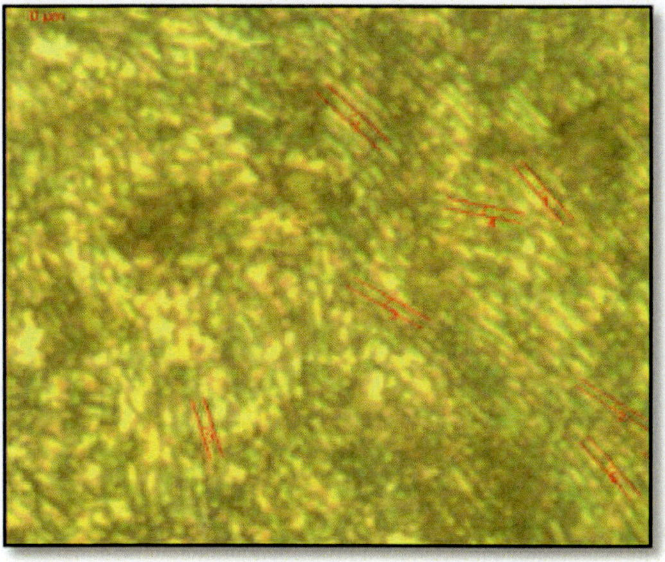

Fig. 7 Interlamellar spacing of sample, T1 (Fine interlamellar spacing—0.23 μm)

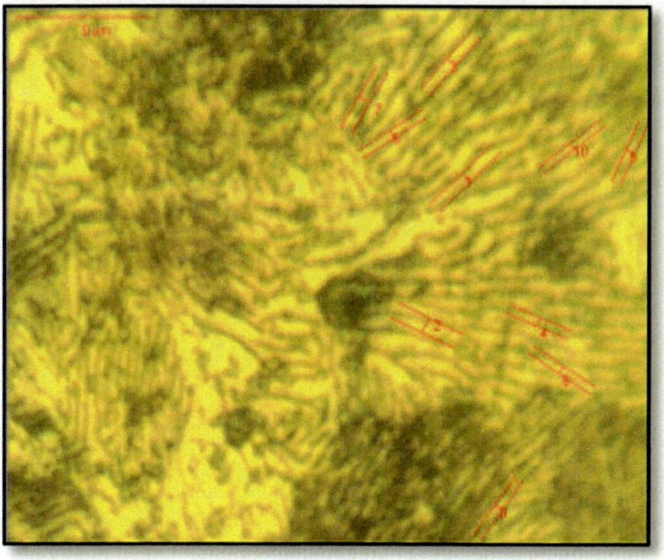

Fig. 8 Interlamellar spacing of sample, P1 (Coarse interlamellar spacing—0.51 μm)

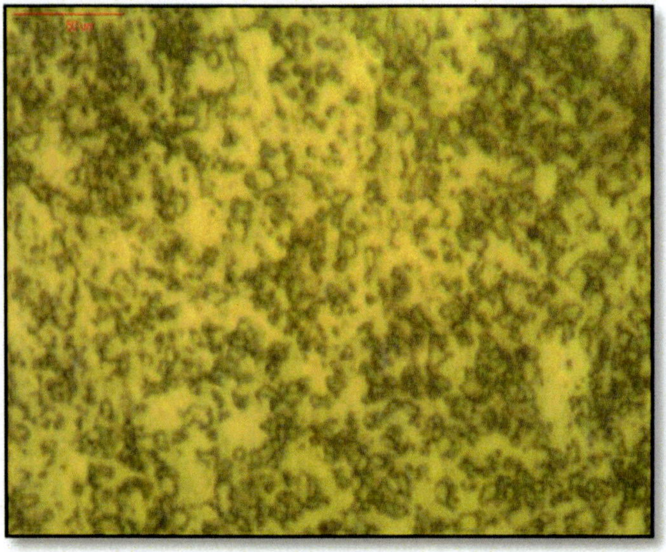

Fig. 9 Spheroidized structure of sample T1

Fig. 10 Spheroidized structure of sample P1

Figures 11 and 12 shows the volume fraction measurement through the gray value phenomena. Figures 13 and 14 shows the ferritic grain size measurement through the circular method.

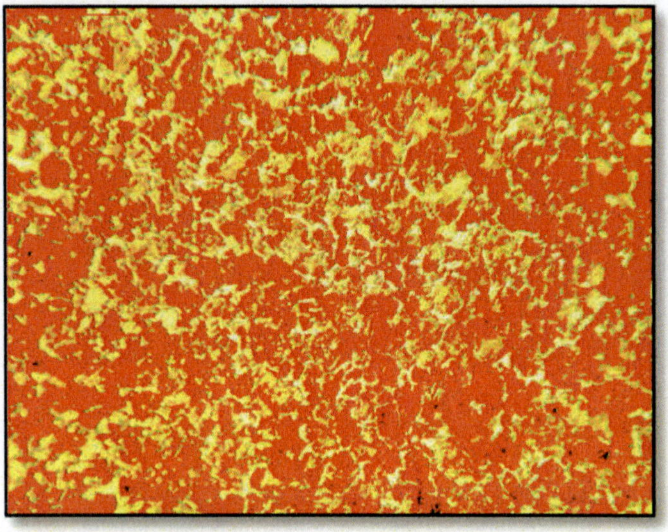

Fig. 11 Volume fraction of sample T1

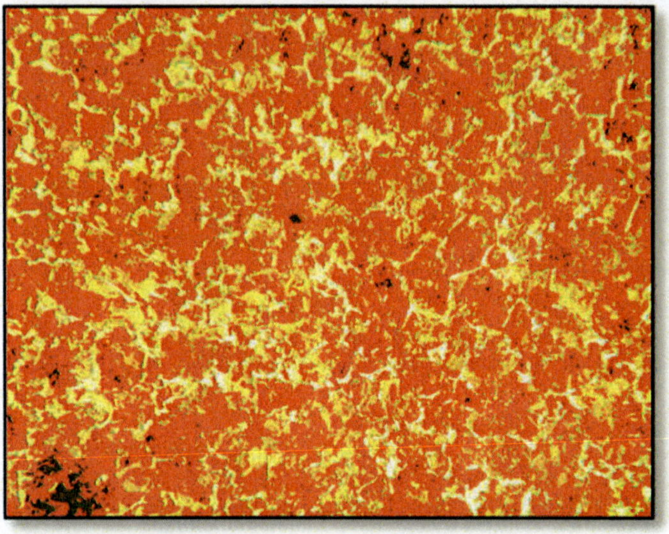

Fig. 12 Volume fraction of sample P1

Fig. 13 Grain size measurement of sample T1

Fig. 14 Grain size measurement of sample P1

Metallographic analysis of samples T1 and P1 was carried out at the raw material stage and after spheroidization annealing. The results are shown in Table 4.

Table 4 Metallurgical of raw material and spheroidized annealed sample

Metallurgical properties	Sample T1	Sample P1	Sample T1	Sample P1
	Properties at RM state		Properties after annealing	
Inclusion	A—0.5 B—0.5 C—0.0 D—0.3	A—0.5 B—1.5 C—0.0 D—0.5	–	–
ASTM grain size No.	8–9	7–8	7–8	6–7
Volume of phases (%)	Pearlite—98.2 Ferrite—1.8	Pearlite—99.1 Ferrite—0.9	–	–
Interlamellar spacing (μm)	0.23	0.51	–	–
Average thickness of Fe_3C (μm)	0.12	0.32	–	–
Spheroidization level (SEP 1520)	–	–	2-1 and 2-2	3-0 and 3-1
Average spheroids size (μm)	–	–	2–4	4–6

3.3 *Summary of Result*

Sample T1 which is having low interlamellar spacing has higher hardness value along with higher yield strength and tensile strength. Elongation percentage increases with an increase in interlamellar spacing, whereas the graphical representation is mentioned in Fig. 15 shows yield ratio decreases with an increase in interlamellar

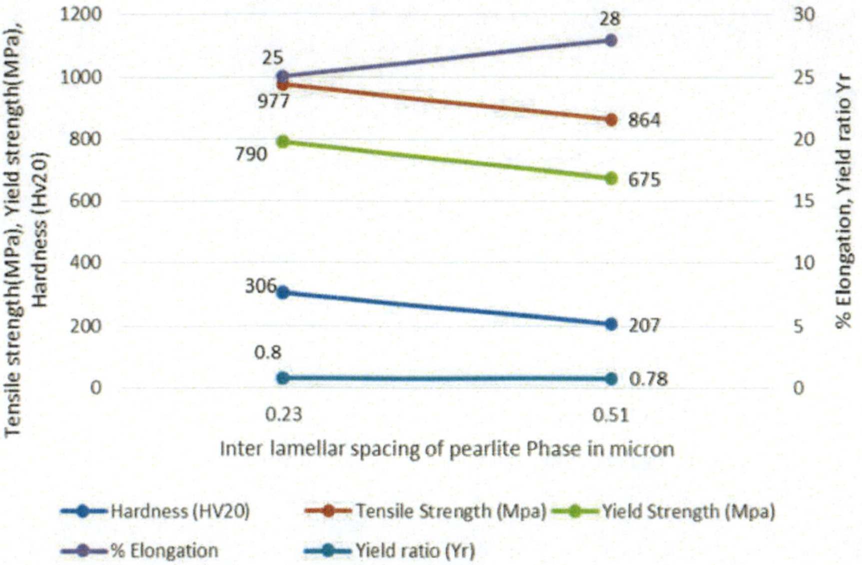

Fig. 15 Properties of samples T1 and P1 at Raw material stage

spacing in pearlite. It is to be mentioned that ductility in the raw material is linearly propositional to interlamellar spacing of pearlite.

Initial Interlamellar spacing in pearlite strongly affects the spheroidization kinetics. In fine pearlite sample T1 (ISP: 0.23 μm), the average spheroid size was observed 2–4 μm after annealing, whereas in coarse pearlite sample P1 (ISP: 0.51 μm), average spheroid size has been found 4–6 μm after annealing.

It is observed that, as the inter-lamellar spacing in the prior micro-structure increases, after spheroidization annealing yield ratio increases. The increase of yield ratio proves that ductility is decreased, which is further endorsed by the reduction of percentage elongation.

For the same annealing time, the diffusion—dissolution time is more in the case of pearlitic structure due to this phenomenon the spheroidized structure sample T1 and P1 have different metallurgical and mechanical properties.

The physical reasons for the scattering of true lamellar spacing have been discussed by a few authors [13, 15, 17–20]. During spheroidization annealing process prior cementite lamellae in pearlite phase is transformed into spheroidal carbides. Micro-structure analysis of the spheroidized sample T1 and P1 shows that cementite is fully spheroidized in case of lower interlamellar spacing, whereas pearlite colonies are found in sample P1 having higher initial interlamellar spacing (Fig. 16).

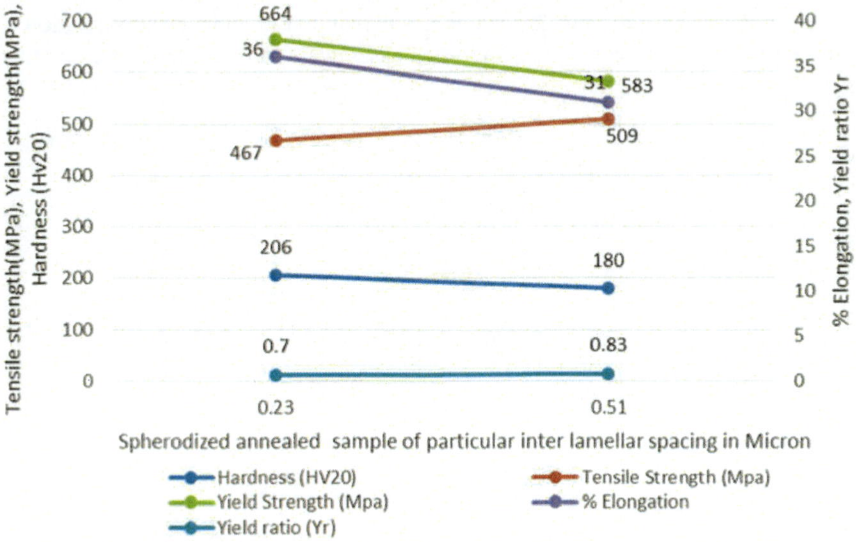

Fig. 16 Properties of T1 and P1 after Spheroidization annealing treatment

4 Conclusion

This work studied the effect of prior interlamellar spacing of pearlite phase in spheroidization annealing process of eutectoid steel. The conclusions are depicted below:

1. In the given spheroidization annealing cycle, the rate of spheroidization is higher in the case of finer interlamellar structure (ISP: 0.23 μm) than coarser once (ISP: 0.51 μm). The spheroidal carbide size in ferrite matrix has been found 2–4 μm in the case of prior interlamellar spacing of 0.23 μm and 4–6 μm in the case of prior interlamellar spacing of 0.51 μm. Since spheroidization is a diffusion-controlled process, therefore less diffusion distance in the fine interlamellar structure accelerates the spheroidization kinetics.
2. Yield ratio (Yr = Yield strength/Tensile strength) of the sample having lower prior interlamellar spacing (ISP: 0.23 μm) has been found 0.70 whereas the same for the coarser one is 0.83. After spheroidization annealing of eutectoid steel, it is found that yield ratio (Yr) decreases with a decrease in interlamellar spacing. The sample having yield ratio 0.70 is more ductile than that of 0.83.
3. Prior interlamellar spacing in the pearlite phase influences the rate of spheroidization, spheroidal carbide size, and the carbide particle distribution in the ferrite matrix.

Acknowledgements The authors would like to thank Mr. Chirag A. Shah, CEO of Theis Precision Steel India Pvt. Ltd., for his continuous support and help.

References

1. Icolas N, Utal N, Edric C, Ommes JG, Ilvia S, Lacher B (2010) Image analysis of pearlite spheroidization based on the morphological characterization of cementite particles. Imaga Anal Stereol 91–98. https://doi.org/10.5566/ias.v29
2. Atlas of Time-Temperature diagrams for iron and steel. ISBN: 0-87170-415-3, p 21
3. Atlas of isothermal transformation and cooling transformation diagrams, pp 23–29
4. Roósz András (1984) Measurement of the interlamellar spacing of pearlite article. Metallography. https://doi.org/10.1016/0026-0800(84)90002-8
5. Belaiew NT (1922) The inter structure of the pearlite grain. J Iron Steel Inst 105:201–239
6. Belaiew NT (1922) On the structure of nodular troosite. J Iron Steel Inst 124:195–214
7. Caballero FG, de Andres CG, Capdevila C (2000) Characterization and morphological analysis of pearlite in a eutectoid steel. Mater Charact 45(2):111–6
8. Cree AM, Faulkner RG, Lyne AT (1995) Cementite particle coarsening during spheroidization of bearing steel SAE 52100. Mater Sci Technol 11:566–571
9. Elwazri AM, Wanjara P, Yue S (2006) Empirical modelling of the isothermal transformation of pearlite in hypereutectoid steel. Mater Sci Technol 22:542–6
10. Krauss G (2005) Steels: processing, structure, and performance. ASM Int, Ohio
11. Yahyaoui H, Sidhom H (2014) Effect of interlamellar spacing on the elastoplastic behavior of C70 pearlitic steel: experimental results and self-consistent modeling. Mater Des. https://doi.org/10.1016/j.matdes.2013.10.062

12. Robert A (2005) McGrew at the colorado school of mines. ASM International: steels: processing, structure, and performance
13. Ai JH, Zhao TC, Gao HJ, Hu YH, Xie XS (2005) Effect of controlled rolling and cooling on the microstructure and mechanical properties of 60Si2MnA spring steel rod. J Mater Process Technol 160(3):390–395
14. Offerman SE, van Dijk NH, Rekveldt MTh, Sietsma J, van der Zwaag S (2002) Ferrite/pearlite band formation in hot rolled medium carbon steel. Mater Sci Technol 297–303
15. Chattopadhyay S, Sellars CM (1977) Quantitative measurements of pearlite spheroidization. Metallography 10:89–105
16. Gensamer M, Pearsall EB, Pellini WS (1942) Trans Am Soc Met 30:983
17. Ridley N (1984) Metall Trans A 15A:1019
18. Boswell PG, Chadwick GA (1977) Acta Metall 25:779
19. Cahn JW, Hagel WC (1963) Acta Metall 11:561
20. Li C, Goto B, Raabe K (2012) Acta mater, vol 60, pp 4005. dierk-raabe.com/pearlitic-steels

Review on Banana Fibre-Reinforced Composites

Chetan Prakash Chaudhari, Kiran Bhole, and Jayram Gholave

1 Introduction

The term "composite" means material made up of more than one discrete materials. Hence, a material consisting of more than one discrete phases or materials may perhaps be referred as composite. The properties of composite are remarkably diverse from the properties of constituent [1]. Composites are combination of two or more materials/constituents providing characteristics not attainable from one separate material. In common, composites consist of two phases, one is fibre and another is matrix. By the reinforcement geometry, the composites are categorized as: flake, particulate and fibres. By the type of matrix, composites are categorized as: carbon, ceramic, metal and polymer. The prime load carrying elements are fibres whereas the matrix helps fibres to retain the desired orientation and location. Furthermore, matrix acts as load transferring medium [2]. Inside the whole configuration, the distinct parts remain separated [3]. Composites are categorized into three categories on basis of the matrix phase as polymer matrix composite (PMC), metal matrix composite (MMC) and ceramic matrix composite (CMC). PMC is a substance made up of a polymer (resin) matrix reinforcing with fibres. Due to their low weight and simple manufacturing methods, PMC is so popular. Usage of unreinforced polymer materials is restricted by their low mechanical characteristics like comparatively less strength and less impact resistance [4]. Even though composites have previously confirmed their value as weight-saving constituents, the existing task is to create them cost effectively [5]. Due to progress in innovation and research, past few years have seen a wide-ranging composite materials development [6].

C. P. Chaudhari (✉) · K. Bhole · J. Gholave
Sardar Patel College of Engineering, Andheri West, Mumbai 400058, India
e-mail: chetanpc117@gmail.com

H. K. Dave and D. Nedelcu (eds.), *Advances in Manufacturing Processes*, Lecture Notes in Mechanical Engineering, https://doi.org/10.1007/978-981-15-9117-4_27

1.1 Study of Natural Fibres

Natural fibres (NFs) are economical, renewable, partly or completely recyclable and eco-friendly. Plants like flax, banana, bamboo, ramie, pineapple, kenaf, sisal, jute, hemp, cotton and wood are used as reinforcement for composites. NFs are an attractive environmental substitute to carbon and glass fibres for fabrication of composites due to their accessibility, price, low density, renewability and acceptable mechanical characteristics. Nevertheless, moisture resistance of NF is very low [7, 8].

Sisal has botanical name of Agave sisalana with origin to southern Mexico. Numerous products are made using a stiff fibre yield by sisal. As for eras hemp was a main source of fibre, sisal occasionally called as "sisal hemp". Widely used agricultural fibre is jute which shows reasonably greater mechanical properties. Jute fibres are mostly cultivated in Thailand, India, Bangladesh and some regions of Latin America. Overall yearly worldwide production of jute is around 2500 thousand tones. The most demandable and essential fibre in Europe is flax. Nearly 80% of entire worldwide flax is harvested in Holland, UK, Spain, Belgium and France. Flax fibres are comparatively stiffer, crisper and stronger. In the course of the past decade, reinforcing constituents for composites flax and jute fibre have gained incredible consideration. Abaca fibre produced chiefly in Philippine is a strong NF. Specific flexural strength, rotting resistant and tensile strength of abaca fibre have been stated similar to that of glass fibre. Therefore, for paper, bags and ropes, abaca has been broadly used as raw material [5, 7]. Some of commonly used NFs are shown in Fig. 1.

The coir fibre production is predominantly vital in some regions of the emerging world. Nearly 60% of entire worldwide supply of white coir fibre is by coastline area of Kerala State. Over half of coir fibre manufactured yearly all over the world is expended in countries of origin [9, 10]. Aligned long coir fibre gives better mechanical properties compared to short randomly orientated coir fibre [11]. Chemical compositions of few NFs can be seen from Fig. 2.

Study on jute fibre at fixed temperatures and determined times under ambient surroundings shows that degradation takes place more quickly at high temperature. It also showcases that degradation takes place rapidly at the low temperatures for prolonged heating [13] (Table 1).

Fig. 1 Coconut, banana and sisal fibres, respectively [5]

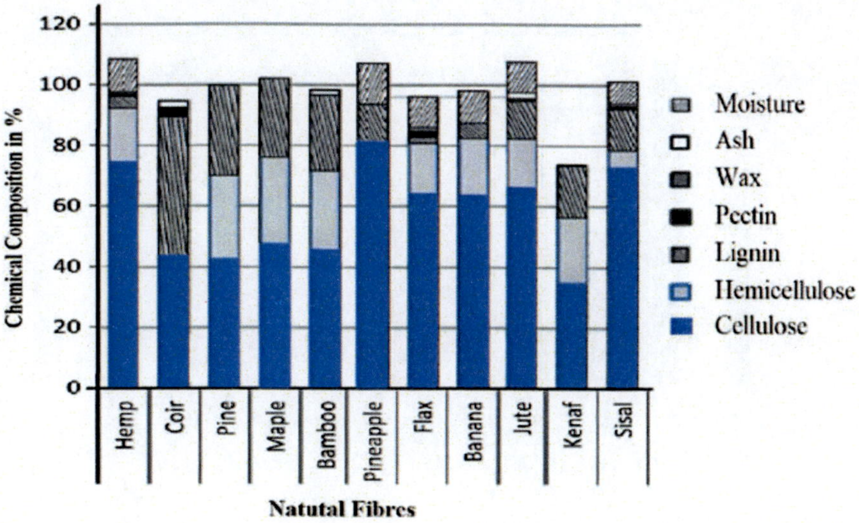

Fig. 2 Chemical compositions some of NFs [12]

Table 1 Comparison of properties of NFs [4, 5, 11, 14, 20]

Fibre type	Density (g/cm^3)	Tensile strength (MPa)	Young's modulus (GPa)	Flexural strength (MPa)
Glass	2.55	2400	73	–
Coir	1.25	175–220	4–6	6
Sisal	1.33	600–700	38	288.6
Hemp	1.48	550–900	70	–
Flax	1.4	800–1500	60–80	165
Jute	1.46	400–800	10–30	45
Cotton	1.51	287–597	5.5–12.6	43.3
Banana	1.35	355–600	17.85–33.8	76.53
Pineapple	1.5	413–1627	34.5–84.5	–
Ramie	1.5	220–938	44–128	–

1.2 Natural Fibre-Reinforced Composites (NFRCs)

NFRCs are a composite material made up of a matrix embedded using high-strength NFs, like flax, kenaf, sisal, palm, jute and banana. The properties of NFRC are diverse with each other according to earlier studies, due to dissimilar moisture surroundings, sources and fibres. The performance of NFRCs depends on aspects like fibre–matrix interaction, chemical properties, physical properties, defects, mechanical composition, temperature, porosity and fibre orientation [14, 15]. To accomplish decent interface, properties in NFs chemical treatments are essential. Figure 3 shows

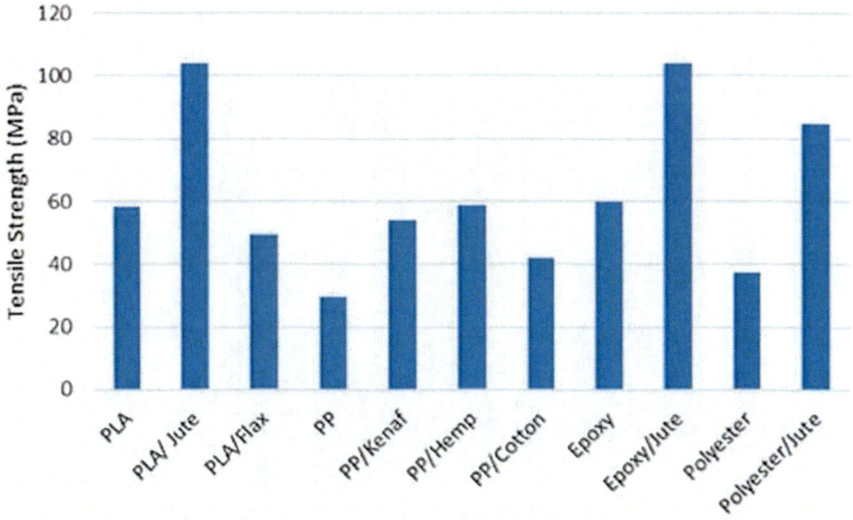

Fig. 3 Mechanical properties of some NFRCs [14]

how mechanical properties of polyester, polypropylene (PP), epoxy and polylactic acid (PLA) matrices can be affected by various NFs. NFRCs show even superior mechanical properties than a matrix alone. Tensile strength of PLAs was improved when jute fibres are added in PLA. Conversely, flax fibres introduction indicated a negative impact on composites as tensile strength gets reduced. Composites of PP were enhanced with the incorporation of cotton, kenaf and hemp. The maximum enhancement can be seen in jute/polyester composites [14].

Usually, much greater strengths are attainable with plant fibres compared to readily existing animal fibres. Matrix choice is restricted by means of the temperature at which NFs degrade. Lots of NFs used in NFRC are thermally unbalanced beyond 200 °C, even though for a small period it is possible for NFs to be treated at greater temperature under special situations. In defining the mechanical characteristics of composites, interfacial bonding among matrix and fibre plays important part. Decent interfacial bonding is necessary for achieving optimum reinforcement as stress is transmitted among fibres and matrix through the interface. For obtaining the best mechanical characteristics of composites, the fibres are aligned parallel to the applied load. Because of inadequate fibre wettability, air addition during processing and the less fibre capability to compact porosity arises in composites. Porosity in NFRCs increases more quickly with content of fibres as soon as the limit of geometrical compaction is surpassed [15, 16].

Experimental study has been performed to define the influence of sisal fibre orientations like 45°, 90° and 0°/90° on mechanical properties of sisal fibre composites. The outcomes of study are shown in Fig. 4 by comparing the tensile and flexural strength for different orientations which shows that the better mechanical properties are obtained with 90° orientation compare to other orientations [5, 17].

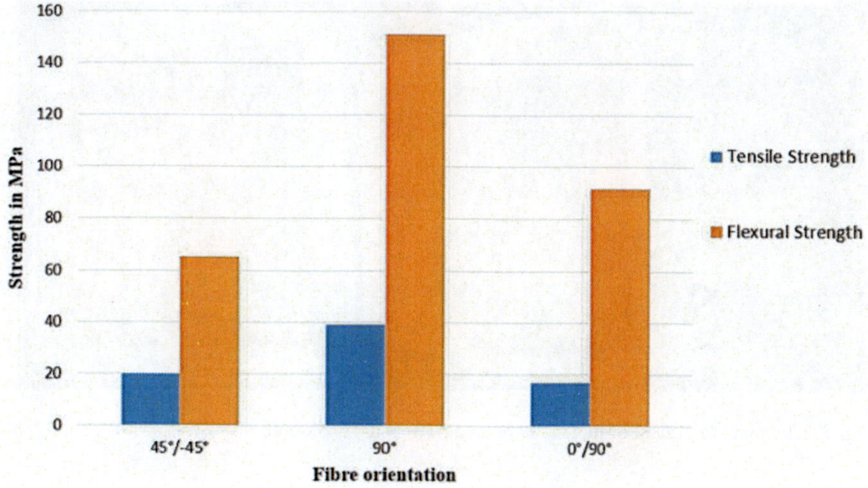

Fig. 4 Tensile and flexural strength for different fibre orientations [17]

NFRCs are environmentally better option compared to glass fibre composites in lots of applications. NF fabrication results in lesser environmental damage in comparison with glass fibre fabrication. For comparable performance, NFRCs have greater fibre content due to which the quantity of extra polluting polymer decreases. Lesser weight of NFRC increases fuel efficiency. However, it has been seen that NFRCs need high fraction of fibre volume to equate the stiffness and strength of synthetic composite [8, 18].

1.3 Banana Fibre-Reinforced Composites (BFRCs)

Banana fibre is achieved from banana stems. Banana fibre is comparatively cheap and plentifully available material. The fibre modification effects can be judged through thermal resistance, morphology, thermal and mechanical characteristics of the resultant composites. Banana crop is principally harvested in tropical nations like India. Banana crop production amount is great in India. Banana fibre is fabricated in lots of famous towns of India like Vishakhapatnam, Chennai and Mumbai. Fabrication rate of banana fibre has been greater than before because of its high usage in industries. Banana fibre is one of the NFs having high specific flexural strength, tensile strength as well as rotting resistance [4, 19].

The term "banana" arises from Arabic which means "finger". Banana fibres are pull out from the dehydrated stalks of banana plant that is not properly utilized remaining product of banana agriculture. BFRC is best appropriately used for agro-industries. It is widely utilized for decorative design items and for more strength requiring certain papers. Production of banana is maximum in India. In entire fruits

Fig. 5 Chemical treatment of banana fibre using 1% NaOH solution

farming of India, cultivation area of banana fruit is second largest. For wood, goods similar to plywood banana board can be likely alternative [20]. Extracted fibres need to be dried as the fibres contains moisture [21]. Appropriate drying of banana fibre is essential as fibre superiority principally rest on moisture content. Artificially dried fibres have usually better grade fibres compared to sun-dried fibres [22].

Banana fibre has worthy-specific strength characteristics as good as that of traditional fibres like glass. Banana fibre has lesser density compared to glass fibre. Alkali treatments are effective for eliminating fibre impurities, reducing moisture captivation in addition permitting mechanical bonding in that way enhancing matrix–fibre interface [16, 21, 23]. Chemically treated banana fibre has found to show enhancement in properties by improving cellulose chains packing after dissolution of bonding reagent lignin [24]. Chemical treatment set-up has been shown in Fig. 5.

2 Banana Fibre-Reinforced Composites Fabrication

2.1 Pre-fabrication Processing

Banana fibres are sun dried for 8 h and untangled by using manual combing. The banana fibres are then chemically treated by using 1% NaOH solution for an hour. After that neutral pH is accomplished by washing the treated fibres in tap water. To eliminate moisture content of the fibre, they are then dried in direct sunlight for couple of hours and later in an oven for 6 h at 55 °C to completely dry out fibres. The dried fibres are then successively cut for required lengths [25–27]. BFRCs are most commonly fabricated by hand lay-up process and compression moulding and injection moulding.

Fig. 6 Fabrication of BFRC using banana fibre and epoxy–hardener mixture

2.2 *Hand Lay-up Process*

The composite can be made by hand lay-up method. The smooth metal surface is achieved by rubbing the cleaned base plate using sand paper. The plate surface is then permitted to dry out once washed by thinner solution. Later the dried surface is covered from the inner side using debonding gel like wax for excellent surface quality. The hardener and epoxy resin are mixed together in ratio of 1:10. Once mixed the curing time for the mixture is 20 min. The caution need to be taken that the resin mixture should not cure itself in container. A persistent guard need to be done on the epoxy–hardener mixture in the container by means of stop watch. The first layer of banana fibre is placed above the base metal plate with epoxy resin. Manually spread the epoxy–hardener mixture with help of roller by ensuring proper reinforcement as well as expelling trapped air. The required thickness of composite can be obtained by adding layers of banana fibre–matrix formulation. The process of layer-by-layer fabrication of BFRC can be visualized from Fig. 6.

The metal plate is then bolted and the arrangement is permit to cure an entire day underneath a loaded circumstance by using compression press. Figure 7 shows the arrangement of composite sheet in compression press. The hardened composite sheet is then ejected and cut [22, 25, 26, 28].

2.3 *Compression Molding (CM)*

CM process is mostly used for big parts in medium amounts. Usual CM process can be either hot or cold besides uses thermosetting matrix. Only pressure is used in case of cold CM. Pressure as well as temperature is used in case of hot CM. Watchful viscosity control is vital for achieving decent bonding among the matrix and fibre. It is very essential to confirm excessive pressure and heat in hot CM do not

Fig. 7 Compression of composite sheet using compression press

break fibres. In the course of CM viscosity, temperature, holding time and pressure of substance to accomplish decent fibre wetting should be accurately controlled. Starting surface to core thicker parts shows huge temperature gradients. Alignment of fibres alongside the flow direction of polymer is significant characteristic of CM process. Processing parameters of CM depends on matrix, fibre and type of resin used in composite. Nevertheless, normally pressing interval can be in between 30 s and 5 min, temperature in between 130 and 250 °C and pressure in between 10 and 20 MPa. CM process have advantages like high reproducibility, cheap, short cycle time and less waste. Below Fig. 8 represents schematic diagram of CM process [16, 29].

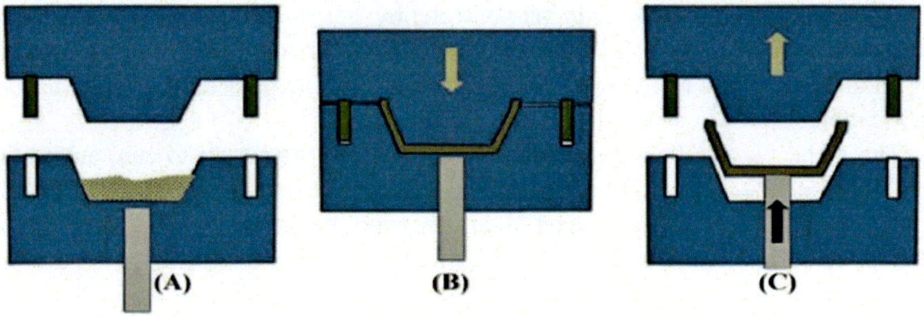

Fig. 8 Schematic diagram of compression moulding process [16]

3 Factors Affecting Performance of Banana Fibre-Reinforced Composites

3.1 *Fibre Configuration*

In a study on BFRC, four kinds of composite samples, namely PP/banana mat, PP/banana yarn, PP/raw banana fibre and unreinforced polypropylene (PP), were produced with 70 wt% matrix and 30 wt% fibre. Unreinforced PP was manufactured aimed at evaluation reasons. The flexural test specimen was tested according to ASTM D790 standard and the tensile test specimen was tested according to ASTM D638 standards. This study validates that the significant part is played by fibre configuration in attaining higher mechanical strength. Diverse mechanical characteristics of composites were recorded like flexural strength and tensile strength. It is established from Fig. 9 that the composite of PP/banana yarn gives the maximum flexural strength and tensile strength in comparison with other configuration as yarn is continuous and unidirectional in tensile load application. The rest two fibre configurations give mixed outcomes [5, 30].

In another study, two kinds of fibre configurations S1 and S2 were produced to know fibre mats positioning in hybrid composites. The S1 has configuration as aloe vera–flax–sisal–flax–aloe vera while S2 has configuration as flax–aloe vera–sisal–aloe vera–flax. Both S1 and S2 are fabricated with epoxy resin between the fibres. A critical identification from outcome of study is that peripheral fibre selection has important part in improving the composite properties. From Fig. 10, it is clear that the peripheral flax hybrid composite (S2) has better flexural and tensile strength compared to (S1) [3].

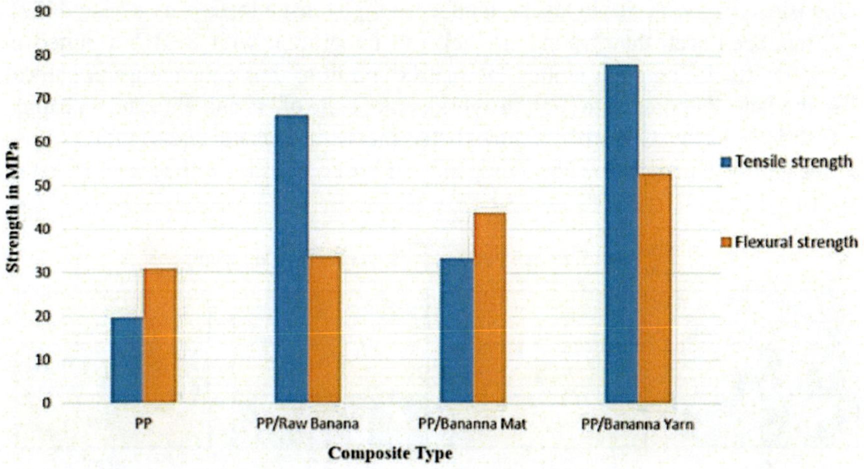

Fig. 9 Bar chart of tensile and flexural strength [5, 30]

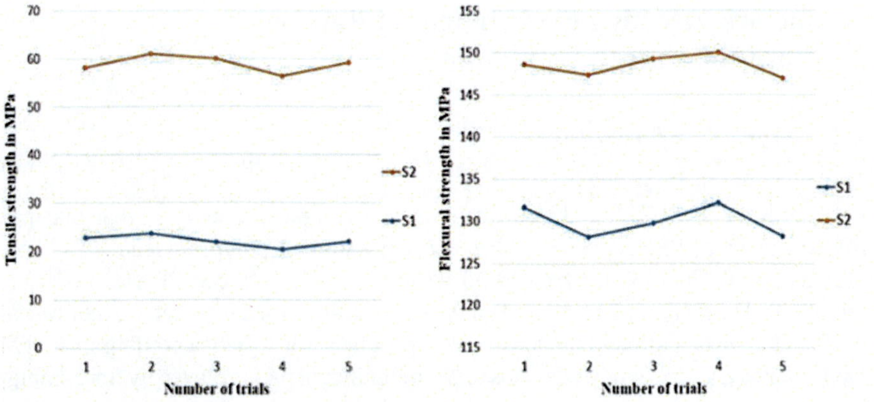

Fig. 10 Tensile and flexural strength of two specimens [3]

In experiment on cross-banana–glass hybrid composites, four different fibre configurations were studied. All four composite samples consist of three layers. Fibre layer arrangement of all four sample composites is shown in Fig. 11 where B stands for banana fibre and G stands for glass fibre. In tension test results, GGG sample composite is found to have higher tensile strength because of more integration of high-strength glass fibre. Best mechanical characteristics can be achieved by incorporating high-strength fibre at outer periphery of composite. However, as a result of poor interfacial bond among matrix and fibre flexural strength of GGG of composite resulted in lower value compare to GBG composite. Greater flexural strength of GBG composites is result of high banana fibre dispersion due to which glass fibres can occupy available void spaces forming dense composite. Figure 12 shows flexural and tensile strength of different fibre configuration [31]. Tensile strength of fibre composites rises with rise in layers' number as high tensile loads can be beared [32].

It has been seen that groundnut shell ash integration with BFRCs resulted in decent interfacial bonding among the matrix and fibre thus enhancing mechanical characteristics of composite [33]. In hybrid, composite of banana fibre and pineapple leaf fabricated using CM process shows better mechanical characteristics for 3:1 ratio of pineapple leaf and banana fibre compared to fibre composites with 1:3 and 1:1 ratio.

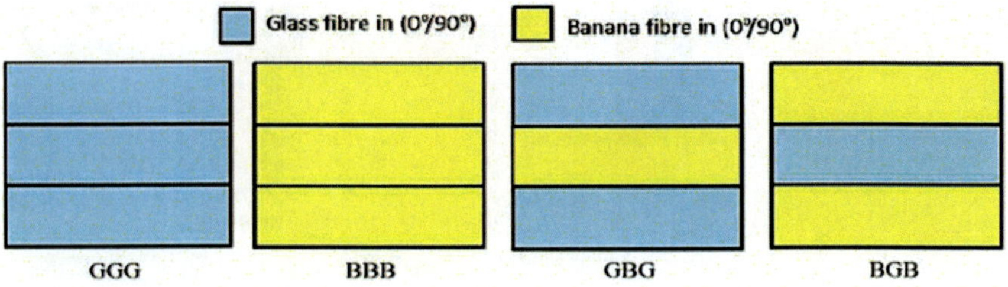

Fig. 11 Fibre layer arrangement in composite samples [31]

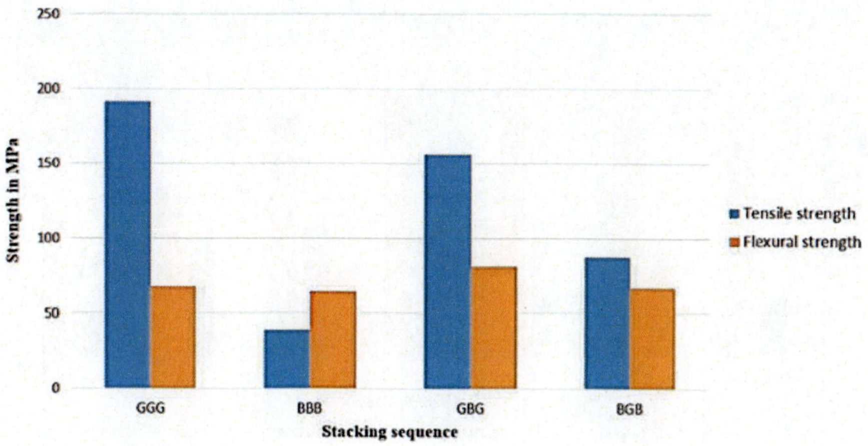

Fig. 12 Tensile and flexural strength of various fibre configuration sample [31]

This enhancement in properties is due to high tensile strength of pineapple leaf fibre compared to banana fibre [21]. The investigational research discovered that fly ash BFRCs absorbed 15.67% fewer moisture in comparison with BFRCs. As a result, fly ash BFRCs delivered enhanced mechanical characteristics like higher value of tensile strength, Young's modulus and toughness by 15.7, 12.38 and 18.39%, respectively, than BFRCs [34]. The outcomes achieved from one of the studies noticeably revealed that the incorporation of coir fibre with banana result in enhancement of thermal stability, toughness, flexural properties and tensile properties of the BFRCs [35].

3.2 Surface Treatment

A study establishes that the NaOH treatment carries a major part in decreasing absorption rate of moisture as well as enhancing the mechanical characteristics of composite. Study was carried out using epoxy (LY556) and banana fibre to fabricate the composite. Hardener and epoxy are in 1:10 ratio by weight. The fibres were dipped for 30 min in the NaOH solution. The tap water-washed fibres are dried by cloth. The dry fibres are then located in an oven for 45 min at 50 °C so as to remove complete moisture content. The composites are fabricated using hand lay-up method. Five samples were tested for each case and average value was obtained. After treating banana fibre with numerous % of NaOH, it is clear that 1% NaOH treated composite delivers superior mechanical properties and can be seen in Fig. 13. Approximately 50% growth in the composite properties in comparison with untreated fibre composite is achieved by 1% NaOH-treated fibre composites. Further increase in NaOH concentration effects in damage to fibre surface, which results in reduction in mechanical characteristics of fibre composite. NaOH treatment eliminates fibre particles such as hemicellulose, lignin, oil and wax from surface of fibre in addition

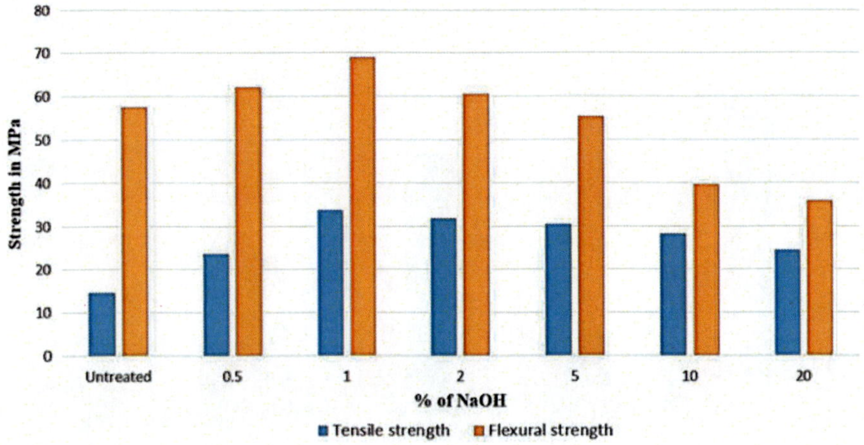

Fig. 13 Tensile and flexural properties of treated and untreated composites [36]

there is improvement in fibre–matrix bonding due to increase in surface roughness [16, 32, 33, 36].

The adhesion among matrix and fibres inside the composites and thermal stability of BFs enhanced when treated with chemical [27, 37]. Poor bonding in BFs leads to moisture absorption and that is why poor bonding among matrix and fibre is principal hurdle to worldwide adoption of BFRCs [38]. The banana fibre treatment with solution of 10% sodium carbonate resulted in better tensile characteristics due to removal of hemicelluloses along with lignin [39]. It has been reported that increment in coagulum quantity result in improved impact characteristics by decreasing water absorption characteristics. Optimum flexural strength can be obtained by replacing 40% polyester resin with coagulum [40]. The compatibilizer use and fibre treatment have shown enhanced mechanical characteristics of low-density polyethylene (LDPE) composites. Without compatibilizer LDPE composites are treated with acrylic acid and alkali has displayed nearly same mechanical characteristics. This shows the improvement in compatibility of banana fibre with LDPE. On the other hand, additional treatment of alkali-treated fibres with acrylic acid result in somewhat poorer tensile properties due to fibre disintegration. However, better flexural strength compares to only alkali-treated fibre composites [19]. Chemical treatments of fibres like formic acid, $KMnO_4$ and alkali that are shown to enhance fibre interfacial bonding with resin were obtained for alkali-treated composites that give maximum glass transition temperature and storage modulus compared to formic acid and $KMnO_4$-treated fibre composites [41]. Plasma treatment is a physical surface treatment which does not modify the bulk characteristics of fibre. It affects only the outermost fibre layer. Plasma treatment cleans the fibre surface along with enhancing fibre polymer adhesion. It also increases the fibre surface roughness by promoting interlocking among matrix and fibre. Plasma technique helps in functionalizing surface of fibre to encourage fibre polymer chemical bonding [32]. Influence of alkali treatment of hybrid pineapple leaf BFRCs on tensile characteristics is presented in Fig. 14.

An experiment on the surface treatment chemicals' effects on tensile strength of BFRCs demonstrates the enhancement in tensile strength when treated with solution of sodium sulphite and aminopropyltriethoxysilane (APTES silane). The better tensile strength of silane-treated BFRCs is due to reduction in hydroxyl groups which result in stable covalent bond [32, 42]. Effect of different surface treatment chemicals on tensile characteristic of BFRC is shown in Fig. 15.

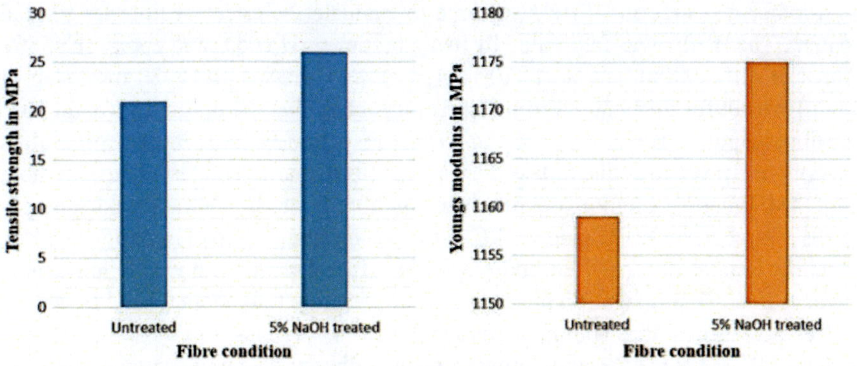

Fig. 14 Tensile strength and Young's modulus of hybrid pineapple leaf BFRCs [21]

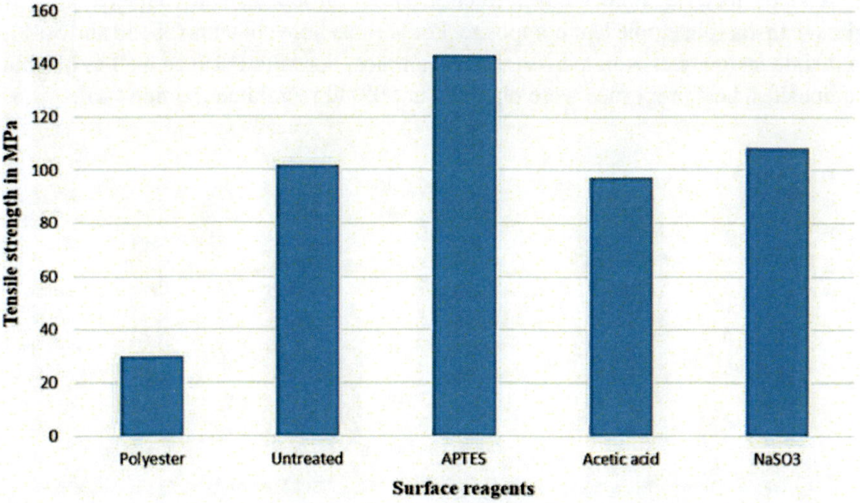

Fig. 15 Tensile of BFRCs treated with surface reagents [42]

3.3 Fibre Length and Fibre Content

A study on BFRC where BFRC was fabricated using the hardener (HY591) and epoxy resin (LY556). ASTM D638 standard tensile test was conducted with 25 mm gauge length. This study shows that composite fabricated from 15 mm fibre length delivers superior mechanical properties [25]. A research on effect of fibre length on the mechanical characteristics of coir fibre-reinforced composite with changing fibre length from 30 to 50 mm in step of 10 mm (30, 40 and 50 mm) delivered best mechanical characteristics for fibre length of 50 mm [11].

Another research on BFRC was done to evaluate the effect of fibre length and content. The BFRC was fabricated by using hardener (HY951) and epoxy (LY556). Fibres of different weight % (8–20%, step of 4%) and length (5–20 mm, step of 5 mm) are cut for this study. From Fig. 16, it is clear that the optimum weight ratio and fibre length for BFRC are 16% and 15 mm, respectively. Extra increment result in lowering the properties due to less adhesion of fibre–matrix and less matrix in more fibre [43]. Fracture toughness experiments on fibre fraction volume for percentages of 13, 17 and 20 on BFRC and GFRC show enhancement in fracture toughness due to reinforcement banana fibre and glass fibre. Increase in volume fraction causes increase in crack propagation load. As a result, fracture toughness GFRC as well as BFRC increases as fibre volume increases. Fracture toughness of GFRCs is superior to that of BFRCs [44]. Accumulation of banana fibre with epoxy enhances physical as well as mechanical characteristics for composite. Likewise, nanosilica addition shows a vital part in reducing wear [6] (Tables 2 and 3).

A study looks at fibre volume fraction effect on mechanical characteristics of glass–banana–pineapple leaf composite. For various fibre contents (30, 40 and 50%), uniaxial tension tests were performed on samples. For both PaLF as well as banana composites, best properties were obtained at 50% fibre volume fraction [45].

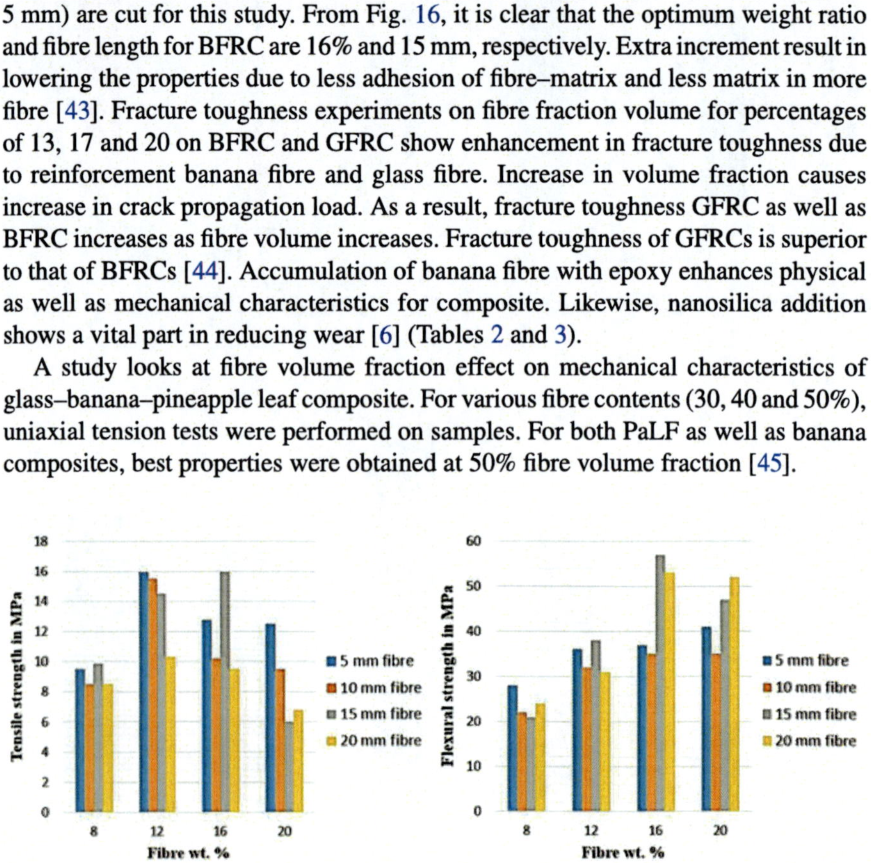

Fig. 16 Effect of fibre length and content on tensile and flexural properties [43]

Table 2 Mechanical characteristics of BFRCs for various fibre volume fraction [24]

Samples	Impact strength (Joules)	Tensile strength (MPa)	Flexural strength (MPa)
40% banana fibre + 60% epoxy resin	8.62	108.42	71.28
Coir 50% banana fibre + 50% epoxy resin	9.48	112.58	76.53
60% banana fibre + 40% epoxy resin	11.22	98.3	66.18

Table 3 Effect of fibre volume and chemical treatment on mechanical behaviour of BFRCs [27]

Property	Flexural strength (MPa)			Tensile strength (MPa)		
Fibre weight (%)	10	20	30	10	20	30
Treated BFRC	59	61	60	28	27	24
Untreated BFRC	56	58	58	27	26	23

3.4 *Temperature*

A study on banana fibre was done to evaluate the effect of temperature on tensile behaviour of fibre using UTM clutch of 20 mm gauge length. The test was performed for changing temperatures of -20, 0, 20, 40, 60, 80, 100, 140, 180 and 220 °C. Tensile strength of banana fibre for different temperature values is represented in Fig. 17 and we can see that banana fibre displayed stable tensile strength in 0 to 100 °C

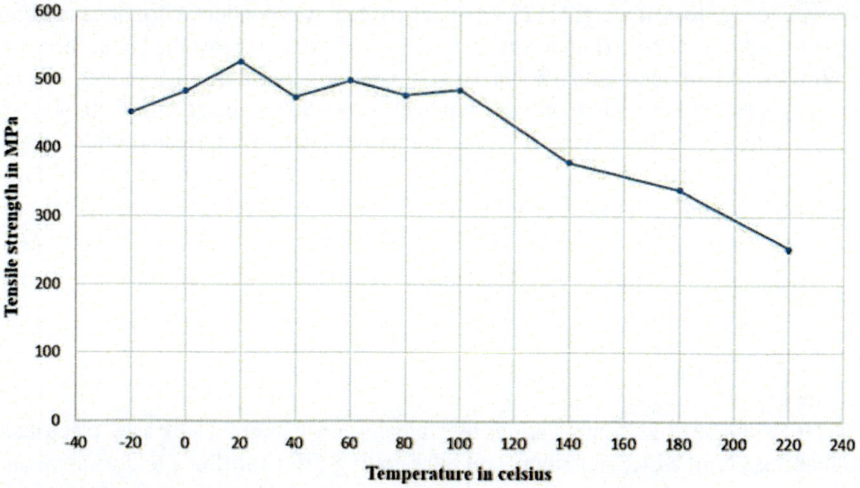

Fig. 17 Tensile strength of banana fibre for different temperatures [46]

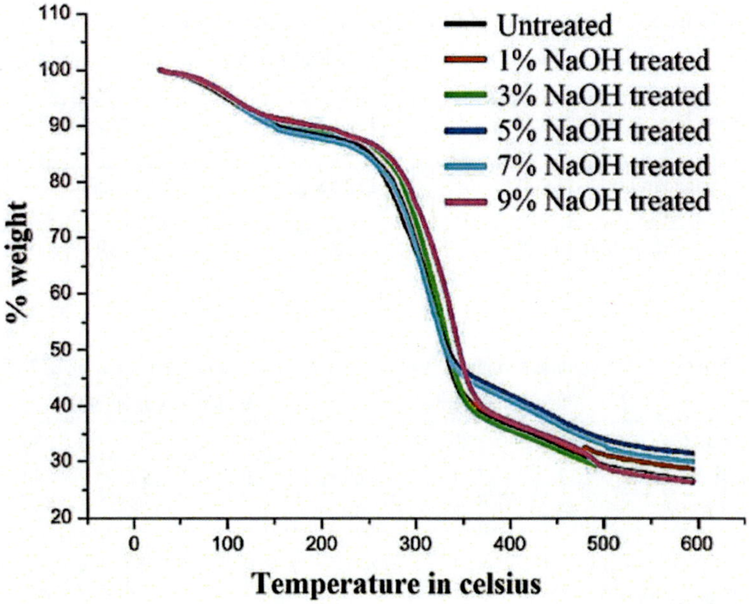

Fig. 18 TGA graph of NaOH-treated and NaOH-untreated banana fibres [47]

temperature range however temperature change beyond the range pointedly reduces tensile strength of banana fibre [46].

Thermogravimetric analysis (TGA) is usually used to check amount of fibre decomposition as well as thermal stability. A distinctive TGA graph for thermal degradability of fibre composite showcases that a specimen exposed to heat will deliberately undergo loss of weight [12]. TGA experiment was conducted for banana fibre treated for different NaOH concentration. Figure 18 demonstrates the better thermal stability of fibre treated with 5% NaOH compared to other fibres. Moreover, it also shows that the all fibres are thermally stable up to 60–80 °C and at 220 °C displayed an adequate weight loss. Among all six samples, banana fibre treated with 5% of NaOH showed smallest loss in weight [47]. Figure 19 showcases a typical TGA graph for different combination of banana and jute hybrid composites. It is seen in TGA analysis that the untreated fibres start losing weight prior to treated fibres [27].

4 Applications

The NFCs can be used in applications like railway coach interior, suitcases, electrical appliances, chair, table, containers and false ceiling [7]. Coir fibres find its applications in paper weights, ropes for nets, helmets, storage tank, padding for mattresses along with building panels [14, 21]. Lots of surfboard makers are nowadays utilizing

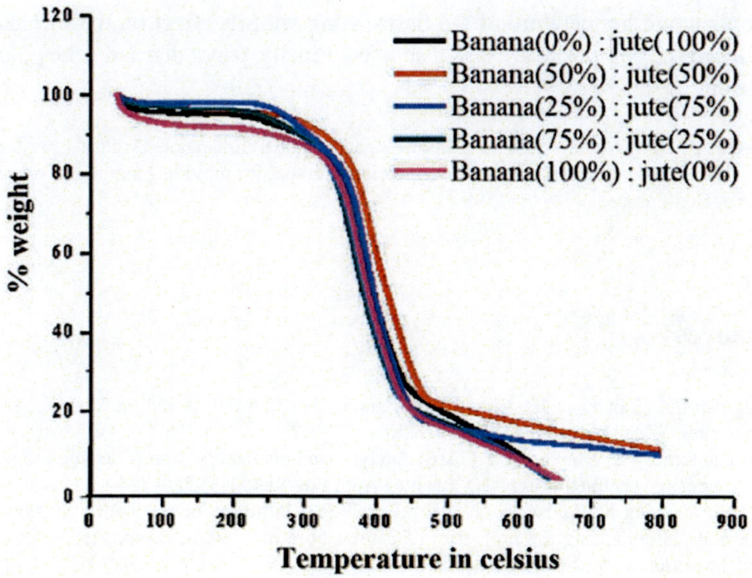

Fig. 19 TGA graph of banana and jute hybrid composites [48]

NFCs for its fabrication and "Ecoboard" manufactured with help of hemp fibre and resin was one of the initial models. The aircraft business is also implementing NFCs in interior panels [15, 22]. A fascinating side of jute composite is its railway coach as well as automobiles applications. Ropes made up of sisal are extensively used for shipping, agricultural as well as marine [16, 20]. NFRCs are developing pretty quickly as the possible alternative of metal-built things in applications consists of construction, electronic, sporting goods, marine, aerospace and automotive industries. In recent times, BFRCs are getting lots of attentions due to its usage in customer cars for under-floor protection [5, 49]. In addition, flax fibre composite can be used in fabrication of automobile hood [50, 51].

5 Conclusion

This paper presents exhaustive literature review of natural fibre, NFRCs and particularly the banana fibre-reinforced composite. The process of fabrication of fibre composite is reported in the paper. The surface treatment process as one of the pre-processing techniques used to reduce the impurities form the fibres is demonstrated in first part of paper. The pre-processing techniques for other natural fibres are also discussed in the paper. Typical hand lay-up technique for development of composite is presented step by step. After development of the composite, the properties of the composite material are presented. The factors like fibre configuration, fibre length, fibre content and temperature of the fibres along with its effect on the mechanical properties of the composite are also discussed. Finally, paper discusses the potential applications of the NFRCs.

Acknowledgements Authors like to acknowledge Technical Education Quality Improvement Program—III of Sardar Patel College of Engineering Mumbai for providing support in presenting the paper at Conference.

References

1. Bhagwan DA, Lawrence JB, Chandrashekhara K (2017) Analysis and performance of fiber composites. Wiley India Pvt Ltd, New Delhi
2. Krishnamoorthy K, Sasikumar T (2016) Analysis and characterization of tensile property of the composite specimen using ANSYS. Int J Appl Eng Res 11(1):380–384
3. Balasubramanian K, Rajeswari N, Vishvak R (2020) Influence of sequential positioning of fibre mats in enhancing the properties of natural fibre mat hybrid composite. Mater Today Proc. https://doi.org/10.1016/j.matpr.2020.01.097
4. Rao PD, Rao DV, Naidu AL, Bahubalendruni R (2017) Mechanical properties of banana fiber reinforced composites and manufacturing techniques: a review. Int J Res Dev Technol 8(5):39–46
5. Chaudhari CP, Bhole K (2020) A review on characteristics of natural fibre composite. In: Paper presented at 2nd international conference on emerging trends in manufacturing, engines and modelling (ICEMEM 2), SVKM NMIMS Shirpur, 23–24 December 2019
6. Rahul K, Shetty MH, Karthik MN, Pavana KB, D'Souza KP, D'Souza L (2017) Processing and characterisation of banana fiber reinforced polymer nano composite. Nanosci Nanotechnol 7(2):34–37
7. Elanchezhian C, Ramnath BV, Ramakrishna G, Rajendrakumar M, Naveenkumar V, Saravanakumar MK (2018) Review on mechanical properties of natural fiber composites. In: Paper presented at international conference on processing of materials, minerals and energy (PMME), Ongole, Andhra Pradesh, 29–30 July 2016
8. James M, Richard E, Stuart RC, Benjamin MW, Kerry K (2012) Natural fibre composite energy absorption structures. Compos Sci Technol. https://doi.org/10.1016/j.compscitech.2011.11.004
9. Harish S, Michael DP, Bensely A, Mohan DL, Rajadurai A (2009) Mechanical property evaluation of natural fiber coir composite. Mater Character. https://doi.org/10.1016/j.matchar.2008.07.001
10. Manjusha K, Kondareddy B, Kumar DP (2016) Effect of fiber length and weight on tensile response of natural fiber reinforced composite. Int J Eng Res Technol 5(4):389–394
11. Walte AB, Bhole K, Gholave J (2020) Mechanical characterization of coir fiber reinforced composite. In: Paper presented at 3rd international conference on advances in materials and manufacturing applications (IConAMMA 3). Amrita School of Engineering Bengaluru, 16–18 August 2018
12. Azwa ZN, Yousif BF, Manalo AC, Karunasena W (2013) A review on the degradability of polymeric composites based on natural fibres. Mater Des. https://doi.org/10.1016/j.matdes.2012.11.025
13. Hart A, Summerscales J (2017) Effect of time at temperature for natural fibres. In: Paper presented at 3rd international conference on natural fibers (ICNF 3), University of Minho Braga, 21–23 June 2017
14. Mohammed L, Ansari MNM, Pua G, Jawaid M, Islam MS (2015) A review on natural fiber reinforced polymer. Int J Polym Sci. https://doi.org/10.1155/2015/243947

15. Pickering KL, Efendy MGA, Le TM (2016) A review of recent developments in natural fibre composites and their mechanical performance. Composites Part A 83. https://dx.doi.org/10.1016/j.compositesa.2015.08.038
16. Balla VK, Kate KH, Satyavolu J, Singh P, Tadimeti JGD (2019) Additive manufacturing of natural fiber reinforced polymer composites processing and prospects. Compos B. https://doi.org/10.1016/j.compositesb.2019.106956
17. Kumaresan M, Sathish S, Karthi N (2015) Effect of fiber orientation on mechanical properties of sisal fiber reinforced epoxy composites. J Appl Sci Eng. https://doi.org/10.6180/jase.2015.18.3.09
18. Joshi SV, Drzal LT, Mohanty AK, Arora S (2004) Are natural fiber composites environmentally superior to glass fiber reinforced composites?. Compos Part A 35. https://doi.org/10.1016/j.compositesa.2003.09.016
19. Prasad N, Agarwal VK, Sinha S (2016) Banana fiber reinforced low density polyethylene composites: effect of chemical treatment and compatibilizer addition. Iran Polym J. https://doi.org/10.1007/s13726-016-0416-x
20. Bavan DS, Kumar GCM (2010) Potential use of natural fiber composite materials in India. J Reinf Plast Compos. https://doi.org/10.1177/0731684410381151
21. Rahman M, Das S, Hasan M (2018) Mechanical properties of chemically treated banana and pineapple leaf fiber reinforced hybrid polypropylene composites. Adv Mater Proces Technol. https://doi.org/10.1080/2374068X.2018.1468972
22. Murugan K, Venkatesh S, Thirumalai R, Nandhakumar S (2020) Fabrication and investigations of kenaf fiber and banana fiber reinforced composite material. Mater Today Proc. https://doi.org/10.1016/j.matpr.2020.04.540
23. Pujari S, Ramakrishna A, Kumar MS (2014) Comparison of jute and banana fiber composites: a review. Int J Current Eng Technol. https://dx.doi.org/10.14741/ijcet/spl.2.2014.22
24. Ramesha M, Atreya TSA, Aswin US, Eashwar H, Deepa C (2014) Processing and mechanical property evaluation of banana fiber reinforced polymer composites. In: Paper presented at 12th global congress on manufacturing and management (GCMM 12), VIT University Vellore, 8–10 December 2014
25. Sumaila M, Amber I, Bawa M (2013) Effect of fiber length on the physical and mechanical properties of random oriented nonwoven short banana fibre/epoxy composite. Asian J Nat Appl Sci 2(1):39–49
26. Bhoopathi R, Ramesh M, Deepa C (2014) Fabrication and property evaluation of banana-hemp-glass fiber reinforced composites. Procedia Eng 97. https://dx.doi.org/10.1016/j.proeng.2014.12.446
27. Komal UK, Verma V, Tarachand A, Verma N, Singh I (2018) Effect of chemical treatment on mechanical behavior of banana fiber reinforced polymer composites. In: Paper presented at advances in materials & processing: challenges and opportunities (AMPCO), IIT Roorkee, 30 November–2 December 2017
28. Gupta SD, Mahato DN, Paswan MK (2020) Differential fabrication and characterization of natural fiber composite laminates—an investigative approach. Mater Today Proc. https://doi.org/10.1016/j.matpr.2020.02.212
29. Kerni L, Singh S, Patnaik A, Kumar N (2020) A review on natural fiber reinforced composites. Mater Today Proc. https://doi.org/10.1016/j.matpr.2020.04.851
30. Amir N, Abidin KAZ, Faizzaty BMS (2017) Effects of fibre configuration on mechanical properties of banana fibre/PP/MAPP natural fibre reinforced polymer composite. In: Paper presented at advances in material and processing technologies conference (AMPTC), VIT Vellore, 11–14 December 2017
31. Zulkafli N, Maligam SD, Fadzullah SHSM, Mustafa Z, Zakaria KA, Subramoni S (2019) Mechanical properties of cross-ply banana-glass fibre reinforced polypropylene composites. Defence S and T Tech Bull 12(1):124–135
32. Verma A, Parashar A, Jain N, Singh VK, Rangappa SM, Siengchin S (2020) Surface modification techniques for the preparation of different novel biofibers for composites. In: Khan A, Rangappa SM, Siengchin S, Asiri AM (eds) Biofibers and biopolymers for biocomposites. Springer, Switzerland, pp 1–34

33. Naidu AL, Kona S (2018) Experimental study of the mechanical properties of banana fiber and groundnut shell ash reinforced epoxy hybrid composite. Int J Eng. https://doi.org/10.5829/ije.2018.31.04a.18
34. Kauser B, Upadhyay V, Chak V, Hasan F (2017) Moisture absorption behavior and mechanical properties of banana fiber-reinforced fly ash/epoxy composites. In: Chattopadhyay J, Singh R, Prakash O (eds) Innov Mater Sci Eng, vol 2. Springer, Singapore, pp 11–16
35. Prasad N, Agarwal VK, Sinha S (2018) Hybridization effect of coir fiber on physico-mechanical properties of polyethylene-banana/coir fiber hybrid composites. Sci Eng Compos Mater. https://doi.org/10.1515/secm-2015-0446
36. Venkateshwaran N, Perumal AE, Arunsundaranayagam D (2013) Fiber surface treatment and its effect on mechanical and visco-elastic behavior of banana/epoxy composite. Mater Des. https://doi.org/10.1016/j.matdes.2012.12.001
37. Reddy RA, Yoganandam K, Mohanavel V (2020) Effect of chemical treatment on natural fiber for use in fiber reinforced composites—review. Mater Today Proc. https://doi.org/10.1016/j.matpr.2020.02.982
38. William J, Patrick C (2017) Improving the properties of banana fiber reinforced polymeric composites by treating the fibers. In: Paper presented at 3rd international conference on natural fibers (ICNF 3), Braga Portugal, 21–23 June 2017
39. Caldas A, Santos JC, Panzera TH, Strecker K (2016) Mechanical properties of epoxy banana fibre composite trated with sodium carbonate. In: Paper presented at Brazilian conference on composite materials (BCCM 3), Gramado Brazil, 28–31 August 2016
40. Kumari S, Rai B, Kumar G (2018) A study on effect of ATH on Euphorbia coagulum modified polyester banana fiber composite. In: Paper presented at 7th International conference on structural analysis of advanced materials (ICSAAM 7), Bucharest Romania, 19–22 September 2017
41. Indira KN, Jyotishkumar P, Thomas S (2014) Viscoelastic behaviour of untreated and chemically treated banana fiber/PF composites. Fibers Polym. https://doi.org/10.1007/s12221-014-0091-5
42. Ike-Eze ICE, Aigbodion VS, Ude SN, Omah AD, Offor PO (2019) Experimental study on the effects of surface treatment reagents on tensile properties of banana fiber reinforced polyester composites. J Mater Environ Sci J 10(5):402–410
43. Venkateshwaran N, ElayaPerumal A, Jagatheeshwaran MS (2011) Effect of fiber length and fiber content on mechanical properties of banana fiber/epoxy composite. J Reinf Plast Compos. https://doi.org/10.1177/0731684411426810
44. Muralikrishna MVV, Suryakumari TSA, Gopi R, Loganathan GB (2020) Development of mechanical properties in banana fiber composite. Mater Today Proc. https://doi.org/10.1016/j.matpr.2019.08.189
45. Hanafee ZM, Khalina A, Norkhairunnisa M, Syams ZE, Liew KE (2017) The effect of different fibre volume fraction on mechanical properties of banana/pineapple leaf (PaLF)/glass hybrid composite. In: Paper presented at 3rd electronic and green materials international conference (EGM 3), Thailand, 29–30 April 2017
46. Cestmir M, David H, Petr H, Abraham K (2017) Effect of temperature and moisture content on tensile behaviour of false banana fibre. Int Agrophys. https://doi.org/10.1515/intag-2016-0067
47. Parre A, Karthikeyan B, Balaji A, Udhayasankar R (2020) Investigation of chemical, thermal and morphological properties of untreated and NaOH treated banana fiber. Mater Today Proc. https://doi.org/10.1016/j.matpr.2019.06.655
48. Boopalan M, Niranjanaa M, Umapathy MJ (2013) Study on the mechanical properties and thermal properties of jute and banana fiber reinforced epoxy hybrid composites. Compos B Eng. https://doi.org/10.1016/j.compositesb.2013.02.033
49. Sanjay MR, Arpitha GR, Naik LL, Gopalakrisha K, Yogesha B (2016) Applications of natural fibers and its composites: an overview. Nat Resour. https://doi.org/10.4236/nr.2016.73011

50. Kong C, Lee H, Park H (2016) Design and manufacturing of automobile hood using natural composite structure. Compos B Eng. https://doi.org/10.1016/j.compositesb.2015.12.033
51. Kenge A, Kulkarni N, Tate N, Shaikh H, Bhole K (2020) Development and mechanical characterization of natural-artificial fiber hybrid composite. In: Paper presented at 3rd Biennial international conference on Nascent Technologies in Engineering (ICNTE 3), FC Rodrigues Institute of Technology Vashi, 4–5 January 2019

Experimental Evaluation of Mechanical Properties of Epoxy Based Composite Material Using Taguchi Method

Vishal Naranje, Ajay Rajan Sankar, Sachin Salunkhe, and Bhanudas D. Bachchhav

1 Introduction

Fiberglass is a common kind of glass fiber enhanced plastic. The fibers can be arranged at random, flattened on a sheet or woven into a cloth. The plastic matrix is usually based on thermosetting polymers such as epoxy, polyester or vinyl ester and a thermoplastic. It is cheaper and more versatile than carbon-fiber and can be moulded in complex shapes, stronger than many metals by weight. Applications are for aircraft, vessels, motor vehicles, bathrooms and bins, swimming pools, hot tubes, septic tanks, water pipes, toilets, pipes, covers, casting and exterior surfboards. Glass enhanced plastics (GRP), glass enhanced plastic (GFP) also known as fiberglass [1]. There are different types of glass fibres are available in the market such as A type, C type, D type, E type, Advantex, ECR Glass, AR Type, R Type, S-2 Type, etc. These Fibre glasses are used for different applications. Type E is used in electrical and thermal insulation, type C (chemical) is having superior resistance to acid; and type T for thermal insulation. The basic substance that forms the glass fiber is actually known glass the other materials are a range of natural minerals and manufactured chemicals. As glass fiber itself is sometimes referred to as "fiberglass,". The key ingredient of the composite material is glass fibre and resin. The fibregalss provides

V. Naranje · A. R. Sankar
Department of Mechanical Engineering, Amity University Dubai, Dubai, UAE
e-mail: vnaranje@gmail.com

S. Salunkhe (✉)
Department of Mechanical Engineering, Vel Tech Rangarajan Dr.Sagunthala R and D Institute of Science and Technology, Avadi, Chennai, India
e-mail: drsalunkhesachin@veltech.edu.in

B. D. Bachchhav
Department of Mechanical Engineering, AISSMS, College of Engineering, Pune, Maharashtra 411 001, India
e-mail: bdbachchhav@aissmscoe.com

H. K. Dave and D. Nedelcu (eds.), *Advances in Manufacturing Processes*, Lecture Notes in Mechanical Engineering, https://doi.org/10.1007/978-981-15-9117-4_28

the reinforcement and the resin provides the binding to the body and required resistance to corrosion. Pigment chemicals, flame retardants, U.V. absorbers, etc. may also be used according to product specifications and manufacturing process. Surface tissue/veil is also used as another material. In the last decades, a significant research effort has been made to study the impact behaviour of composite materials with textile reinforcements. Several researchers studied the effect of carbon nanotube and its influence on the mechanical properties of nanocomposites [2, 3]. Most of the researchers have studied the mechanical properties of the composites [4–7]. This article aims to evaluate the optimum input parameter values under different constraints (for example, temperature, time and hardener), which will improve the mechanical characteristics of the composite material.

2 Materials and Methods

2.1 Material Preparation

Below is the procedure for the manufacture of the composite material and the same is shown in Fig. 1.

1. Take clean flat base sheet of plywood or metal
2. To check the hardness of the composite material, CSM-300 gm/m^2 mat cut into 8 pieces of 3 mm thickness.
3. Measure the weight of mat using weighing machine and take the double amount for resin. i.e. the ratio 1:2 for mat to resin. It is mixed in a ratio of 10:1 by weight as recommended.
4. Measure the amount of hardener/catalyst MEKP (methyl ethyl ketone peroxide) as the given standard table to the resin weight and keep aside.

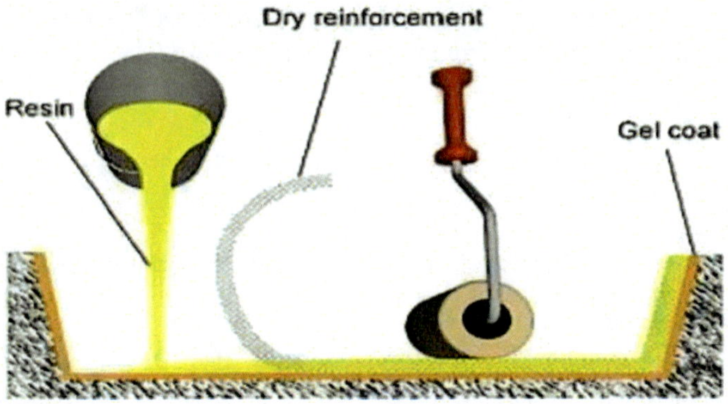

Fig. 1 Schematic diagram of the experimental set up for preparation of composite material [2]

5. Take a mixing bowl and add general-purpose orthophthalic polyesterresin and catalyst one by one.
6. Mix well without formation of any bubbles.
7. The uniformly mixed dough (SGM-filled epoxy) is then slowly poured into glass mould, pre-coated with wax and a uniform thin film of silicone release agent.
8. Apply releasing mould wax on top of the sheet metal.
9. Rub the wax again with a clean polishing cloth in order to remove any excess wax.
10. Put one layer of mat on the sheet and add few drops of resin and roll with a steel roller.
11. Make sure all sides are equally filled.
12. Add another layer of mat on top of the resin and roll again, then add more resin.
13. Continue this process until all the layers are filled.
14. It's then kept for drying at room temperature for specific time.
15. After hardening the sample is kept in a thermal dying oven at specific temperature range and for time to harden.
16. Follow the same procedure to make nine different compositions with varying SGM content.
17. Later the samples are removed from the oven.

2.2 *Design of Experimentation*

The experiment is planned to study the effect of the process parameters on the mechanical properties of the composite. In order to identify critical process parameters and their levels, a thorough literary review was conducted. After conducting extensive initial screening experiments, the process parameters considered for the study are temperature, curing time, hardener percentage. Table 1 shows the range of these process parameters.

An experiment design based on well-defined guidelines uses a Taguchi method. This procedure uses a special array known as orthogonal arrays. Such standard sets provide the means to carry out the minimum number of tests that could provide all the factors that affect the output parameter with complete information. For testing the L9 orthogonal array is chosen. The number of the experiments and process parameter values are indicated in Table 2.

Table 1 Selected process parameters and their levels

Sr. No.	Factor symbol	Process parameters	Unit	Level		
				1	2	3
1	A	Temperature	°C	25	35	45
2	B	Hardener	%	1	1.5	2
3	C	Curing time	minutes	60	75	90

Table 2 Layout of the experiment using L9 orthogonal array with parameters values

Sr. No.	Hardener	Time	Temp
1	1	60	25
2	1	75	35
3	1	90	45
4	1.5	60	45
5	1.5	75	35
6	1.5	90	25
7	2	60	45
8	2	75	25
9	2	90	35

2.3 Testing of Test Specimens

The test was performed in accordance with the plan shown in Table 3. The tensile strength and hardness were selected as the response parameters. The hardened samples of composite were cut into the required shape using hack saw and file. Total of nine samples are made and tested for tensile strength and hardness. For tensile testing Ultimate tensile testing machine (UTM) was used, shown in Fig. 2. Tests were carried out in accordance with the ASTM E1309 standard. The hardness of the composite is tested using Brinell hardness tester, shown in Fig. 3. The tested samples are shown in Fig. 4.

Table 3 Experimental results

S. No.	Hardener	Time	Temp	Brinell hardness BHN	Tesile test KN
1	1	1	25	134	7.45
2	1	1:15	35	124	7.14
3	1	1:30	45	141.5	8.81
4	1.5	1	45	144	9.21
5	1.5	1:15	35	125.25	9.03
6	1.5	1:30	25	130	7.68
7	2	1	45	147	9.26
8	2	1:15	25	125.85	9.15
9	2	1:30	35	133	7.075

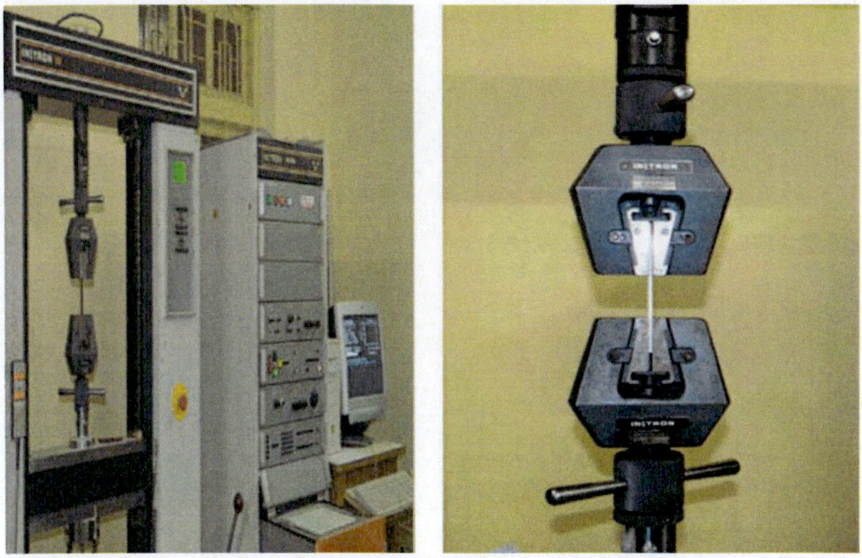

Fig. 2 Experimental setup for tensile testing on UTM

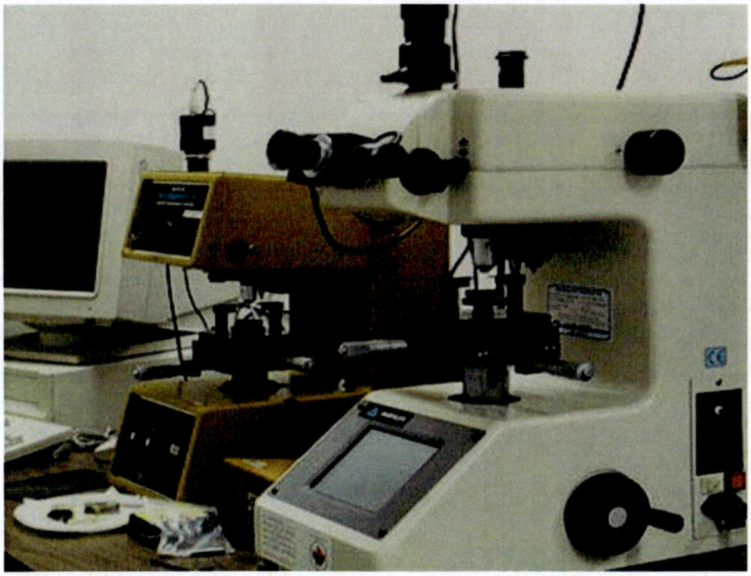

Fig. 3 Brinell hardness tester

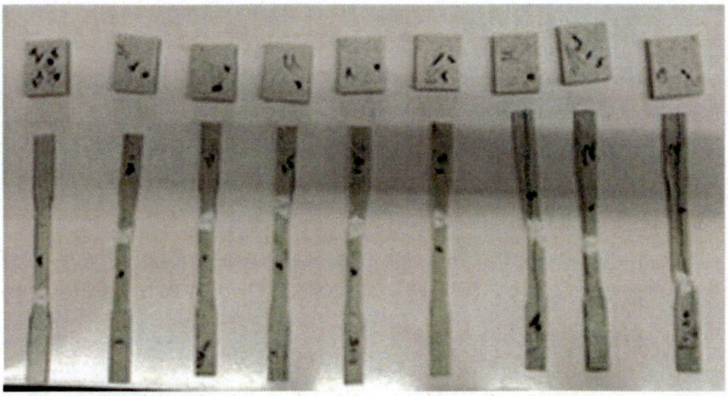

Fig. 4 Tensile testing samples

3 Result and Discussion

The experimental results for tensile strength are shown in the following figure and values of the tensile strength and hardness is shown in Table 3. The observation from the experimental investigation are discussed in the below section.

Figure 5 represent the variation in the tensile strength due to the change in the percentage of catalyst in the composite and change of curing time to make the composite. It is observed that the change in the catalyst percentage is not playing much significant role in the tensile strength of the composite. On the other hand, the curing time and curing temperature played a significant role in the tensile strength of material. Both these parameters are directly proportional to the tensile strength of the material. Further, the investigation shows that all the process parameters are significant for the hardness of the composite material. The individual effects of catalyst show that the hardness is increasing with increase in catalyst percentage, but it starts decreasing once the percentage reached to 1.5% in the mixture. High percentage of catalyst with low curing temperature and time produced insignificant effect on the mechanical properties. The time is inversely proportional to the hardness which is clearly depicted in Fig. 6. The temperature is directly proportional to the hardness, as the time and curing temperature are increasing the hardness of the material is also increasing. Figure 6 shows the effect of the induvial parameters on the tensile and hardness of the material. The interaction between individual parameters are also shown in Fig. 7. The main effect plot for tensile strength, hardness is shown in Figs. 8 and 9 respectively. The interaction plot for tensile strength is shown in Fig. 10. The Pareto Chart for process parameters are shown in Fig. 11.

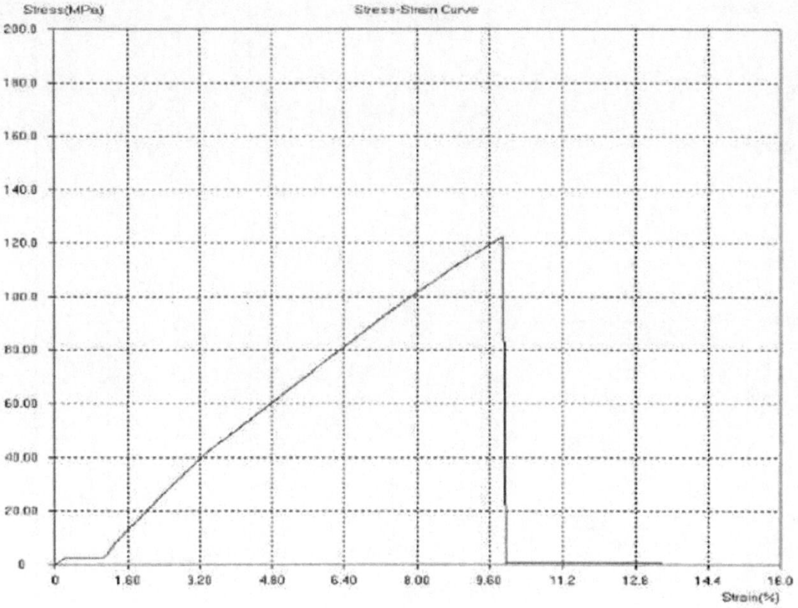

Sample 1

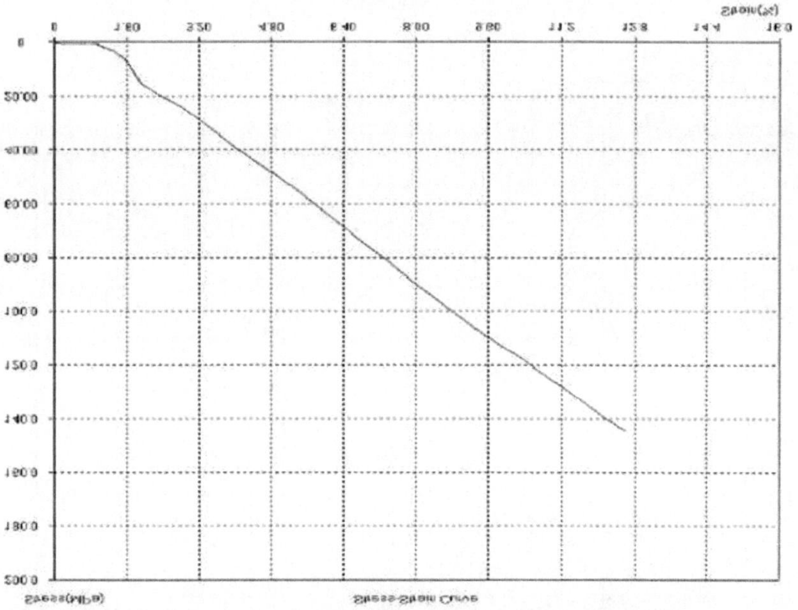

Sample 2

Fig. 5 Response from tensile testing experiments

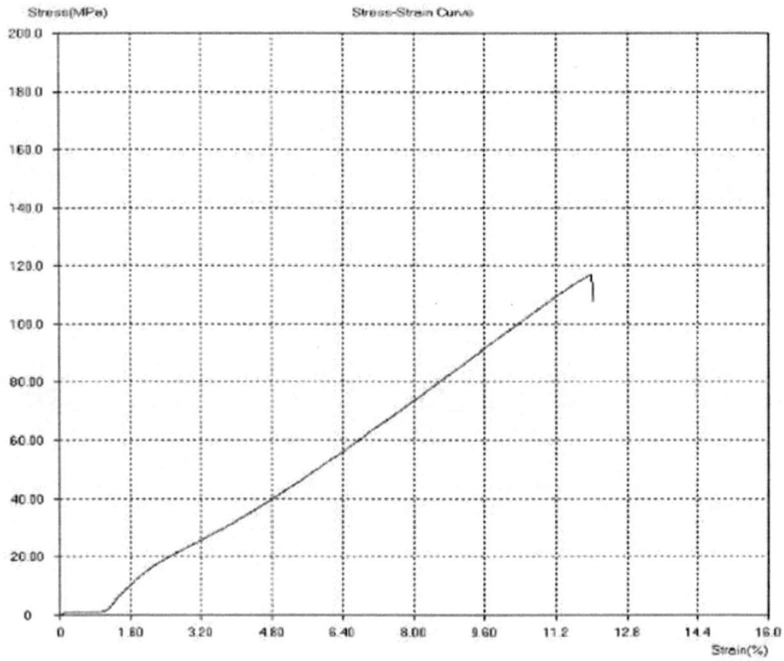

Sample 3

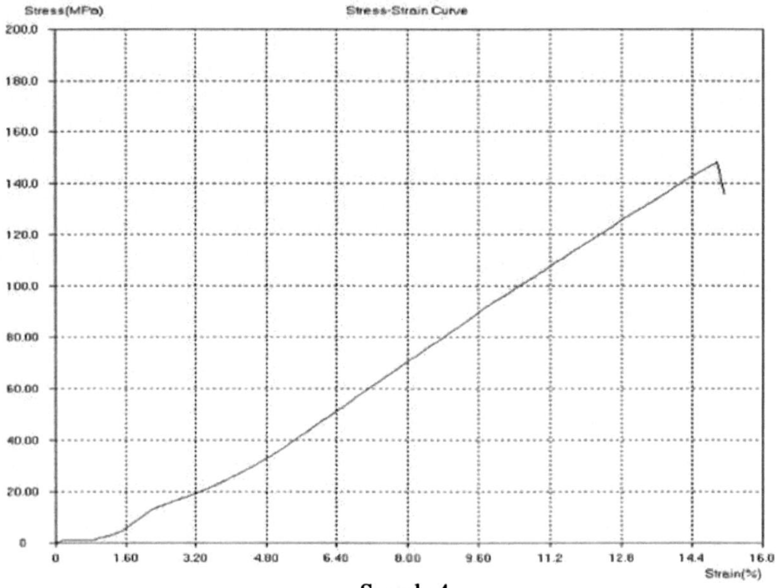

Sample 4

Fig. 5 (continued)

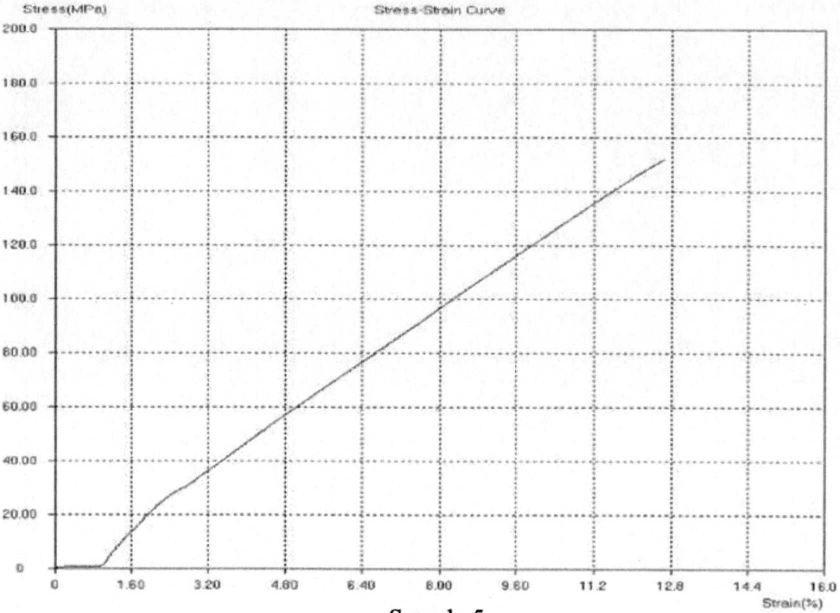

Sample 5

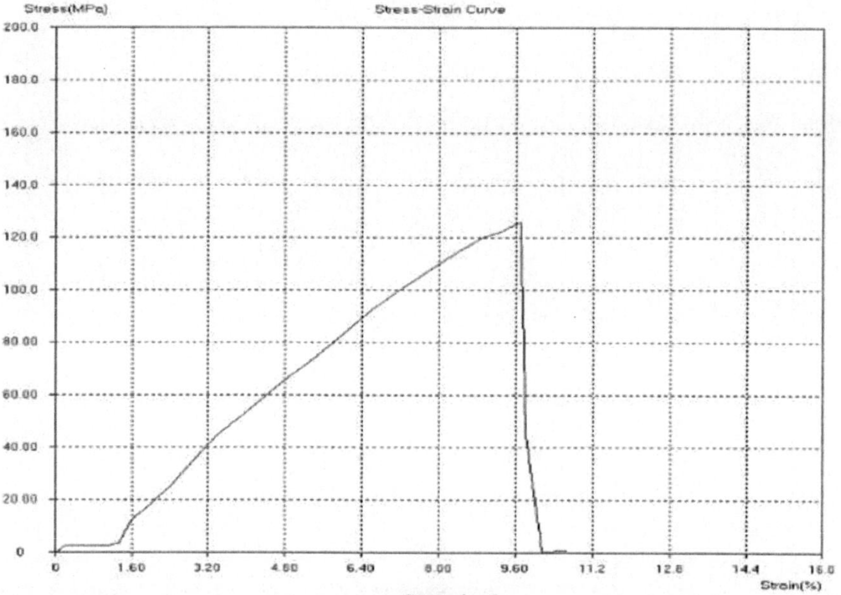

Sample 6

Fig. 5 (continued)

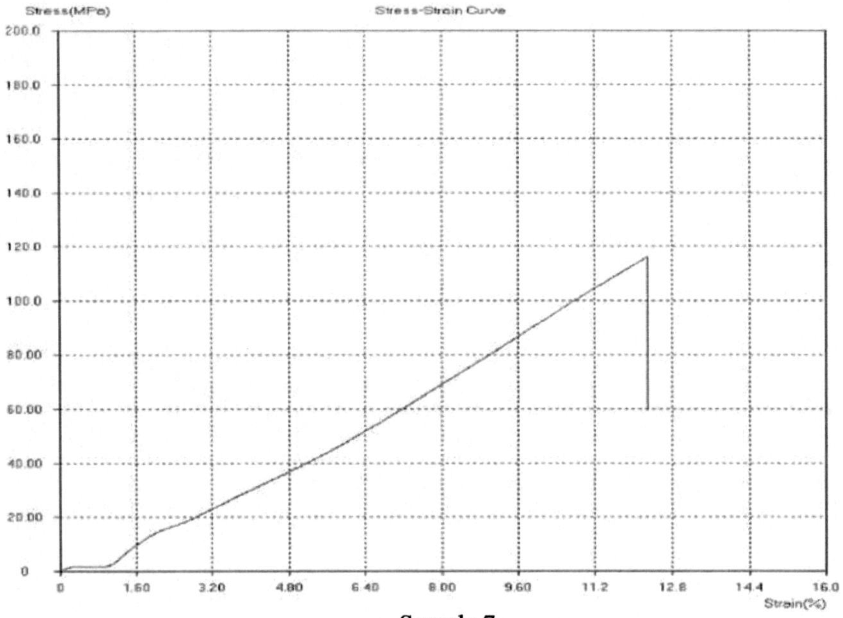

Sample 7

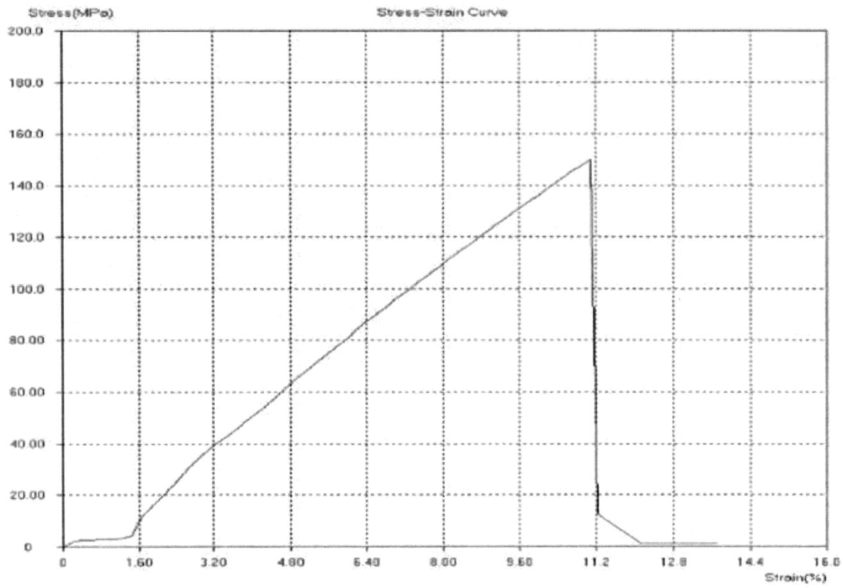

Sample 8

Fig. 5 (continued)

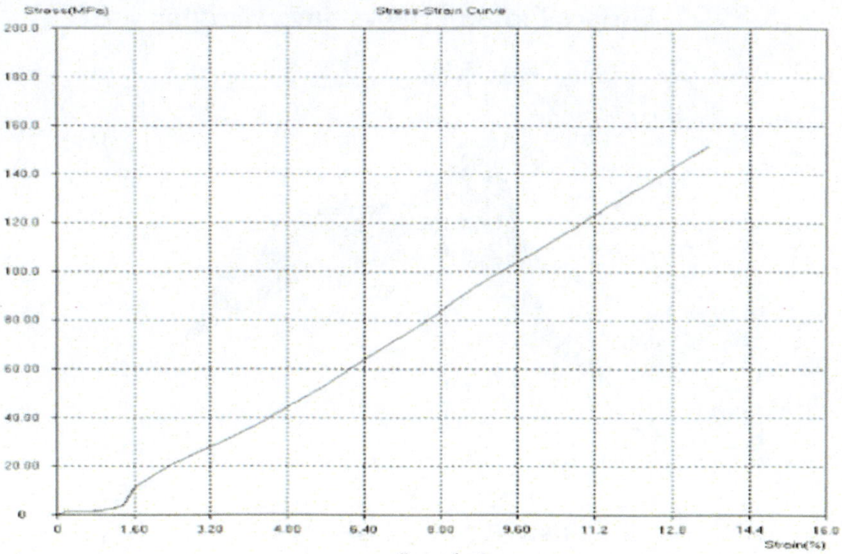

Sample 9

Fig. 5 (continued)

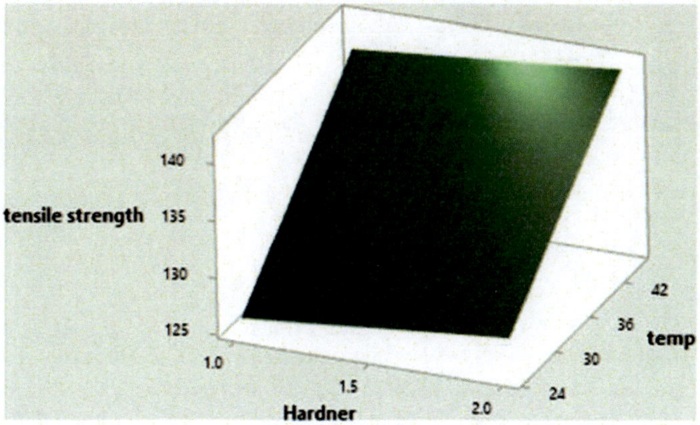

Fig. 6 Surface plot of tensile strength versus temp, harder

4 ANOVA Analysis

In order to determine significant differences between factors, ANOVA was performed. Through measuring the F-ratio (variance ratio) and the percentage contribution by each design parameter and error, it determines the importance of control factors. From the ANOVA table, it is concluded that the temperature is the most significant factor compared to curing time. In the mechanical properties of the material,

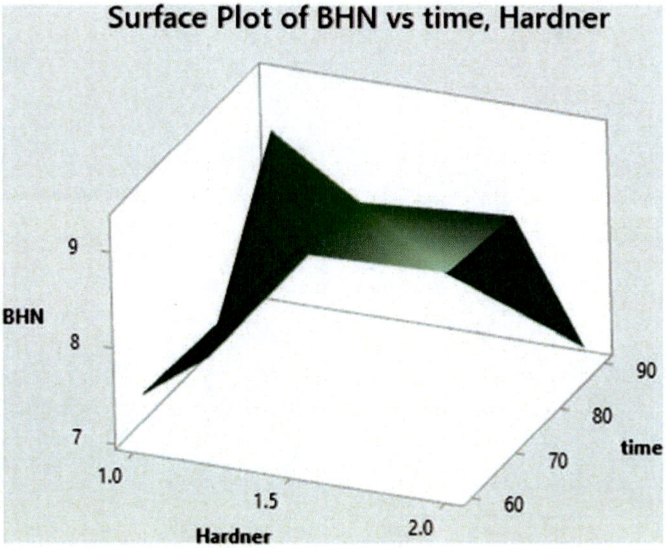

Fig. 7 Surface plot of hardness versus temp, harder

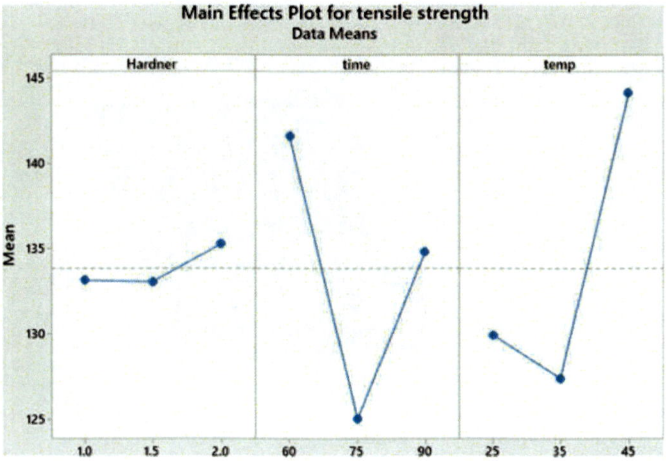

Fig. 8 Main plot for tensile strength

the catalyst percentage did not have a significant role. Below is the linear equation for composite tensile strength (Table 4).

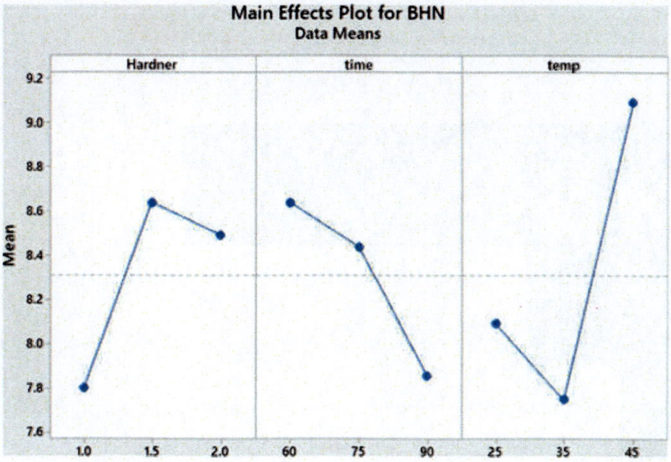

Fig. 9 Main plot for hardness

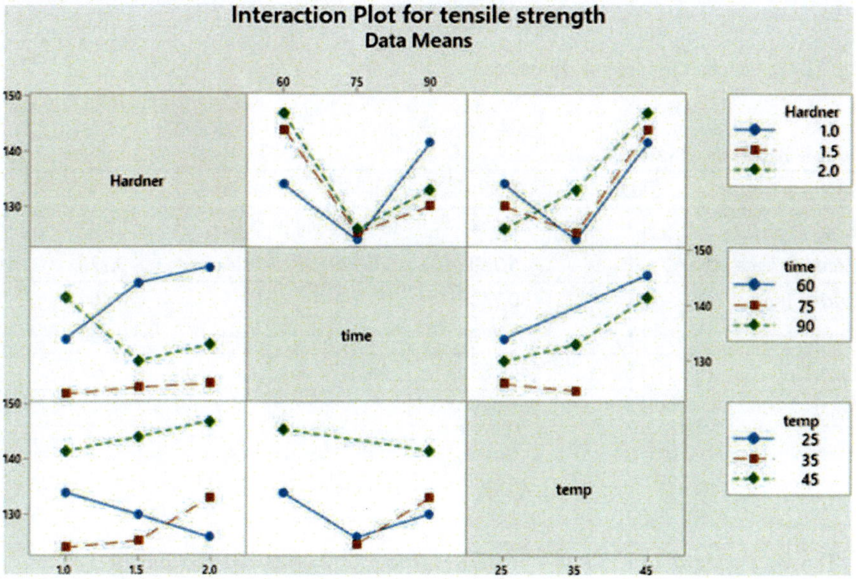

Fig. 10 Interaction plot for tensile strength

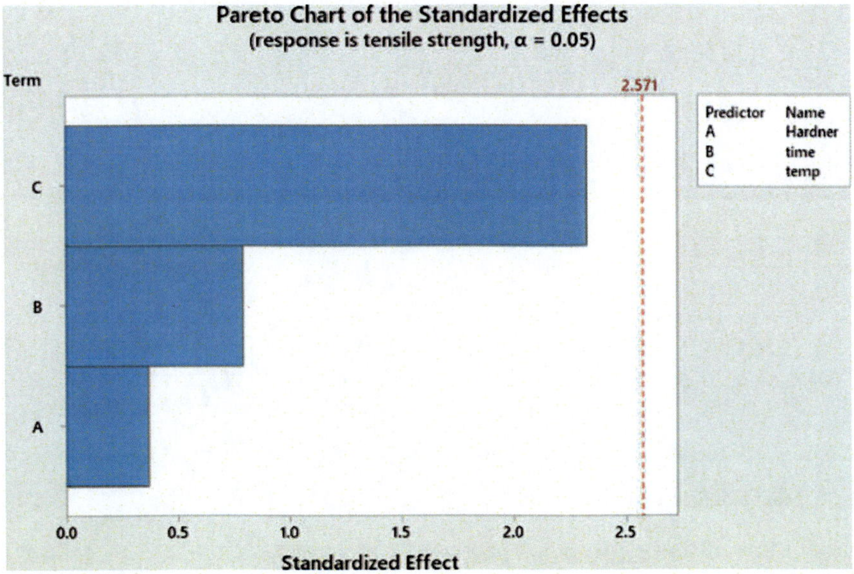

Fig. 11 Pareto chart for process parameters

Table 4 Analysis of variance

Source	DF	Adj SS	Adj MS	F-value	P-value
Regression	3	340.634	113.545	2.32	0.192
Hardener/catalyst	1	6.720	6.720	0.14	0.726
Curing time	1	30.743	30.743	0.63	0.464
Curing temp	1	263.872	263.872	5.40	0.068
Error	5	244.383	48.877		
Total	8	585.017			

$$\begin{aligned}\text{Tensile Strength} = {} & 133.844 - 0.678\,\text{Hardner\_1.0} - 0.761\,\text{Hardner\_1.5} \\ & + 1.439\,\text{Hardn}er\text{\_2.0} + 4.426\,\text{time\_60} - 5.414\,\text{time\_75} \\ & + 0.989\,\text{time\_90} - 3.894\,\text{temp\_25} - 3.148\,\text{temp\_35} \\ & + 7.042\,\text{temp\_45}\end{aligned}$$

$$\begin{aligned}\text{Hardness} = {} & 8.312 - 0.512\,\text{Hardner\_1.0} + 0.328\,\text{Hardner\_1.5} \\ & + 0.183\,\text{Hardner\_2.0} - 0.399\,\text{time\_60} + 0.855\,\text{time\_75} \\ & - 0.457\,\text{time\_90} - 0.218\,\text{temp\_25} - 0.981\,\text{temp\_35} \\ & + 1.200\,\text{temp\_45}\end{aligned}$$

5 Conclusion

The produced epoxy-based composite displays the best mechanical properties at 2 percent catalyst level, with 45 °C curing temperature at 1 h curing period. The effect of catalyst, curing time and curing temperature has been determined using Taguchi L8 experimental design. Curing temperature and time have been found to play a major role in the composite's mechanical properties. At the same time, hardener is also a significant factor, providing best effect at a high curing temperature. It is thus found that all of these parameters have a major interaction effect on mechanical properties of composite.

References

1. Kotsilkova R (ed) (2007) In: Thermoset nanocomposites for engineering applications, Smithers Rapra Technology, pp 8–15
2. Ahmada KZK, SHj Ahmada,Tarawneha MA, Apte PR (2012) Evaluation of mechanical properties of epoxy/nanoclay/multi-walled carbon nanotube nanocomposites using taguchi method. Procedia Chem 4:80–86
3. Dhanasekar D, Chidhambaram K, Anil Kumar S, Mohan Kumar T (2019) Experimental evaluation of glass fiber reinforced composites subjected to different loads. Int Res J Eng Technol (IRJET) 6(5)
4. Thirumalai R, Prakash R, Ragunath R, SenthilKumar KM (2019) Experimental investigation of mechanical properties of epoxy-based composites. Mater Res Exp 6(7)
5. Dandekar CR, Shin YC (2012) Modeling of machining of composite materials: a review. Int J Mach Tools Manuf 57:102–121
6. Jacobs IS, Bean CP (1963) Fine particles, thin films and exchange anisotropy. In: Rado GT, Suhl H (eds) Magnetism, vol 3. Academic, New York, pp 271–350
7. Gujjala R, Ojha S, Acharya SK, Pal SK (2014) Mechanical properties of woven jute–glass hybrid-reinforced epoxy composite. J Compos Mater 48:3445–3455

Experimental and Simulation Study on Permeability of Hybrid Composite

Prasanth Kottapalli, Rahul Narkhede, Harshit K. Dave, Himanshu V. Patel, and Rajesh kumar Verma

1 Introduction

The fiber-reinforced composites are gaining attention due to its numerous advantages such as high strength-to-weight ratio, corrosion resistance, and durability. The selection of the composite manufacturing process is based on the application of composites Puglia et al. [1]. The liquid composite molding (LCM) is widely used for composite manufacturing. The process consists of liquid matrix material that flows through porous reinforcement material. The resin transfer molding (RTM) and vacuum-assisted resin transfer molding (VARTM) processes are known as LCM processes. The vacuum-assisted resin transfer molding (VARTM) process is the advancement of conventional resin transfer molding (RTM) process. In VARTM process, one of the rigid molds is replaced by a flexible vacuum bag. In recent days, natural fiber-reinforced composites are gaining attention due to its lightweight, economical, and eco-friendliness in nature. Some standard natural fibers are cotton, jute, flax, hemp, kneaf, ramie, banana, pineapple leaves, etc. The permeability is an essential parameter which is affected by the mold size, pressure, and gating arrangement. The mold filling time is influenced by reinforcement permeability. The simulation is an effective tool to reduce trial and error approach during the fabrication of composite using LCM process.

P. Kottapalli · R. Narkhede · H. K. Dave · H. V. Patel (✉)
Department of Mechanical Engineering, S. V. National Institute of Technology, Surat 395 007, India
e-mail: hpsvnit07@gmail.com

R. Verma
Mechanical Engineering Department, Madan Mohan Malaviya University of Technology, Gorakhpur, India

H. K. Dave and D. Nedelcu (eds.), *Advances in Manufacturing Processes*, Lecture Notes in Mechanical Engineering, https://doi.org/10.1007/978-981-15-9117-4_29

Various software packages, viz. RTM-WORKS, PAM-RTM, ANSYS CFX ANSYS FLUENT, and LIMS are popular for simulations of RTM and VARTM resin flow fronts Luz et al. [2].

Li et al. [3] studied the permeability in hybrid composites using modified Darcy's equation. From their study, they found that capillary pressure is significant hybrid composites. Amico and Lekakou [4] studied the relationship between the capillary pressure and permeability in the woven fiber. The experiments were performed using silicone oil as fluid. Rajkumar et al. [5] studied the permeability of sisal fiber of both untreated and treated in vacuum infusion process. They compared the flow front characteristics with permeability measurement models like Kozeny–Carman and contact model. Kim et al. [6] studied the permeability and interfacial properties on different glass fiber pattern. VARTM is used to measure the permeability. Ameri et al. [7] performed the in-plane permeability study on flax fibers. From their study, they revealed that the permeability across the fiber direction was higher. Subbiah et al. [8] studied the permeability of sisal and glass fiber during vacuum infusion process. They found that fiber swelling affects the permeability of natural fibers when bio-epoxy was used as a matrix material. Jhan et al. [9] investigated the resin flow in the sandwich composites for VARTM process and the in-plane permeability is calculated with the Darcy equation. They applied the identical conditions for predicting the commercial boat deck and the results were compromising with experimental data. Wu and Larsson [10] had studied the resin flow in the thin-walled preforms in the VARTM. He compared the variations in the thickness between the simulation and digital image correlation (DIC). The experimental results were delighted with the DIC images. Yun et al. [11] had studied the void formation due to heterogeneity by using the manifold embedding method. The void can be characterized by the gamma distribution parameter and Moran's. The VARTM flow in fabric is simulated, and the factors were measured on the image. Schmidt et al. [12] investigated the in-plane permeability of glass, PP foam, sisal, and their hybrids with different stacking sequence. They found that the permeability for all three fibers with same fiber volume fraction ($V_f = 20\%$) by using Darcy's equation is in the RTM process. The results verified that the permeability as sisal is higher than other two.

Many authors investigated the permeability by using VARTM with single fabric but less literature found for hybrid reinforcements. The present work is aimed to study the effect of stacking sequence and substacking sequences on the permeability of natural fibers. The value of permeability is obtained using VARTM experiments. The value obtained permeability is used for mold filling simulations. The flow front simulations were carried out to study the effect of stacking sequence and substacking sequence on mold filling time.

2 Experimental Methodology

The natural fibers jute and flax are used for fabrication of composite. The unsaturated polyester resin is used as matrix and methyl ethyl ketone peroxide (MEKP) catalyst

Table 1 Properties of reinforcement materials

Material	GSM (g/m^2)	Density (g/cc)
Jute fiber	215	1.3
Flax fiber	210	1.5
Infusion mesh	265	0.95
Peel ply	95	1.14

as harder. The composite fabrication was carried out at ambient temperature using VARTM Process. Table 1 represents the properties of the reinforcement materials and other materials that are used during infusions.

2.1 *Fabrication of Laminates*

For the permeability measurement, VARTM setup was used as shown in Fig. 1. On the flat mold, reinforcement materials were stacked as per the mold dimensions. The stacked reinforcement materials were closely enclosed by a flexible vacuum bag. The polyester resin was mixed with the hardener methyl ethyl ketone peroxide (MKEP) and cobalt accelerator. The weight ratio of the following resin, hardener, and accelerator is 100:10:1.04, respectively. The liquid resin was infused under vacuum, and the final composite laminates were demolded after the curing of 24 h.

For the fabrication of the composite, the mold dimension was taken 400 × 300 mm. The reinforcement material was cut as per the mold dimensions. For current study, hybridization carried out using jute and flax reinforcements. For the in-plane permeability measurement, hybridization was done using a total of eight layers of reinforcements. These hybridizations consist of different stacking sequence and substacking

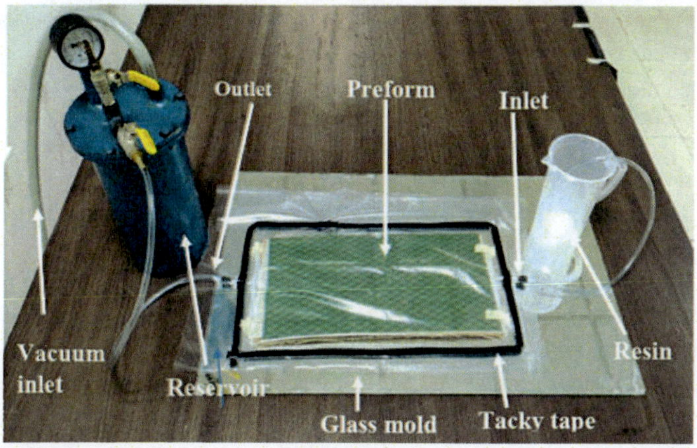

Fig. 1 VARTM setup

Table 2 Hybridization using jute and flax fibers

Stacking sequence	Hybridization ratio	Fiber placement
F8	8:0	FFFFFFFF
F6J2	6:2	FJFFFFJF
F4J4	4:4	FJJFFJJF
F2J6	2:6	FJJJJJJF
J8	0:8	JJJJJJJJ
Substacking sequences	Hybridization ratio	Fiber placement
(FJ)4	4:4	FJFJFJFJ
(FFJJ)S	4:4	FFJJJJFF
(JJFF)S	4:4	JJFFFFJJ
(F4/J4)	4:4	FFFFJJJJ
(FJJF)S	4:4	FJJFFJJF

sequences that are shown in Table 2. The uniform hybridization was done to identify the effect of reinforcement material on permeability. For the substacking sequences, a fiber volume fraction was taken constant and placement of the fibers was kept different to make a hybrid composite. Total of five stacking sequences and four substacking sequences were formed using jute and flax fiber reinforcements.

2.2 Permeability

The permeability is an essential property of porous reinforcements. The measurement of permeability is a crucial aspect in the liquid composite molding process. The permeability measurement required two known parameters, i.e., porosity of reinforcement and viscosity of resin. The viscosity of the resin was measured using Anton Paar rotational viscometer at room temperature. The porosity of the reinforcement was calculated using the theoretical fiber volume fraction (FVF) approach. Equation 1 indicates the theoretical fiber volume fraction.

$$FVF = \frac{v_1 + v_2 + \ldots v_8}{\text{Total Volume of Composite}} \tag{1}$$

$$\text{Porosity}(\varepsilon) = 1 - FVF \tag{2}$$

where V_1 is a volume of fiber 1, V_2= is the volume of fiber 2. The porosity of the laminate is determined by Eq. 2. The in-plane permeability of the reinforcement is calculated by the Darcy's equation of flow through porous medium shown in Eq. 3.

$$K = \left(\frac{x^2}{t}\right)\frac{\mu\varepsilon}{2\Delta P_i} \tag{3}$$

P_i is the vacuum pressure applied, x is the distance traveled by resin and t is the time. ε is porosity of the reinforcement and μ is viscosity of the resin and K is in-plane permeability.

In VARTM process, resin was infused in reinforcements by vacuum pressure. During the resin infusion, flow front was recorded using a digital camera that was placed normal to the VARTM setup. The flow front progression was captured with respect to time. The flow front was captured till complete mold filling. The flow front was tracked using captured flow front progression with respect to time. The resin flow distance with respect to time is a required data in the calculation of the permeability. The slope of squared distance versus time was utilized for the permeability calculation as per Eq. 3. Figure 2 shows the squared distance versus time.

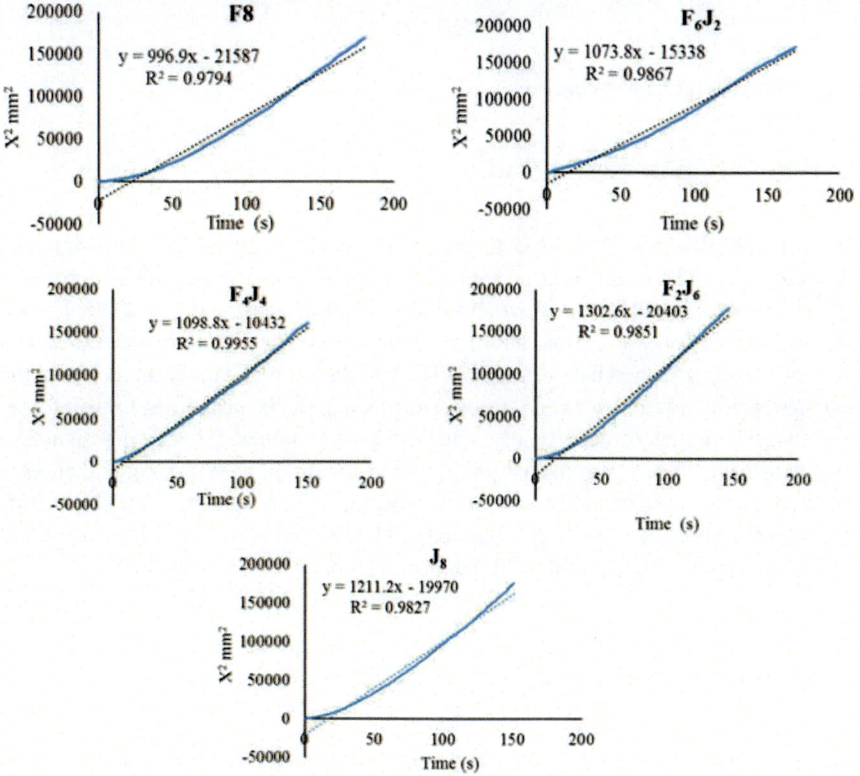

Fig. 2 Squared distance versus time graphs of hybrid laminates

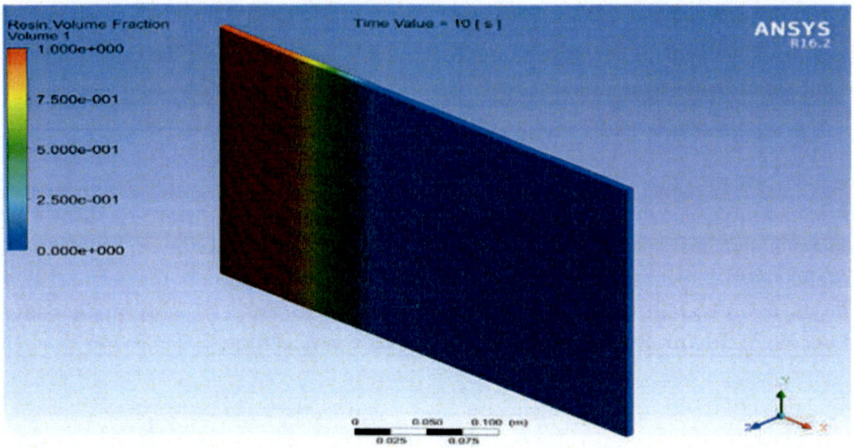

Fig. 3 Interface of resin flow simulation

3 Simulation of Resin Flow

The simulation approach is used to reduce the cost of fabrication with trial and error approach. The trial and error approach is not feasible for the manufacturing of larger components. The resin flow simulation is carried out using ANSYS Fluent. The flow simulation was carried out in transient mode. The geometry generation and messing were performed using ANSYS. The simulation was carried out between the two phases, i.e., liquid resin and porous reinforcement. The properties for reinforcements were calculated theoretically, and few were obtained through the manufactures datasheets. The resin viscosity is required for the flow experiments that were obtained using experimentally at room temperature. Through the VARTM experiments, permeability values were obtained, and these values were utilized for mold filling simulation. Figure 3 shows the interface of resin flow Simulation.

4 Result and Discussion

The resin flow front was tracked and infusion process till mold filling was recorded during the VARTM process. The permeability values were obtained through the experiments. The values of permeability for different stacking and substacking sequences of jute and flax fibers are shown in Table 3.

The calculated permeability values were used for mold filling simulations. The simulated and experimental mold filling time is compared. Table 4 shows the comparison of experimental and simulated mold filing time. Figure 4 shows good the agreement between simulation and experimental mold filling for $(FFJJ)_s$ at 10 s and 120 s.

Table 3 Permeability values of laminates

	FVF	Porosity (ε)	Slope $\left(\frac{x^2}{t}\right)$	Mold filling time (s)	Permeability (m^2)
Stacking sequences					
F_8	0.22	0.780	996.9	173	1.764×10^{-9}
F_6J_2	0.225	0.775	1073	158	1.886×10^{-9}
F_4J_4	0.224	0.776	1098	151	1.933×10^{-9}
F_2J_6	0.229	0.771	1302	133	2.277×10^{-9}
J_8	0.219	0.781	1211	142	2.145×10^{-9}
Substacking Sequences					
$(FJ)_4$	0.22	0.780	1132	155	2.005×10^{-9}
$(FFJJ)_S$	0.223	0.777	1380	122	2.445×10^{-9}
$(JJFF)_S$	0.212	0.788	993	169	1.775×10^{-9}
F_4/J_4	0.204	0.796	1116	152	2.015×10^{-9}
$(FJJF)_S$	0.224	0.776	1098	151	1.933×10^{-9}

Table 4 Experimental and simulated mold filling time

	Experimental mold filling time E_t (s)	Simulated mold filling time S_t (s)	Error (%)
Stacking sequences			
F8	173	179	3.5
F6J2	158	163	3.2
F4J4	151	159	5.3
F2J6	133	136	2.2
J8	142	146	2.8
Substacking sequences			
$(FJ)_4$	155	159	2.6
$(FFJJ)_S$	122	130	6.6
$(JJFF)_S$	169	177	4.7
F4/J4	152	160	5.3
$(FJJF)_S$	151	159	5.3

Table 4 shows the comparison of all the experimental and simulated mold filling times during the resin infusions. During the experiments, mold filling time was measured using a stopwatch for the identical interval of time simulation process compared and good agreement found between the experimental and simulation interface.

Figure 5 shows the experimental and simulated mold filling time for all stacking sequences and substacking sequences. From Fig. 5, it can be seen that there is good

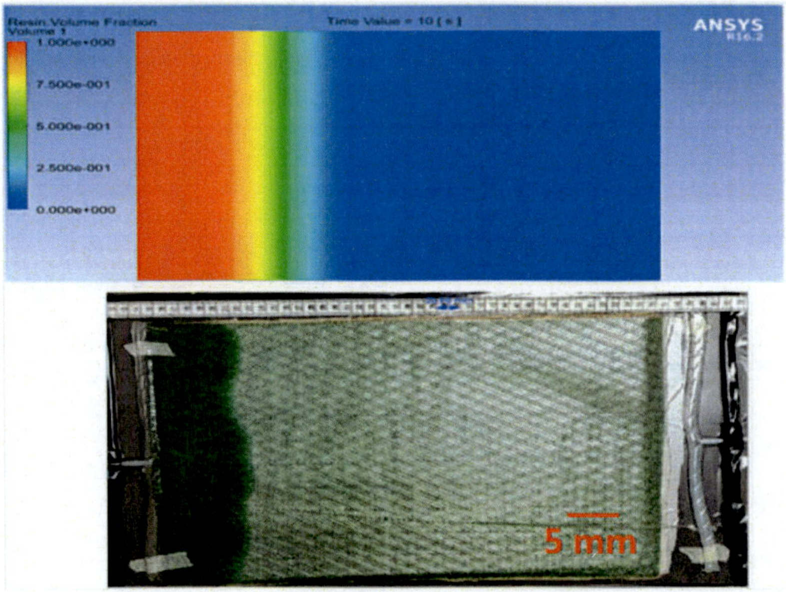

a) (FFJJ)s at 10 sec

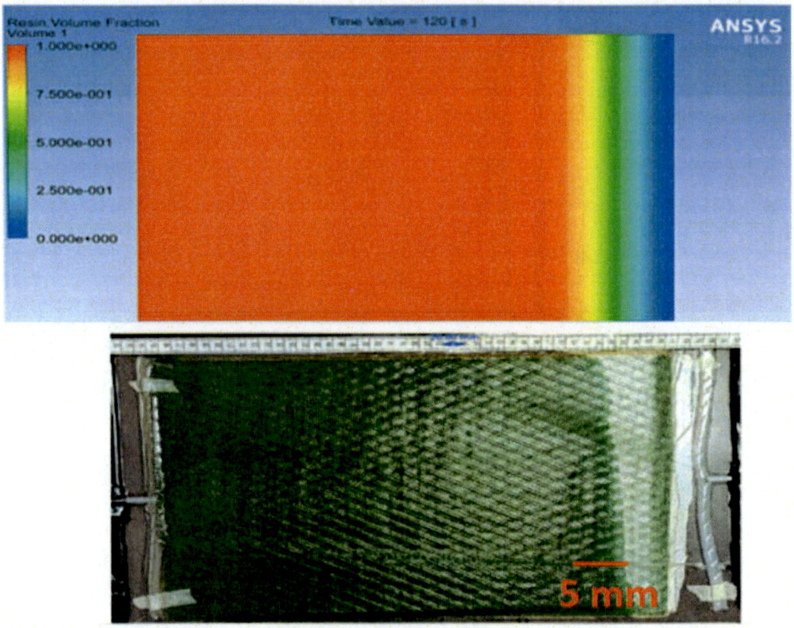

b) (FFJJ)s at 120 sec

Fig. 4 Comparison of experimental and simulation process

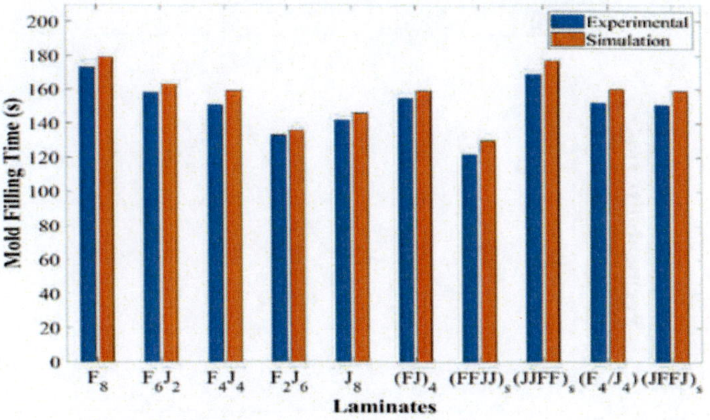

Fig. 5 Comparison between experimental and simulated mold filling time

agreement between experimental and simulated mold filling time. During the experiments, the highest and the lowest mold filling time were observed for F8 and (FFJJ)S among all stacking sequences and substacking sequences. The liquid matrix flows faster in inter-tow space during resin infusion under vacuum. The reinforcement gets wet due to the capillary action of the resin over the porous medium. During resin infusion, capillary pressure allows fiber tows to absorb resin until complete saturation of the fiber tow is achieved.

As the arrangement of the reinforcement changed, the flow pattern would also change that will affect the permeability of the reinforcements. The less flow resistance is offered between the layers when reinforcement consists of more inter-tow spaces. The less intra-tow provides fast flow inside the fabric due to less surface tension. From Fig. 6, it is observed that the permeability is increasing with an increase in amount of jute. Pan and Zhong [11] reported that fluid flow depended on inter-tow and intra-tow space of the reinforcements. The jute has more inter-tow spaces compared to the flax, so this enables faster flow which results in more permeability than flax. Permeability values are increasing gradually from F_8 to F_2J_6, but decrease in the J_8 shown in Fig. 6.

From Fig. 7, it can be observed that jute fiber consist of the larger intra-tow spaces and poor porous connectivity. The air entrapment can be observed due to the larger intra-tow space.

Jpati and Sengupta [13] reported that wetting in flax reinforcement is faster than jute. Faster wetting means resin flows faster inside inter-tows. F_8 was found less permeable due to it consists of less inter-tow space. Form F_6J_2 to F_2J_6 the fluid flow is increasing, because of the flow of fluid between the jute layers that reduces the flow resistance. The flax fibers carry the resin faster and help in filling of inside jute in F_2J_6, so faster filling time. In case of J_8, where flax is absent, which shows fluid flow inside the jute is slow, so it takes more time to fill the mold. Hence permeability is less than F_2J_6. Figure 8 shows the inter-tow space between hybrid composites.

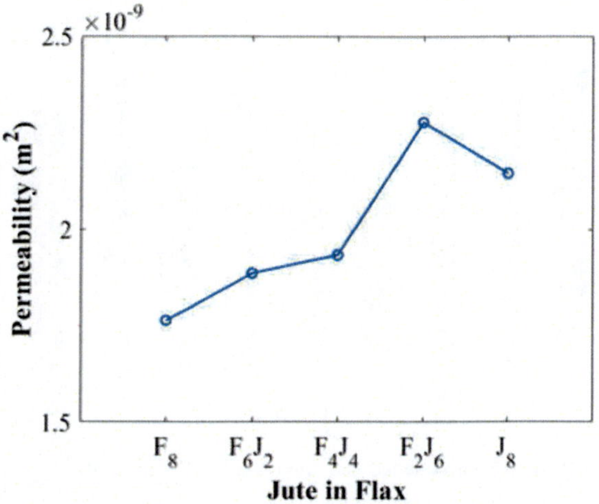

Fig. 6 Permeability in hybridization

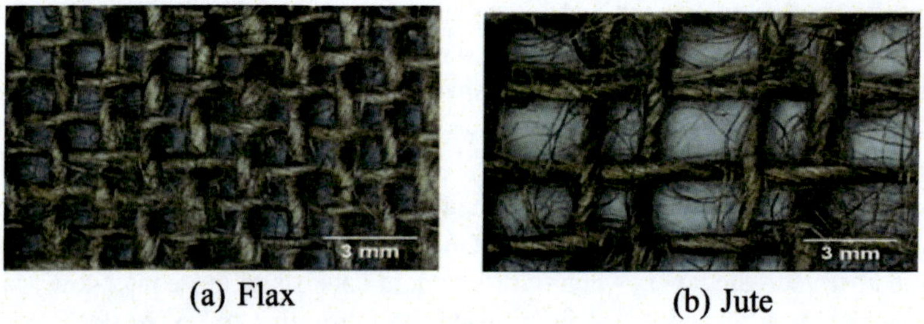

Fig. 7 Intra-tow spaces of reinforcements

From Fig. 8, it can be seen that narrow space is observed inside the flax inter-tows, while wider space is observed inside the jute.

For the substacking sequence $(FFJJ)_S$, it is observed that the arrangement of jute fibers is placed in the mid position that leads to a faster flow. The quicker flow gives higher permeability value. The lowest permeability is in $(JJFF)_S$, where flax fibers placed at mid position. The change in permeability for substacking sequence can be seen in Fig. 9. The placement of fibers plays an important role for the change in permeability for substacking sequences. (F_4/J_4) the four flax layers were placed at lower half that reduces viscous resistance offered by the mold. In same substacking sequence, on the upper half, four jute offers more inter-tow space, which enables fast mold filling. The major issue of jute layers is, due to fast flow between the tows results in intra-tow voids. These voids reduce the strength of the composite. The flax fibers will take more filling time, but it saturates fibers with the resin. The $(FJ)_4$ arrangement provides the lesser voids with fast filling time, because of flax reduce the void issue and jute improves the flow, when both fibers placed alternative.

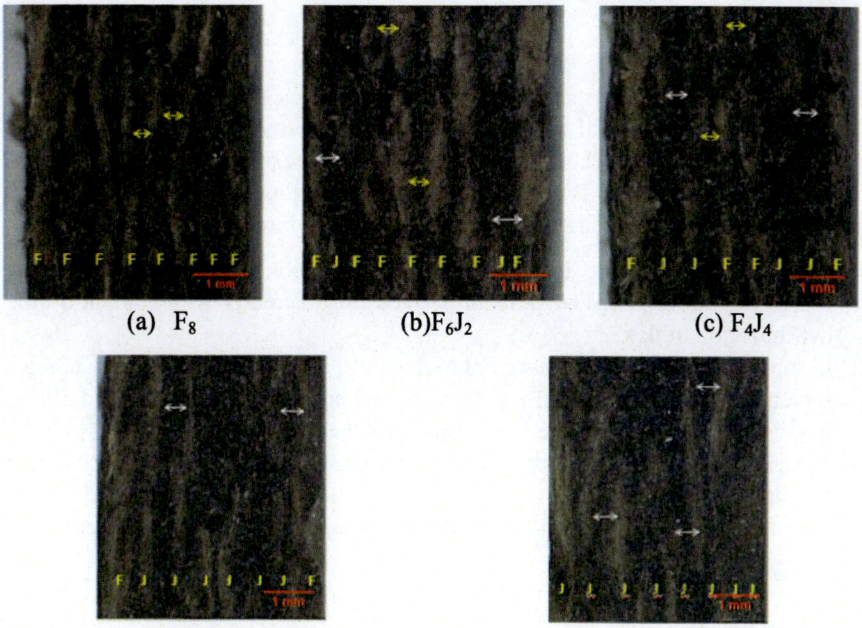

(a) F_8 (b)F_6J_2 (c) F_4J_4

(d)F_2J_6 (e) J_8

Fig. 8 Inter-tow spaces of hybrid composites

Fig. 9 Permeability in substacking of F_4J_4

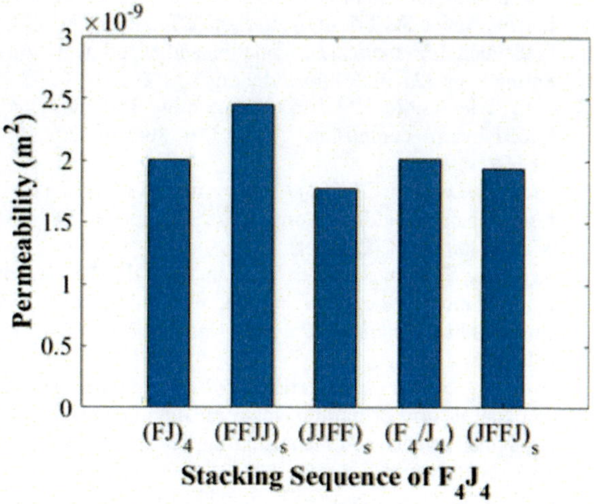

5 Conclusions

In this work, the effect of stacking sequence and substacking sequence on permeability during fabrication of the hybrid composites is studied, using jute and flax reinforcements. The permeability is determined by the experiments, and the calculated value of permeability is used as input in transient flow simulation for VARTM process. Form the study, following conclusions can be drawn:

1. The permeability of the jute fiber is higher than the flax, due to more inter-tow spaces present in jute. The value of permeability increases with amount of jute hybridization in flax.
2. In substacking sequence, placement of fiber also affects the permeability of reinforcement. The placement of jute fibers at mid position gives faster mold filling. In $(FJ)_4$, arrangement of jute fibers generates the lesser voids.
3. The placement of flax fiber at bottom position results less flow resistance due to the mold surface.
4. The simulation study of resin flow gives good agreement with experiments.

References

1. Puglia D, Biagiotti J, Kenny JM (2005).A review on natural fibre-based composites—part II: application of natural reinforcements in composite materials for automotive industry. J Nat Fibers 1. Taylor and Francis, pp 23–65. https://doi.org/10.1300/J395v01n03_03.
2. Luz FF, Amico SC, Souza JÁ, Barbosa ES, de Lima AG (2012). Resin transfer molding process: fundamentals numerical computation and experiments. Advanced structured materials vol 27. Springer, pp 121–151. https://doi.org/10.1007/978-3-642-30532-0_5
3. Li Y, Xie L, Ma H (2015) Permeability and mechanical properties of plant fiber reinforced hybrid composites. Mater Des 86:313–320. https://doi.org/10.1016/j.matdes.2015.06.164 (Elsevier)
4. Amico S, Lekakou C (2001) An experimental study of the permeability and capillary pressure in resin-transfer moulding. Compos Sci Technol 61:1945–1959. https://doi.org/10.1016/S0266-3538(01)00104-X (Elsevier)
5. Rajkumar S, Tjong J, Nayak SK, Sain M (2017) Permeability and mechanical property correlation of bio based epoxy reinforced with unidirectional sisal fiber mat through vacuum infusion molding technique. Polymer Compos 38:2192–2200. https://doi.org/10.1002/pc.23797 (Elsevier)
6. Ameri E, Lebrun G, Laperrière L (2016) In-plane permeability characterization of a unidirectional flax/paper reinforcement for liquid composite molding processes. Composites Part A: applied science and manufacturing 85:52–64. https://doi.org/10.1016/j.compositesa.2016.03.002 (Elsevier)
7. Subbiah R, Tjong J, Nayak SK, Sain M (2015) Studies on permeability of sisal fibre mat during thermoset resin filling in vacuum infusion process. Can J Chem Eng 93:1364–1370. https://doi.org/10.1002/cjce.22228 (Elsevier)
8. Jhan YT, Lee YJ, Chung CH (2011) Resin flowing analysis in sandwich laminates under VARTM process. J Reinf Plast Compos 30:533–545. https://doi.org/10.1177/0731684411399142 (Elsevier)
9. Wu D, Larsson R (2019). A shell model for resin flow and preform deformation in thin-walled composite manufacturing processes. Int J Mater Forming. Springer, pp 1–15. https://doi.org/10.1007/s12289-019-01517-z.

10. Yun MY, Lopez E, Chinesta F, Advani S (2018) Manifold embedding of heterogeneity in permeability of a woven fabric for optimization of the VARTM process. Compos Sci Technol 168:238–45. https://doi.org/10.1016/j.compscitech.2018.10.006 (Elsevier)
11. Schmidt TM, Goss TM, Amico SC, Lekakou C (2009) Permeability of hybrid reinforcements and mechanical properties of their composites molded by resin transfer molding. J Reinf Plast Compos 28:2839–2850. https://doi.org/10.1177/0731684408093974 (Elsevier)
12. Pan N, Zhong W (2006) Fluid transport phenomena in fibrous materials. Textile Prog 38:1–93. https://doi.org/10.1533/tepr.2006.0002
13. Jpati SS, Sengupta S (2006) Wetting characteristics of long vegetable fibres. Indian J Fiber Textile Res NISCAIR 31:262–266

Impact and Flexural Testing of Jute and Flax Fiber Reinforced Composites Fabricated by VARTM Process

Chandresh B. Kumbhani, Harshit K. Dave, and Himanshu V. Patel

1 Introduction

The composite material is finding a broad area of application due to their superior property like increased tensile strength, good wear rate, high corrosion resistance, low density, etc. Development of new material is trending in today's time for penetrating more and more field of application. Glass fiber and carbon fiber has the highest strength to weight ratio but usage is limited due to non-bio degradability and high cost of synthetic fibers. So the interest of the researcher is rapidly growing towards natural fiber-reinforced composites (NFRC) in terms of their fundamental research and industrial applications. Their availability, price, recyclability, and strength make them attractive and alternative to glass, carbon fibers used for the manufacturing of composites. The advantage of NFRC in the automobile is a lightweight and high efficiency of the automobile [1, 2].

The orientation of fiber plays an important role in the mechanical properties of polymer composites like tensile strength, Compressive strength, impact strength, flexural strength, etc. [3]. If fibers are placed parallel to the loading direction then it gives the highest strength compared to other combinations. Properties of NFRC can be improved by providing some surface treatment like coating [4–8].

The addition of nanoparticle can improve the mechanical properties but the uniform dispersion is the main issue in addition to the nanoparticle. Nanoparticle develops a good fiber-matrix interface and due to this the tensile, flexural, and impact strength increases [9]. Hybrid composite gives a combination of different reinforcement and results in an improvement in properties. It can also improve the moisture

C. B. Kumbhani (✉)
School of Engineering, P P Savani University, Surat, Gujarat, India
e-mail: chandresh.kumbhani@ppsu.ac.in

H. K. Dave · H. V. Patel
Mechanical Engineering Department, S V National Institute of Technology, Surat, Gujarat, India

H. K. Dave and D. Nedelcu (eds.), *Advances in Manufacturing Processes*, Lecture Notes in Mechanical Engineering, https://doi.org/10.1007/978-981-15-9117-4_30

absorption capacity of the composites [10, 11]. The water absorption tendency is the main drawback of NFRC. This property varies for different natural fibers. Fiber volume fraction (FVF) has a significant effect on the mechanical properties of NFRC [12, 13]. In an automobile, if we are able to reduce the weight of the system then efficiency and mileage can be increased. This is possible by the use of polymer composite using glass and carbon fiber as a recent trend but usage is limited due to environmental issues. So natural fiber is another option provided that NFRC should resist different types of load such as impact, bending, etc. Here, two different NFRC is prepared and mechanical testing (Impact and flexural) is carried out to see the effect of a number of layers on impact and flexural strength [13–16]. Fabrication of NFRC is done with the VARTM process as it gives good quality of components [17–19].

The reason behind these two testings is to check the suitability of natural fiber in the application where components are subjected to impact and flexural load like leaf spring in the automobile. A lot of researcher tries to implement synthetic composite material in leaf spring but the problem is with cost and we have to add number of leaf as compared to steel leaf. So natural fiber can reduce the cost of spring because they are easily available and reduce the weight of spring but number of leaf will be highest as compared to other all materials.

2 Materials and Methods

2.1 Materials

Details of material requirements and its supplier are given in Table 1 below. The Density and GSM are taken from the technical data sheet of the material provided by the supplier. Jute fiber and flax fiber are supplied by Basu jutex, Kolkata, and Vrukhsha Composites, Tenali respectively. The Mixing ratio is 100:30 by weight percentage of resin and hardener as per technical specification given by the supplier. The chemical Name of epoxy resin is bisphenol A diglycidyl ether and amines is used as hardener (Fig. 1).

Table 1 Properties of reinforcement material

Material	GSM	Density (g/cc)	Thickness (mm)
Jute fiber	215	1.3	0.70
Flax fiber	210	1.5	0.52

(a) Jute Fiber Fabric (b) Flax fiber Fabric

Fig. 1 Reinforcement materials

2.2 *Fabrication of NFRC*

For the fabrication of jute and flax fiber-reinforced composites, the experimental setup of the VARTM process is prepared as shown in Fig. 2. VARTM process gives good quality than hand layup and RTM process, especially in complex geometry. The strength obtained from the VARTM process-based composites is more due to high FVF (up to 60%) compared to other processes. Two types of mould are used one is glass mould and the other is flexible mould. The mould release agent is used for easy removal of the final part from the glass mould. The preform is placed over the glass mould and peel ply and infusion mesh is placed over it.

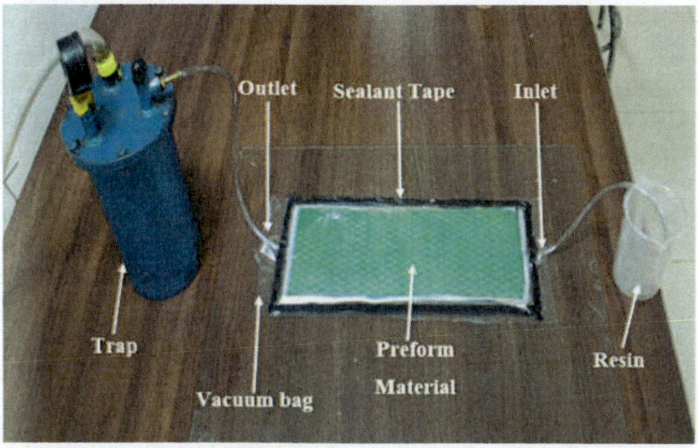

Fig. 2 Complete mould setup for VARTM process

Fig. 3 Impact testing specimens

The preform is covered with sealant tape with the provision of inlet and outlet and the vacuum bag is covered on the mould. After mould preparation, the connection is given to the pump, and a vacuum is applied to check the leakage from the mould and ensure a pure vacuum environment inside the mould. Now by giving the pressure difference, the resin will flow from the preform and converge towards the outlet. After complete mould filling of resin, the mould is kept for 24 h for the curing of the resin.

2.3 *Impact Test*

Impact testing of flax and jute fiber-reinforced composites is carried out as per the ASTM D 256 standard. The specimen length is 63.5 mm, width is 12.7 mm and thickness varies from 3.5 mm to 5 mm. 2.5 mm deep V type notch with 45^0 angles and 0.25 mm radius is cut by AWJM. Five specimens from each specification are tested under the same conditions as shown in Figs. 3 and 4.

2.4 *Flexural Test*

3 point flexural testing of flax and jute fiber-reinforced composites is carried out as per ASTM D 7264 standard. The span to thickness ratio of 20:1 (20 × Thickness) is maintained in all specimens. The thickness of the specimen varies from 3.5 mm to 5 mm and the width of the specimen is 13 mm. the support span is 80 mm for flax 3 layer, flax 5 layer, and jute 5 layer and 100 mm for jute 7 layer. Three specimens from each specification are tested under the same conditions as shown in the figure. Specimens for flexural testing are shown in Fig. 5.

The maximum load that specimen can withstand up to failure is recorded and flexural strength is calculated using 3 points bending equation as shown below [ASTM D 7264].

$$\text{Flexural stress}\ \sigma = \frac{3PL}{2bh^2} \tag{1}$$

where P is load in N, L is supported span, b is the width of a specimen, and h is the thickness of the specimen.

Flexural testing of jute and flax fiber-reinforced composite is carried out using a tensometer of 20KN load cell supplied by kudale instruments private limited, Pune. The testing speed is 1 mm/min.

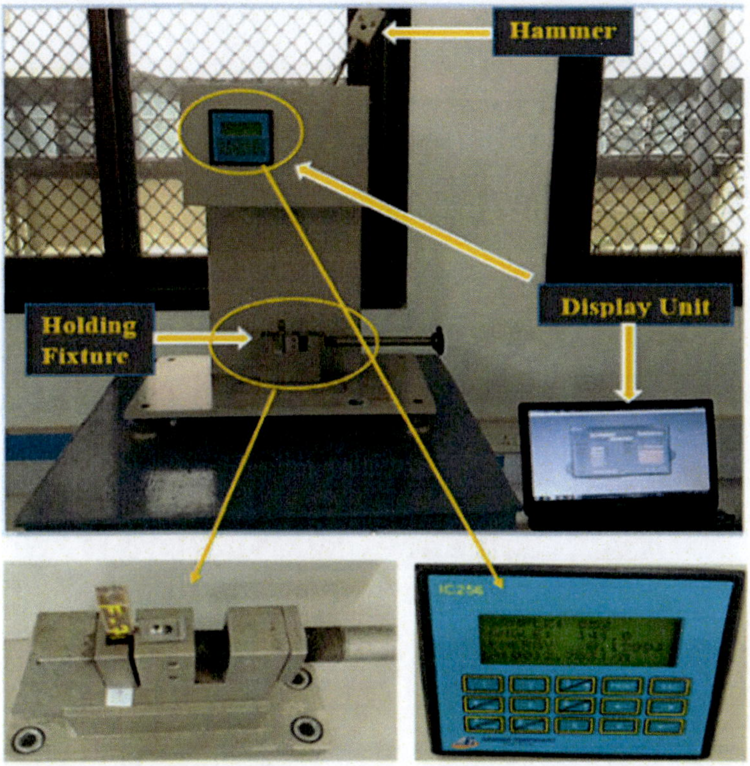

Fig. 4 Izod impact testing machine

3 Results and Discussion

3.1 Impact Test

The energy absorbed by the material is recorded for all specimens and impact energy calculated in J/m.

The reason for lower energy absorption is defects and notching of the specimen. The unnotched impact energy will more as compared to notched impact energy. The mean of all 5 specimens is taken and it is shown in Fig. 6.

The highest impact strength is found to be 44 J/m for flax 7 layer and lowest impact strength is found to be 37 J/m for jute 5 layer. The impact strength increases with an increase in a number of layers for both jute and flax fiber-reinforced composites. The FVF varies from 0.19 to 0.24 and this is the main reason for the small increase in impact strength. Energy absorbed by material increases with an increase in a number of the layer due to thickness variation.

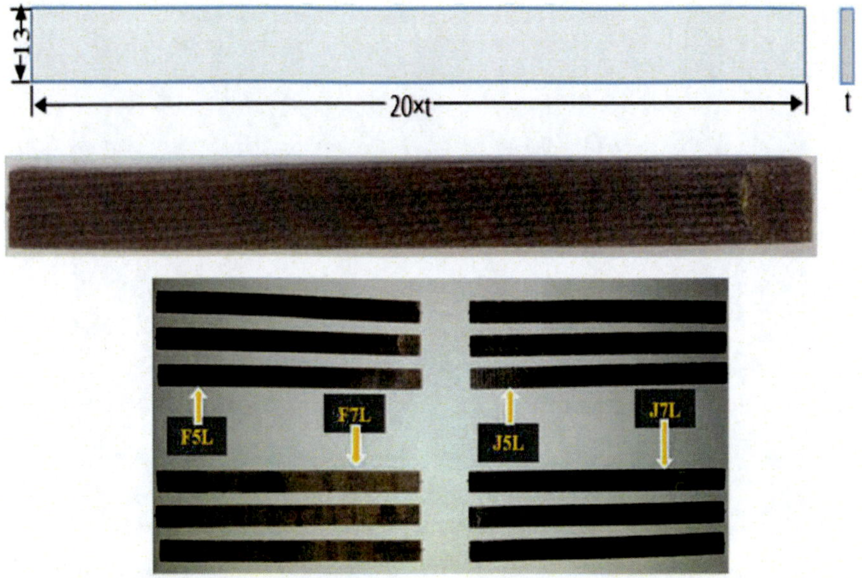

Fig. 5 Flexural testing specimens

Fig. 6 Impact strength of jute and flax fiber-reinforced composites

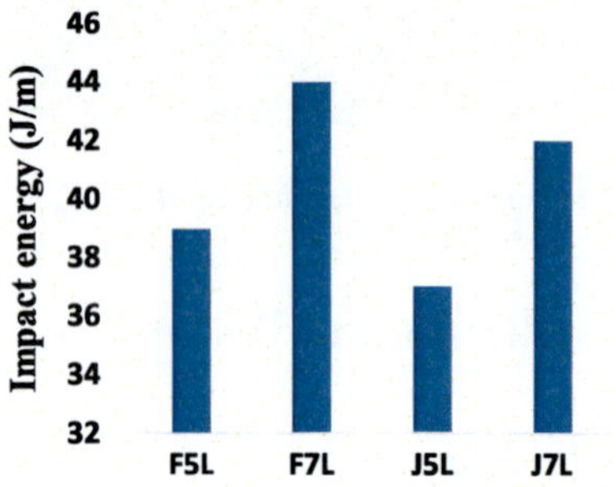

As we know that impact strength is low for both fiber as compared to synthetic fiber, but by this strength, we can calculate the approximate number of leaf required as per design procedure (Fig. 7).

Some images taken by a vision measuring system is shown in Fig. 8. The notch defect and voids are the main reasons for the low strength of the composites. The failure of the impact specimen is also shown in Fig. 8.

The failure in the notched specimen will always start from the notch and it will propagate into thickness direction. So fibers that are in a parallel direction to loading will not break but it will slip to adjacent layers and fibers which are perpendicular loading direction will first break as shown in the figure. First crack initiation will start from resin and after it suddenly breaks. Due to the brittle nature of this composite, the capacity of absorbing impact energy is less as compared to ductile materials.

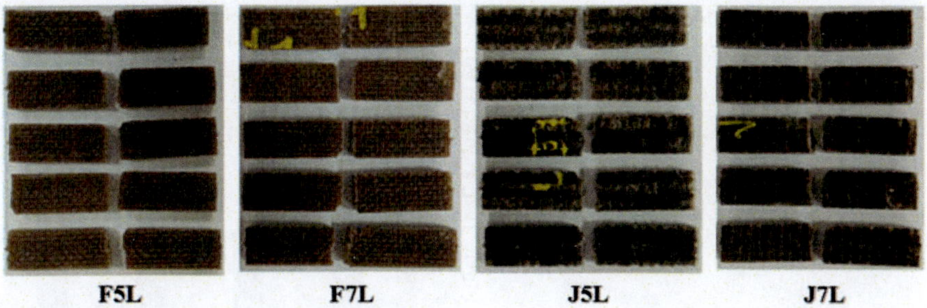

Fig. 7 Break sample after impact strength

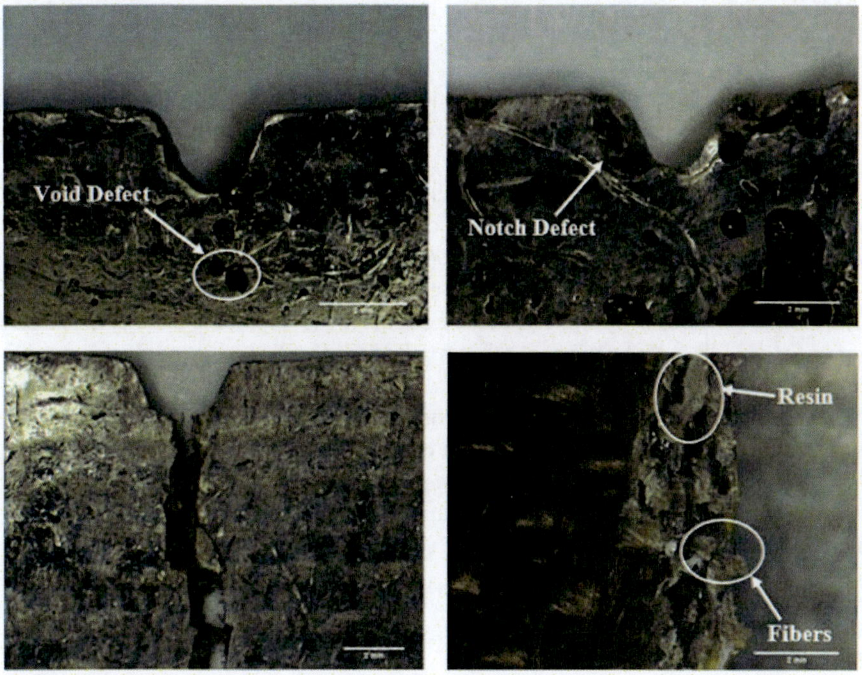

Fig. 8 Images captured by the vision measuring system

So the impact strength greatly depends on the amount of fiber presents in perpendicular to the loading direction. If we have more fibers in perpendicular to loading direction then impact strength will be high and vice versa.

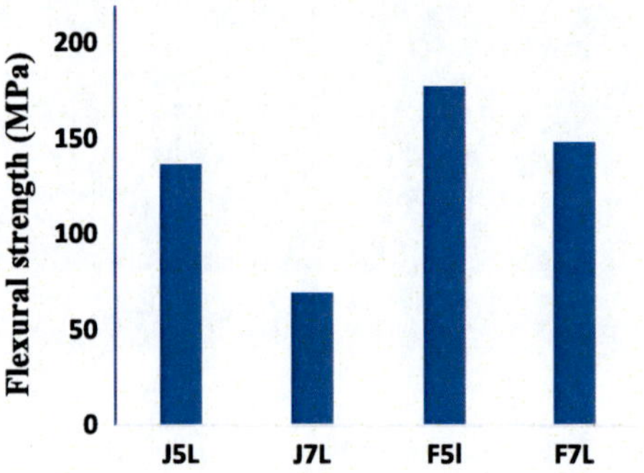

Fig. 9 Flexural strength of jute and flax fiber-reinforced composites

3.2 Flexural Testing

Total 4 combinations two for jute fiber and two for flax fiber with three replication is tested in tensometer. The results obtained from the flexural test is shown below (Fig. 9).

If we increase the number of the layer then the load required to break the specimen will increase but at the same time, the thickness of the specimen will also increase. The flexural strength is directly proportional to the load required to break the specimen and inversely proportional to the square of the thickness of the specimen. The denominator increases by square but the nominator will increases linearly as we can see in Eq. 5.1. That's why the flexural strength of the flax 5 layer is more as compared to the flax 7 layer.

Due to bending, the portion above the neutral layer will experience a tensile load, and the portion below will experience a compressive load. So failure will always happen at the top surface due to maximum stress and fiber is pulling out from the upper surface but here we cannot see fracture at the upper layer due to brittle fracture. The fracture of the flexural component is shown in Fig. 10.

4 Conclusion

The impact and flexural test for 5 and 7 layers of jute and flax fiber-reinforced composites are carried out and the following conclusion can be made.

1. Impact strength is found to be high for flax fiber-reinforced composites as compared to jute fiber-reinforced composites. The impact strength increases by increasing the number of layers.
2. Impact strength and flexural strength can be improved by increasing the fiber volume fraction of material.
3. Flexural strength of flax fiber-reinforced composite is high as compared to jute fiber-reinforced composite but it is decreasing by increasing number of layers.

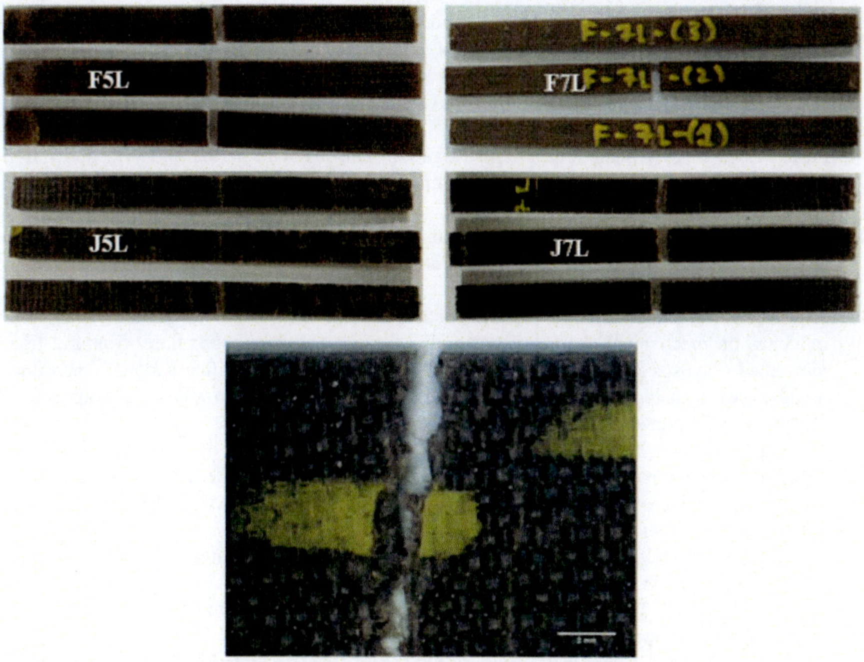

Fig. 10 Fracture behavior during flexural strength

4. By using natural fiber reinforced composite in leaf spring material, the number of leaf will be highest as it shows lower impact and flexural strength. It can minimize cost and weight of leaf spring.

References

1. Elanchezhian C, Ramnath BV, Ramakrishnan G, Rajendrakumar M, Naveenkumar V, Saravanakumar MK (2018) Review on mechanical properties of natural fiber composites. Mat Today: Proc 5(1):1785–90
2. Holbery J, Houston D (2006) Natural-fiber-reinforced polymer composites in automotive applications. Jom 58(11):80–86
3. Nasir AA, Azmi AI, Chye LT, Khalil AN, Kim KN (2018) Tensile behaviour of open hole flax/epoxy composites: influence of fibre lay-up and drilling parameters. J Compos Mater 10(5):228–236
4. Dong C (2018) Review of natural fibre-reinforced hybrid composites. J Reinf Plast Compos 37(5):331–348
5. Xia C, Ren H, Shi SQ, Zhang H, Cheng J, Cai L, Chen K, Tan HS (2016) Natural fiber composites with EMI shielding function fabricated using VARTM and Cu film magnetron sputtering. Appl Surf Sci 362:335–340
6. Xu C, Gu Y, Yang Z, Li M, Li Y, Zhang Z (2016) Mechanical properties of surface-treated ramie fiber fabric epoxy resin composite fabricated by vacuum-assisted resin infusion molding with hot compaction. J Compos Mater 50(9):1189–1198

7. Li Y, Xie L, Ma H (2015) Permeability and mechanical properties of plant fiber reinforced hybrid composites. Mater Des 86:313–320
8. Cavalcanti DK, Banea MD, Neto JS, Lima RA, da Silva LF, Carbas RJ (2019) Mechanical characterization of intralaminar natural fibre-reinforced hybrid composites. Compos B Eng 175:107149
9. Prasad V, Joseph MA, Sekar K (2018) Investigation of mechanical, thermal and water absorption properties of flax fibre reinforced epoxy composite with nano TiO2 addition. Compos Part A: Appl Sci Manuf 115:360–370
10. Ibrahim MI, Sapuan SM, Zainudin ES, Zuhri MY (2020) Preparation and characterization of cornhusk/sugar palm fiber reinforced Cornstarch-based hybrid composites. J Mater Res Technol 9(1):200–211
11. Arputhabalan J, Palanikumar K, Adaikalaraj SR, Priyan MS (2019) Investigation of glass fiber influence on mechanical characteristics and resistance to water absorption of natural fiber reinforced polyester composites. Materials Today: Proc 16; In: 4th international conference on advances in materials, manufacturing and applied sciences, Amrita viswa vidyapeetham, Coimbatore, Tamilnadu, pp 843–852
12. Muktha K, Gowda BK (2017) Development and characterization of sisal fiber reinforced polypropylene composite materials. Mater Today: Proc 4(8); In: 1st International conference on advancements in aeromechanical materials for manufacturing (ICAAMM 2016), MLR Institute of technology, Hyderabad, 2016, pp 8307–8312.
13. Jang BP, Kowbel W, Jang BZ (1992) Impact behavior and impact-fatigue testing of polymer composites. Compos Sci Technol 44(2):107–118
14. Boey FY, Chia LH, Teoh SH (1985) Compression, bend, and impact testing of some tropical wood-polymer composites. Radiat Phys Chem 26(4):415–421
15. Sahu R, Gupta MK, Chaturvedi R, Tripaliya SS, Pappu A (2020) Moisture resistant stones waste based polymer composites with enhanced dielectric constant and flexural strength. Compos B Eng 182:107656
16. Pani PR, Nayak RK, Routara BC, Sekhar PC (2019) Flexural and specific wear rate of seawater aged bamboo, jute and glass fiber reinforced polymer hybrid composites. In: Singh SK, Davim JP, Kumar K (eds), Mater Today: Proc 18; In: 9th International conference of materials processing and characterization (ICMPC-2019), Gokaraju Rangaraju Institute of Engineering and Technology, Hyderabad, pp 3409–3414
17. Srivastava P, Garg CK, Sinha S (2018) The influence of chemical treatment on the mechanical behaviour of hair fibre-reinforced composites. Mater Today: Proc 5(11); In: 5th International conference on advances in energy research (ICAER 2015) IIT Bombay, pp 22922–22930
18. Kumbhani CB, Dave HK, Patel HV (2019) Fabrication and tensile testing of jute and flax fiber reinforced composites using VARTM process. In: Paper presented at 11th international conference on precision, Meso, Micro and nano engineering (COPEN 11), IIT Indore, 12–14 Dec
19. Bayldon JM, Daniel IM (2009) Flow modeling of the VARTM process including progressive saturation effects. Compos a Appl Sci Manuf 40(8):1044–1052

Influence of Fiber Orientation and Number of Layer on Tensile and Flexural Strength of Carbon Fiber-Reinforced Composites Fabricated by VARTM Process

Himanshu V. Patel, Shripal M. Patel, and Harshit K. Dave

1 Introduction

Carbon fiber-reinforced composites are gaining attraction by scientific and industrial interests due to their high specific modulus and strength. Carbon fiber-reinforced composites extensively used in many structural applications including aerospace, automobile, civil, and marine structures where high performance and lightweight of structures are essential [1]. Epoxy resins as a matrix material are extensively used with fiber-reinforced composites due to their better thermal, mechanical, and electrical properties. The high adhesive strength, physical properties, chemical resistance, heat resistance, and electrical resistance are possible to achieve based on chemical composition and curing kinetics [2]. The metal matrix composite with continuous fibers is also gaining attention. These type of composites are widely used because of it consist of better mechanical performance over an alloy material [3]. The use of recycled materials is also becoming a trend. The attempts were made to replace plastic by the recycled material called liquid Wood. The material shows better mechanical properties, hydrophobicity, fire retardance, and durability with reusability up to five times without compromising their properties [4].

Fu and Lauke [5] developed a predictive analytical model as a function of fiber length and fiber orientation to predict tensile strength of short fiber-reinforced polymers. They found fiber length and fiber orientation is the significant parameters for the tensile strength.

H. V. Patel · H. K. Dave (✉)
Department of Mechanical Engineering, S. V. National Institute of Technology, Surat 395007, India
e-mail: hpsvnit07@gmail.com

S. M. Patel
Mechanical Engineering Department, Silver Oak College of Engineering and Technology, Ahmedabad 382 481, India

H. K. Dave and D. Nedelcu (eds.), *Advances in Manufacturing Processes*, Lecture Notes in Mechanical Engineering, https://doi.org/10.1007/978-981-15-9117-4_31

Davis et al. [6] studied effect of orientation on molded fiber-reinforced composite parts. During the simulation, they found fiber orientation has an important role during the mold filling. The fiber orientation could affect the final structural performance of the part.

Vallons et al. [7] investigated effect of fiber orientations on tensile properties of non-crimp fibers (NCF)-reinforced composites. They found that small misorientation affects the static stiffness. Due to the larger off-axis angle, the strength sharply decreased and change in static stiffness was found.

Bora et al. [8] performed a study to investigate the effect of fiber orientation on surface scratch deformation of laminated composites. They found scratch hardness, scratch velocity, penetration depth, and coefficient of friction were the governing parameters. They also found that surface deformation was affected by fiber orientations.

Hossain et al. [9] performed an experimental study using natural fiber reinforcement. They studied the effect of stacking sequence and fiber orientations on tensile strength. They found longitudinal tensile strength was higher than the transverse direction.

Patel and Dave [10] performed experimental and simulation study to determine permeability of fiber reinforcement. Permeability of reinforcement is depended parameter of the fiber architecture. From the study, mold filling time has a good agreement between the experimental and simulation results.

Rahmani et al. [11] fabricated carbon fiber/epoxy multi-axial multi-ply fabric (MMF) composites. During fabrication, they studied the effect of fiber contents, numbers of plies, and layer orientations on tensile, flexural, and impact properties. They found layer orientation was most significant than other parameters.

Subagia et al. [12] fabricated hybrid composite using VARTM process. They fabricated hybrid composite using interply hybridization approach. They performed flexural testing of hybrid composites and they found proper stacking sequence can improve the performance of the composites.

Tseng et al. [13] validated the accuracy of the orientation prediction using the improved anisotropic rotary diffusion and retarding principal rate (iARD-RPR) model. They found good agreement between simulation and prediction model. The fiber orientation is a critical parameter for automotive components that consist of complex geometry, thickness variations, and direction of the matrix flow.

In this work, carbon fiber-reinforced composites were prepared with unidirectional carbon fiber as reinforcements and epoxy resin as the matrix. The composite fabrication was carried out using VARTM process at the level of laboratory scale. The objective of this study was to investigate the effect of fiber orientation of fabric and the number of layers on the tensile and flexural strength of the fabricated composite.

Table 1 Properties of reinforcement and matrix material

Properties of carbon fiber		Properties of resin and hardener		
Density	1.8 g/cm^3		Epoxy	Hardener
Filament diameter	7 μm	Viscosity @ 25° C	22,000–30,000 mPas	< 50 mPas
Tensile strength	4000 MPa	Density @ 25° C	1.2–1.25 g/cc	0.94–0.95 g/cc
Tensile modulus	240 GPa	Flash point	> 200 ° C	> 123 ° C
Elongation	1.7%	Epoxy content	180–200 g/eq	-

2 Experimental Procedure

2.1 Materials

Unidirectional woven carbon fabric with tow size of 3 K and unit weight of 200 g per square meter was used as reinforcement material in a composite laminate. The bisphenol A-based liquid epoxy with a hardener in the ratio of 100:35 (resin: hardener) was used as matrix material. Table 1 shows the properties of reinforcement and material.

2.2 Composite Fabrication

The carbon fiber-reinforced epoxy composite laminates were fabricated for 3, 5, and 7 ply using VARTM process. The VARTM process consists of an infusion of liquid matrix inside the dry form of reinforcement under vacuum. The dry reinforcements are stacked on a rigid mold surface and covered with a flexible vacuum bag using tacky tape. The dry reinforcements were stacked in such a way that forms a stacking sequence of fiber orientations. Through the unidirectional carbon fiber fabric total, four fiber orientations were obtained, the arrangement of the fibers was as shown in Fig. 1. The obtained fiber orientations were [0°], [90°], [0°/90°], [±45°].

Figure 2 illustrates the VARTM process flowchart used for composite fabrication and Fig. 3 schematic diagram of VARTM process setup. After the completion of the molding process, resin infusion was performed inside the dry preform under vacuum.

Figure 4 shows the resin infusion process during composite fabrication. The infused reinforcement allowed to cure under vacuum and at room temperature for 24 h duration.

The cured laminates were demolded after the curing period. For the mechanical testing, these fiber-oriented specimens were cut using abrasive water jet machining (AWJM) to avoid thermal alteration and to obtain a smooth finish of the specimen. Table 2 shows the orientation obtained to investigate the effect on tensile and flexural strength.

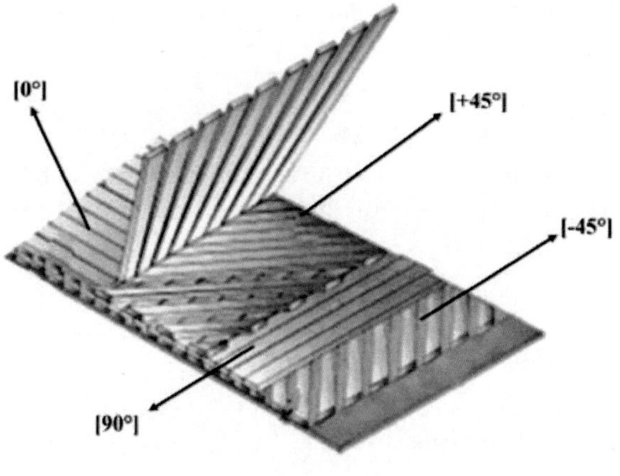

Fig. 1 Schematic of fiber orientations

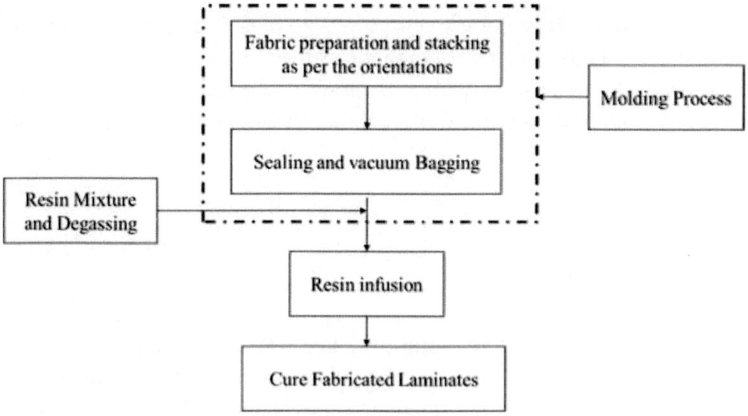

Fig. 2 VARTM process flowchart

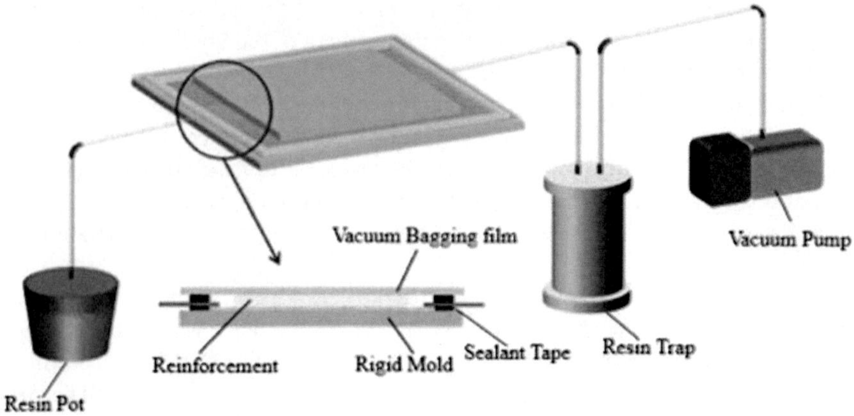

Fig. 3 Schematic of VARTM setup

Fig. 4 Resin infusion during composite fabrication

Table 2 Fiber orientation w.r.t. loading axis and no. of layers

Orientation, θ	3 layers	Five layers	7 layers
[0°]	0°/0°/0°	0°/0°/0°/0°/0°	0°/0°/0°/0°/0°/0°/0°
[90°]	90°/90°/90°	90°/90°/90°/90°/90°	90°/90°/90°/90°/90°/90°/90°
[0°/90°]	0°/90°/0°	0°/90°/0°/90°/0°	0°/90°/0°/90°/0°/90°/0°
[±45°]	+ 45°/-45°/ + 45°	+ 45°/-45°/ + 45°/ -45°/ + 45°	+ 45°/-45°/ + 45°/-45°/ + 45°/-45°/ + 45°

2.3 Mechanical Testing of Fabricated Laminates

Specimens were prepared and tested for measuring the tensile and flexural strength as per the ASTM D3039 and ASTM D7264 standards, respectively. Tensile and flexural tests were performed on a universal testing machine (Instron make model 5980). The machine has a capacity of 100 kN load with a measurement accuracy of ± 0.5% of reading. The experimental values of tensile and flexural strength are shown in Table 3.

3 Results and Discussion

Tensile and flexural strength have been measured by varying fiber orientation and the number of layers. Further, the fracture pattern of the fractured specimen is also observed by vision measurement system. The observations and results are discussed in the following sections.

Table 3 Results of tensile and flexural testing

Sr. No.	Orientation, θ	No. of layers	Mean tensile strength, σ_t, MPa	Standard deviation, S_T	Mean flexural strength, σ_f, MPa	Standard deviation, S_F
1	0°	3	894.78	29.86	873.50	108.54
2	90°	3	23.38	3.20	42.46	5.17
3	(0°/90°)	3	633.68	96.65	910.11	3.46
4	(±45°)	3	83.78	38.86	119.09	2.61
5	0°	5	903.26	47.97	721.29	71.62
6	90°	5	12.55	4.64	19.64	5.29
7	(0°/90°)	5	613.11	56.72	737.16	43.99
8	(±45°)	5	124.09	6.89	213.37	20.76
9	0°	7	968.77	96.36	642.86	41.10
10	90°	7	6.76	0.50	12.86	4.27
11	(0°/90°)	7	612.56	35.93	404.47	22.45
12	(±45°)	7	85.52	2.75	151.68	17.92

3.1 *Tensile Behavior*

The experimental results of tensile strength are compared between different orientation and number of layers. Figure 5 shows the mean value of ultimate tensile strength for by varying fiber orientation and the number of layers of reinforcement ply.

It can be seen that irrespective of the number of layers, strength for specimens with [0°] orientation has the highest strength and [90°] has shown least strength.

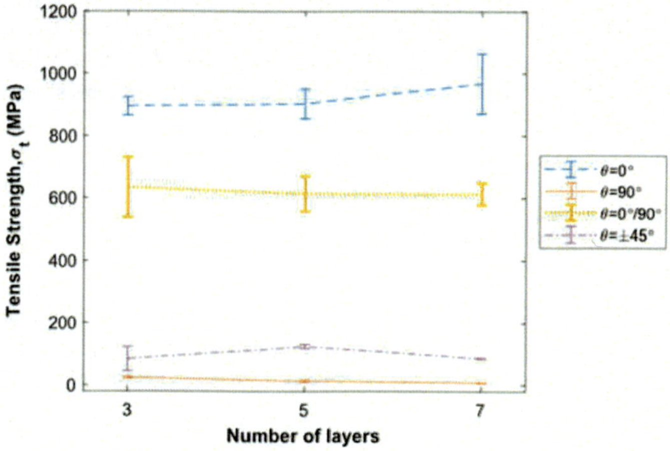

Fig. 5 Tensile behavior of composites

Fig. 6 Fracture pattern of tensile test specimens

The intermediate values of strength are observed on specimens with [0°/90°] orientation. The strength values on the specimen with [±45°] orientation are found to be significantly less but slightly higher than that observed at [90°] orientation. Tensile strength increased from an average of 6.76 MPa for [90°] orientation and seven layers to 968.77 MPa for [0°] orientation and seven layers. From the above results, it can be seen that fiber orientation is the dominant factor. The laminates formed with [0°] orientation fibers are most robust along their length and weakest in the lateral direction. Any crack first starts from the matrix–fiber interface that further propagates and results into failure of the component. Here, unidirectional woven fabric is used in which only 2% fibers that are running in a lateral direction. Thus, off-axis pulling of fibers easily separates the two fibers in the vicinity. Hence, higher the angle between pull axis and fiber axis, higher will be the amount of off-axis pulling.

The fracture pattern of the specimens was observed and analyzed using vision measurement system. The specimen having [0°] orientation failed longitudinally. Specimen with [90°] orientation failed along the axis perpendicular to the loading direction.[0°/90°] specimens have mixed fracture effect of both [0°] and [90°] laminates, i.e., zigzag pattern was obtained in most of the specimen. [45°] orientation specimens have failed at an angle of approximately [45°] to the loading axis. Figure 6 shows the fracture pattern during tensile test of specimen of different fiber orientations. The different fracture patterns in different orientations had one similarity, i.e., all had separated from the interface of two fibers. Two fiber tows are prone to separate from each other when the bonding element, i.e., the matrix cracking.

3.2 *Flexural Behavior*

The experimental results of flexural strength are compared between different orientation and number of layers. Figure 7 shows the flexural strength for all the orientations and number of layers of ply in the present study. Similar to tensile strength, flexural strength obtained for [90°] was the lowest for all layer numbers and slightly higher for [±45°]. The laminate is subjected to bending load, at that time fibers in topmost

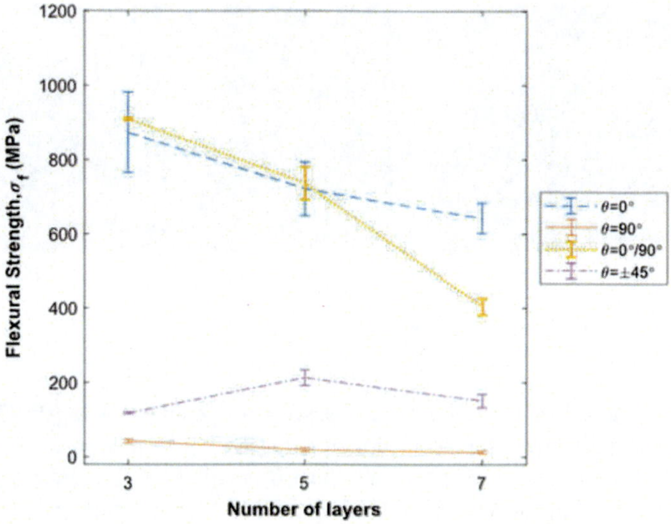

Fig. 7 Flexural behavior of composites

layers are in the highest compression stress, and bottom fibers are in the highest tensile stress.

The fibers placed at [0°] orientation have continuity while those in [90°] orientation have fiber–matrix interface present in the structure. Due to discontinuity at fiber–matrix interface, crack may be initiated. It is also observed that flexural strength is decreasing with an increase in the number of layers. Flexural strength obtained for [0º] and [0º/90º] laminates were almost similar due to identical fiber arrangement at uppermost and lowermost. Figure 8 shows the fracture pattern for [0°] and [0°/90°] orientation during the flexural loading. The uppermost layer experiences the highest compression stress and lowermost layer experience the highest tensile stress. Thus,

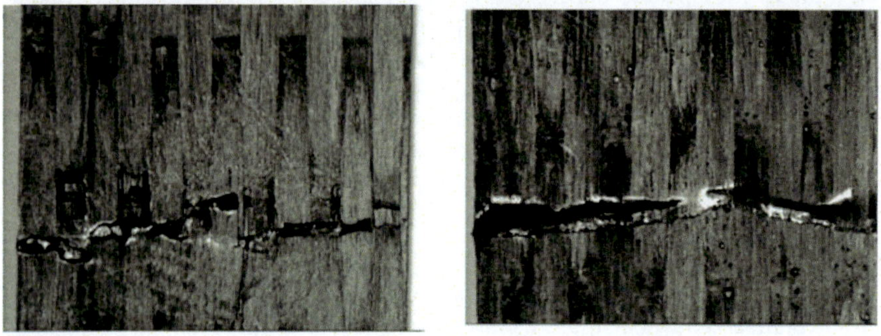

0º 0º/90º

Fig. 8 Fracture pattern for flexural test specimens

if the tensile strength is lower than compressive strength, the specimen would have failed from lowermost layer, and if compressive strength is lower than the tensile strength, the specimen would have failed from the uppermost layer. It can also be observed that specimen fractured from the uppermost surface had lower compression strength.

4 Conclusions

In the present work, composite laminates were fabricated using VARTM method with different fiber orientations and number of layers. The effect of fiber orientations and number of layers on tensile strength and flexural strength were observed. The fracture surfaces were used to explain the mode of failure in different orientations. From the above study, following conclusions can be drawn:

1. The results shows that the laminates having 0° orientation has the highest tensile strength. The laminates having 90° orientation consists of the lowest tensile and flexural strengths.
2. The significant flexural strength was achieved for 0° laminates for three layers.
3. ± 45° laminates can deflect more than any other laminates and necking phenomena had been observed during tensile testing that suggests laminates with ± 45° orientation can be preferred for applications where strength has lesser significance than deflection.
4. Higher tensile strength was observed when number of layers are increased because fibers are more aligned with the loading direction.

References

1. Ning H, Li J, Hu N, Yan C, Liu Y, Wu L, Liu F, Zhang J (2015) Interlaminar mechanical properties of carbon fiber reinforced plastic laminates modified with graphene oxide interleaf. Carbon 91:224–33. https://doi.org/10.1016/j.carbon.2015.04.054 (Elsevier)
2. Njuguna J, Pielichowski K, Alcock JR (2007) Epoxy-based fibre reinforced nanocomposites. Adv Eng Mater 9:835–847. https://doi.org/10.1002/adem.200700118 (Wiley)
3. Nedelcu D, Milosevic O, Chelariu R, Roman C (2010) Some experimental aspects concerning the stratified composite materials with metallic matrix. Int J Mod Manuf Technol 2:65–71
4. Nedelcu D, Stefan A, Mindru T, Plavanescu S (2012). Flexural properties of samples obtained from "liquid wood". Selected engineering problems, pp 151–154.
5. Fu SY, Lauke B (1996) Effects of fiber length and fiber orientation distributions on the tensile strength of short-fiber-reinforced polymers. Compos Sci Technol 56:1179–1190 (Elsevier)
6. Davis B, Gramann P, Rios A (2002). Effect of fiber orientation anisotropies on the structural performance of molded FRP composite parts. The Madison Group of Polymer Processing Research Corporation.
7. Vallons K, Duque I, Lomov SV, Verpoest I (2009). Fibre orientation effects on the tensile properties of biaxial carbon/epoxy NCF composites. In: Proceeding of international conference on composite materials (ICCM) vol 17

8. Bora MO, Coban ON, Sinmazcelik T, Gunay V (2010) Effect of fiber orientation on scratch resistance in uni-directional carbon-fiber-reinforced polymer matrix composites. J. Reinf Plast Compos 29:1476–1490. https://doi.org/10.1177/0731684409103953 (SAGE)
9. Hossain RM, Islam A, Van Vuure AW, Ignaas V (2013) Effect of fiber orientation on the tensile properties of jute epoxy laminated composite. J Sci Res 5:43–54 (JSR)
10. Patel HV, Dave HK (2020) Study of permeability and resin flow front during the fabrication of thin composite using VARTM process. Int J Mod Manuf Technol 12:125–130
11. Rahmani H, Najafi SH, Saffarzadeh-Matin S, Ashori A (2014) Mechanical properties of carbon fiber/epoxy composites: effects of number of plies, fiber contents, and angle-ply layers. Polymer Eng Sci 54:2676–2682. https://doi.org/10.1002/pen.23820 (Wiley)
12. Subagia IA, Kim Y, Tijing LD, Kim CS, Shon HK (2014) Effect of stacking sequence on the flexural properties of hybrid composites reinforced with carbon and basalt fibers. Compos Part B: Eng 58:251–258. https://doi.org/10.1016/j.compositesb.2013.10.027 (Elsevier)
13. Tseng HC, Goto M, Chang RY, Hsu CH (2018). Accurate predictions of fiber orientation and mechanical properties in long-fiber-reinforced composite with experimental validation. Polymer Composites vol 39. Wiley, pp 3434–3445. https://doi.org/10.1002/pc.24361